Horticulture: Plants for People and Places, Volume 3

Geoffrey R. Dixon • David E. Aldous
Editors

# Horticulture: Plants for People and Places, Volume 3

## Social Horticulture

Springer

*Editors*
Geoffrey R. Dixon
GreenGene International
Hill Rising, Sherborne, Dorset
United Kingdom

David E. Aldous
School of Agriculture and Food Sciences
The University of Queensland
Lawes
Queensland
Australia

ISBN 978-94-017-8559-4 ISBN 978-94-017-8560-0 (eBook)
ISBN 978-94-017-8640-9 (set)
DOI 10.1007/978-94-017-8560-0
Springer Dordrecht Heidelberg New York London

Library of Congress Control Number: 2014936300



Printed on acid-free paper

Springer is part of Springer Science+Business Media (www.springer.com)

*We dedicate these Books to our wives;*
*Mrs. Kathy Dixon and Mrs. Kaye Aldous*
*in gratitude for their lifetimes of unstinting support, forbearance and understanding*

*Professor David E. Aldous – deceased 1st November 2013.*

*The concepts underlying the Trilogy Horticulture—Plants People and Places were developed by David Aldous and me during the International Horticultural Congress 2010 in Lisbon, Portugal. These Books celebrate our common views of the scholastic and intellectual depth and breadth of our discipline and the manner by which it is evolving in response to the economic, environmental and social challenges of the*

*21st Century. Jointly, over more than two years we enlisted international authorities as lead authors, reviewed drafts and edited final texts. Despite there being half the world between us we formed a deep rapport. His sudden and wholly unexpected death from a brain aneurysm came as a shattering blow leaving me with the tasks of seeing our work through to completion. This Trilogy stands as a legacy to horticulture as "the first of all the Arts and Sciences" from an internationally acclaimed, respected and much loved: scientist, educator and author. It is appropriate that the Trilogy should be dedicated to David united with our original intention of paying tribute to our respective wives.*

*Professor Geoffrey R. Dixon*

# Preface

This Trilogy of books answers the question "What is Horticulture?". Their contents span from tropical plantations growing exotics crops such as cocoa, pineapples and rubber through to the interior landscaping of high-rise office tower blocks, to applications which encourage physical and mental health. The common thread uniting this Discipline is the identification, breeding, manipulation of growth and stimulation of flowering and fruiting in plants either for food or environmental and social improvement. Understanding the scientific principles of why plant productivity increases following physical, chemical and biological stimuli has fascinated horticulturists for several millennia.

Epicurus (341BC–270BC) the Athenian philosopher of the 3rd century BC believed that plants achieved "the highest good was calmness of mind". Calmness comes to some Horticulturists with the satisfaction of entering vast hectarages of bountiful orchards, to others from well designed and carefully maintained landscapes while others are entranced by participation in conserving components of the Earth's fragile biodiversity. Horticulture while being a scientific discipline has much wider and deeper dimensions. There are historic, artistic and cultural facets which are shared with the Humanities and these aspects are included within this Trilogy. Wherever Horticulturists gather together they share a common language which interprets useful scientific knowledge and cultural understanding for the common benefit of mankind. For while Horticulture is about achieving an intensity of growth and development, flowering and fruiting, it is wholly conscious that this must be achieved sustainably such that the resources used are matched by those passed on for use by future generations.

The structure of this Trilogy is such that it traces the evolution in emphasis which has developed in Horticultural philosophy across the second half of the 20th and into the 21st century. Following the worldwide conflicts of the 1940s the key aspirations were the achievement of food sufficiency and the eradication of hunger from the planet. In an increasingly affluent and developed world there is food sufficiency *par excellence*. Never before has such an array of plenty been made available year-round. This plenty is nowhere more evident that in the fresh fruit and vegetable aisles of our supermarkets. Horticulture has given retail shoppers the gift of high quality and diversity of produce by manipulating plant growth and reproduction and post harvest care across the globe.

This third volume illustrates in considerable depth the scientific and technological concepts interacting with the arts and humanities which now underpin the rapidly evolving subject of Social Horticulture. This covers considerations of: Horticulture and Society, Diet and Health, Psychological Health, Wildlife, Horticulture and Public Welfare, Education, Extension, Economics, Exports and Biosecurity, Scholarship and Art, Scholarship and Literature, Scholarship and History and the relationship between Horticulture and Gardening. This volume firmly brings the evolution of the Discipline into the 21st Century. It breaks new ground by providing a detailed analysis of the value of Horticulture as a force for enhancing society in the form of social welfare, health and well-being, how this knowledge is transferred within and between generations, and the place of Horticulture in the Arts and Humanities. The social domain which describes peoples' behaviour shows how dependent mankind is on nature and green open space, not just for material requirements, but also for our physical, psychological, emotional and spiritual needs. Green open spaces, associated with urban communities, are usually the only possible source of connection with the natural world and so contribute to improving the quality of life in our communities. These green open spaces have been shown to provide considerable social capital in terms of employment, education, and recreational benefit. Physical activity, such as walking or cycling to work or for pleasure, or being directly involved with a natural green space activities has been shown to alleviate stress, reduce mental fatigue and potential for anger, increase wellbeing and self-esteem and aid in a more rapid recovery and spending less time in a hospital. Research has also shown by association patients working in a natural environment have improved health, visited their general practitioner less, taken fewer prescription drugs, felt safer in their community, experienced less pain and discomfort, had more opportunities to use their skills, and culminating in reduced community health costs. The lack of plants in an environment can lead to reduced mental and physical development, poor performance at school and in the workforce and the continuation of poverty for generations Other social benefits of an association with plants include reducing the potential for domestic violence, vandalism, ethnic conflict and crime by building interpersonal relationships. Consequently, this volume places substantial emphasis on the relationship between health, well-being and plants. The success of the Eden Project as described in this volume identifies how keen members of the general public are for an association with plants and the pleasures that they derive from being near, or working with plants. This concept is taken further when examining the importance of plants for psychological well being. It is now well accepted that diets which are rich in fruit and vegetables can contribute hugely towards the reduction in the incidence of diseases of affluence such as cancers, coronary heart disease and strokes. The relationship between working with plants forming part of the wild flora is examined in this volume. Adding areas of wilderness into people's lives can be highly beneficial to their psychological and physical health and well-being. The transfer of knowledge either to students or to practitioners of horticulture is of major significance. Two chapters examine these aspects looking at means by which knowledge is delivered and the wider contexts within which Horticultural Education is provided. Understanding the Economics of Horticulture is of paramount importance in justifying public and private financial provisions for the discipline. Biosecurity is difficult to achieve since global air travel takes new plants, microbes and animals around the world at ever increasing speed. This has lead to some considerable problems with the growth of alien species which have exploited environments devoid of predators. Of particular importance in this volume are the three chapters dealing with aspects of the relationship between Horticulture and scholarship. Here the aspects of Horticulture which integrate with the Humanities are explored in considerable depth. This is an ancient relationship and one where Horticulture demonstrates the intensity of its connections with man's cultural spirit. Horticulture here becomes far more than an attempt to understand the science of plant growth and reproduction and takes on roles which pertain to artistic and historical significance. Finally, there is an examination of the relationship between Horticulture and Gardening. This is an area of thought deserving of much deeper analysis. Recently, regard for gardening has become levelled down through the activities of the media which wish to reduce topics to sound-byte levels. But culturally gardening has much to do with the relationship between man, plants and the human spirit. It is a truism that "Horticulture is to English Literature as Gardening is to Theatre". In other words Gardening is a physical process whereas Horticulture is cerebral. Some would contend that this does not do justice to gardening which itself may be cerebral. Certainly in history there were political aspects to gardening whereby choosing an incorrect style of landscaping could spell serious even fatal penalties for the owner.

The first volume in this Trilogy covers Crop Production Horticulture (volume 1) and the second volume is devoted to Environmental Horticulture (volume 2). Volume 1 illustrates in considerable depth the science and technology which underpin the continuous production of Horticultural Fresh Produce. Firstly there is a consideration of aspects of industrial development based on basic scientific discoveries. This is followed by chapters written by acknowledged world experts covering the production of: Field Vegetables, Temperate Fruit, Tropical Fruit, Citrus, Plantation Crops, Berry Crops, Viticulture, Protected Crops, Flower Crops, Developing New Crops, Post-harvest Handling, Supply Chain Management and the Environmental Impact of Production. Production Horticulture may now be found supporting the economies of less developed nations, consequently the final Chapter focuses especially on the impact of Production Horticulture in Africa.

The second volume Environmental Horticulture covers considerations of: Horticulture and the Environment, Woody Ornamentals, Herbs and Pharmaceuticals, Urban Greening, Rural Trees, Urban Trees, Turfgrass Science, Interior Landscaping, Biodiversity, Climate Change and Organic Production. These subjects are united by consideration of the need for sustainable use of resources and careful conservation applied of all points where Horticulture and the environment coincide. Horticulture plays an enormous role in aiding environmental care and support. Indeed this discipline could be considered as having founded much of the basis for is now considered to be ecological and environmental science.

The value of Horticulture for human development was emphasised by Jorge Sampaio (United Nations High Representative for the Alliance of Civilisations and previously the President of the Republic of Portugal) in his opening address to the 28th International Horticultural Congress in Lisbon, 2010. He stated that Horticulture can achieve "a lot to overcome hunger and ensure food security". In the face of estimates that the world's population, particularly in developing counties, will reach 9.1 billion by 2050 much does need to be achieved, and in this Horticulture has an especially important roles. Intensive plant production has much to offer as urbanization continues at an accelerating pace. Shortly about 70 per cent of the world's population will choose to live in urban and peri-urban areas of many countries. In the developing World many millions of the world's population continue to be undernourished and in poor health. Climatic change, over population, soil degradation, water and energy shortages, pollution and crippling destruction of biodiversity are the challenges facing all of humanity. Horticulture in its Production, Environmental and Social dimensions offers important knowledge and expertise in these areas. This has been well explained in "Harvesting the Sun", a digest recently published by the International Society for Horticultural Science. In summary form the international interactions between horticultural science, technology, business and management are explained. This offers pointers as to how over the early part of the 21st Century world food production must rise by at least some 110 per cent in order to meet the demands of a expanding populations in countries such as China, India, parts of Asia and in South America.

Considerable breadth and depth of intellect are demanded of those who seek an understanding of horticulture. This is not a discipline for the faint hearted since the true disciple needs a considerable base in the physical, chemical, and the biological sciences and natural resource studies linked with an understanding of the application of economics, engineering and the social sciences. Added to this should also comes an appreciation of the artistic, historic and cultural dimensions of the Discipline. The teaching of fully comprehensive horticultural science courses in higher educational institutions has regrettably been diminished worldwide. It is to be hoped that this Trilogy may go some small way in providing an insight into the scale, scope and excitement of the Discipline and the intellectual rigour demanded of those who seek a properly proportioned understanding of it.

Enormous thanks go to all those who have contributed to these three volumes. Their devotion, hard work and understanding of the Editors' requests are greatly appreciated. Thanks are also due to our colleagues in Springer for all their continuing help, guidance and understanding. In particular we would like to thank Dr Maryse Walsh, Commissioning Editor and Ir Melanie Van Overbeek, Senior Publishing Assistant.

Professor Geoffrey R. Dixon affectionately records his thanks to his mentor Professor Herbert Miles, then Head of the Horticulture Department of Wye College, University of London (now Imperial College, London) who challenged him to "define Horticulture". Regrettably, it has taken half a century of enquiry to respond effectively.

Sherborne, Dorset, United Kingdom
Queensland, Australia
August 2013

Geoffrey R. Dixon
David E. Aldous

# Contents

# Contributors

**Manuel Agustí** Instituto Agroforestal Mediterráneo, Universitat Politècnica de València, València, Spain

**David E. Aldous** School of Land, Crop and Food Science, The University of Queensland, Lawes, Queensland, Australia

**Neil O. Anderson** Flower Breeding & Genetics Department of Horticultural Science, University of Minnesota, Saint Paul, MN, USA

**Vera Andjic** Department of Agriculture, Perth, Australia

**Paul Arens** Wageningen University and Research Centre, Plant Breeding, Wageningen, The Netherlands

**Peter Batt** School of Management, Curtin University, Perth, WA, Australia

**Christina M. Bavougian** Department of Agronomy and Horticulture, University of Nebraska, Lincoln, NE, USA

**Andreas Bettin** Faculty of Agricultural Sciences and Landscape Architecture, University of Applied Sciences Osnabruck, Osnabrück, Germany

**Indrabrata Bhattacharya** Department of Plant Pathology, Bidhan Chandra Krishi Viswavidyalaya, Nadia, West Bengal, India

**John Blackman** National Wine and Grape Industry Centre, Wagga Wagga, NSW, Australia

**P. N. Brás de Oliveira** Instituto Nacional de Investigacao Agraria e Veterinaria, Oeiras, Portugal

**R. M. Brennan** Fruit Breeding and Genetics Group, James Hutton Institute, Invergowrie, Dundee, Scotland, UK

**Robin G. Brumfield** Department of Agricultural, Food and Resource Economics, Rutgers, The State University of New Jersey, New Brunswick, NJ, USA

**G. Buck-Sorlin** AGROCAMPUS WEST Angers Centre, National Institute of Horticulture and Landscape, UMR1345 Research Institute of Horticulture and Seeds (IRHS), Angers, France

**P. D. S. Caligari** Instituto de Biología Vegetal y Biotecnología, Universidad de Talca, Talca, Chile

**Ross W. F. Cameron** Department of Landscape, The University of Sheffield, Sheffield, South Yorkshire, UK

**Andrew Clark** National Wine and Grape Industry Centre, Wagga Wagga, NSW, Australia

**J. R. Clark** University of Arkansas, Fayetteville, AR, USA

**Rosemary H. Collier** Warwick Crop Centre, The University of Warwick, Warwick, UK

**Sylvia Concepcion** School of Management, University of the Philippines Mindanao, Mintal, Davao, The Philippines

**Guglielmo Costa** Department of Agricultural Sciences—DipSA, Alma Mater Studiorum, University of Bologna, Bologna, Italy

**Lyle E. Craker** Medical Plant Program, University of Massachusetts, Amherst, MA, USA

**Kevin M. Crosby** Department of Horticultural Sciences, Vegetable and Fruit Improvement Center, Texas A&M University, College Station, US

**Susanna Curtin** School of Tourism, Bournemouth University, Poole, Dorset, UK

**Jeff Daniells** Department of Agriculture, Fisheries and Forestry, Centre for Wet Tropics Agriculture, South Johnstone, Queensland, Australia

**Rebecca L. Darnell** Horticultural Sciences Department, Gainsville, FL, USA

**Harry Day** Juneau, USA

**Yves Desjardins** Institute of Nutrition and Functional Foods/Horticulture Research Center, Laval University, Québec City, QC, Canada

**Yan Diczbalis** Department of Agriculture, Fisheries and Forestry, Centre for Wet Tropics Agriculture, South Johnstone, Queensland, Australia

**Geoffrey R. Dixon** School of Agriculture, University of Reading, Earley Gate, Berkshire, United Kingdom

GreenGene International, Hill Rising, Sherborne, Dorset, United Kingdom

**R. Drew** School of Biomolecular and Physical Sciences, Griffith University, Nathan, Queensland, Australia

**Menahem Edelstein** Department of Vegetable Crops, Agricultural Research Organization, Newe Ya'ar Research Center, Ramat Yishay, Israel

**René Eschen** CABI, Delémont, Switzerland

**C. E. Finn** USDA-ARS, HCRL, Corvallis, OR, USA

**Dorothy Fox** School of Tourism, Bournemouth University, Poole, Dorset, UK

**Victor Galán Saúco** Departamento de Fruticultura Tropical, Instituto Canario de Investigaciones Agrarias, La Laguna, Tenerife, Spain

**Gert Groening** Forschungsstelle Gartenkultur und Freiraumentwicklung, Institut für Geschichte und Theorie der Gestaltung, Universität der Künste, Berlin, Germany

**Nazim Gruda** Division of Horticultural Sciences, University of Bonn, Bonn, Germany

Department for Innovation Promotion, Federal Office for Agriculture and Food, Bonn, Germany

**Heiner Grüneberg** Department of Horticultural Plant Systems, Humboldt-Universität zu Berlin, Berlin, Germany

**Yann Guisard** National Wine and Grape Industry Centre, Orange, NSW, Australia

**Luke Hailey** Bartlett Tree Research Laboratory, Bartlett Tree Research Laboratory, Reading, UK

**J. F. Hancock** Department of Horticulture, Michigan State University, East Lansing, MI, USA

**Stefanie Hennecke** Fachgebiet Freiraumplanung, Universität Kassel, Kassel, Germany

**Maria Herrero** Department of Pomology, Estación Experimental de Aula Dei—CSIC, Zaragoza, Spain

**Andrew Hirons** Lecturer in Arboriculture, Myerscough College, Lancashire, UK

**Bruno Holzapfel** National Wine and Grape Industry Centre, Wagga Wagga, NSW, Australia

**Jose I. Hormaza** Instituto de Hortofruticultura Subtropical y Mediterránea La Mayora (IHSM-CSIC-UMA), Algarrobo-Costa, Malaga, Spain

**Luis Hualda** School of Management, Curtin University, Perth, WA, Australia

**Alan Hunter** College of Life Sciences, School of Agriculture and Food Science, Agriculture & Food Science Centre, University College Dublin, Belfield, Dublin 4, Ireland

**Maiko Inoue** Medical Plant Program, University of Massachusetts, Amherst, MA, USA

**Jules Janick** Department of Horticulture and Landscape Architecture, Purdue University, West Lafayette, IN, USA

**D. Jarret** Fruit Breeding and Genetics Group, James Hutton Institute, Invergowrie, Dundee, Scotland, UK

**Mark Johnston** Arboriculture and Urban Forestry, Myerscough College, Lancashire, UK

**Charlotte Jordan** Redhood City, USA

**Tony Kendle** Eden Projcct, Cornwall, UK

**Olaf van Kooten** Horticultural Production Chains Group, Wageningen University, Wageningen, The Netherlands

**Smilja Lambert** Cocoa Sustainability Research Manager Asia/Pacific, Mars Chocolate—Cocoa Sustainability, Cairns, Queensland, Australia

**Daniel I. Leskovar** Texas A&M AgriLife Research Center, Texas A&M University, Uvalde, TX, US

**Robert Lillywhite** Warwick Crop Centre, University of Warwick, Warwickshire, UK

**G. A. Lobos** Plant Breeding and Phenomic Center, Faculty of Agricultural Sciences, Universidad de Talca, Talca, Chile

**Virginia I. Lohr** Department of Horticulture, Washington State University, Pullman, WA, United States of America

**Pieternel A. Luning** Food Quality and Design Group, Wageningen University, Wageningen, The Netherlands

**Peter M. Martin** Amenity Horticulture Research Unit, University of Sydney Plant Breeding Institute, Cobbitty, NSW, Australia

**Amparo Martínez-Fuentes** Instituto Agroforestal Mediterráneo, Universitat Politècnica de València, València, Spain

**Aaron Maxwell** School of Veterinary and Life Sciences, Murdoch University, Perth, Australia

**Keith W. McAuliffe** Sports Turf Research Institute, Ormiston, QLD, Australia

**Peter F. McSweeney** Department of Agriculture and Food Systems, Melbourne School of Land and Environment, The University of Melbourne, Melbourne, VIC, Australia

**Carlos Mesejo** Instituto Agroforestal Mediterráneo, Universitat Politècnica de València, València, Spain

**William B. Miller** Department of Horticulture, Cornell University, Ithaca, NY, USA

**Elizabeth J. Mitcham** Horticulture Collaborative Research Support Program, University of California, Davis, CA, USA

**Roy Murray-Prior** School of Management, Curtin University, Perth, WA, Australia

**Panayiotis A. Nektarios** Department of Crop Science, Lab. of Floriculture and Landscape Architecture, Agricultural University of Athens, Athens, Greece

**Ruth A. Nettle** Department of Agriculture and Food Systems, Melbourne School of Land and Environment, The University of Melbourne, Melbourne, VIC, Australia

**Marco A. Palma** Department of Agricultural Economics, Texas A&M University, College Station, US

**David Pearson** Faculty of Arts and Design, University of Canberra, Australian Capital Territory, Australia

**Glynn Percival** Bartlett Tree Research Laboratory, Bartlett Tree Research Laboratory, Reading, UK

**Ghillean T. Prance** The Old Vicarage, Lyme Regis, UK

**James E. Pratley** Charles Sturt University, School of Agricultural and Wine Sciences, Wagga Wagga, NSW, Australia

**David Rae** Royal Botanic Garden Edinburgh, Edinburgh, Scotland

**S. Raffle** Horticultural Development Company, Agriculture and Horticulture Development Board, Kenilworth, Warwickshire, UK

**Angelo Ramina** Department of Agronomy, Food, Natural resources, Animals and Environment—DAFNAE, University of Padova, Padova, Legnaro, Italy

**Andrew Rawson** School of Agricultural and wine Sciences, Orange, NSW, Australia

**John P. Rayner** Department of Resource Management and Geography, Melbourne School of Land and Environment, University of Melbourne, Richmond, VIC, Australia

**Paul E. Read** Department of Agronomy and Horticulture, University of Nebraska, Lincoln, NE, USA

**Carmina Reig** Instituto Agroforestal Mediterráneo, Universitat Politècnica de València, València, Spain

**P. Diane Relf** Department of Horticulture, Virginia Tech, VPI & SU, Blacksburg, VA, United States of America

**Suzy Rogiers** National Wine and Grape Industry Centre, Wagga Wagga, NSW, Australia

**Maria Fay Rola-Rubzen** CBS—Research & Development, Curtin University, Perth, WA, Australia

**Pia Rowe** Faculty of Arts and Design, University of Canberra, Australian Capital Territory, Australia

**Silviero Sansavini** Dipartimento di Scienze Agrarie, University of Bologna, Bologna, Italy

**Emma Schaffert** Bartlett Tree Research Laboratory, Bartlett Tree Research Laboratory, Reading, UK

**Leigh Schmidtke** National Wine and Grape Industry Centre, Wagga Wagga, NSW, Australia

**Rob E. Schouten** Horticultural Production Chains Group, Wageningen University, Wageningen, The Netherlands

**Chris Searle** Suncoast Gold Macadamias, Bundaberg, Queensland, Australia

**Kevin Seaton** Department of Agriculture and Food Western Australia, South Perth, WA, Australia

**D. Simpson** East Malling Research, East Malling, Kent, UK

**Jason Smith** National Wine and Grape Industry Centre, Wagga Wagga, NSW, Australia

**Christopher Steel** National Wine and Grape Industry Centre, Wagga Wagga, NSW, Australia

**Jane Stoneham** Sensory Trust, Cornwall, UK

**Josef Tanny** Institute of Soil, Water & Environmental Sciences, Agricultural Research Organization, Bet Dagan, Israel

**Leon A. Terry** Plant Science Laboratory, Vincent Building, Cranfield University, Bedfordshire, UK

**Peter M. A. Toivonen** Agriculture and Agri-Food Canada, Pacific Agri-Food Research Centre, Summerland, Canada

**Jaap M. van Tuyl** Wageningen University and Research Center, Plant Breeding, Wageningen, The Netherlands

**Anna Maria Vettraino** DIBAF, University of Tuscia-Viterbo, Viterbo, Italy

**Jack G. A. J. van der Vorst** Operations Research and Logistics Group, Wageningen University, Wageningen, The Netherlands

**Henry Wainwright** The Real IPM Company (K) Ltd, Thika, Kenya

**Ian Warrington** Massey University, Palmerston North, New Zealand

**Chris C. Williams** Department of Resource Management and Geography, Melbourne School of Land and Environment, University of Melbourne, Richmond, VIC, Australia

# Chapter 27
# Horticulture and Society

**Tony Kendle and Jane Stoneham**

**Abstract** The interaction between society and horticulture is explored in this chapter. People relate to plants and gardens at different levels of perception and intensity. This can be at individual or group levels. In both instances plants have positive influences on human activity and consciousness. Studies of individual associations with plants demonstrate psychological and physical health benefits of consistent and substantial value. At the community level, plants may act as bridges between people as individuals and as groups. Association with plants can provide bridges between diverse groups in a community bringing cohesion and shared benefits. Each of these facets is discussed and illustrated in this chapter and draws on experience and evidence derived from the Eden Project and Sensory Trust in Great Britain.

**Keywords** Community · Individual · Benefits · Psychology · Health · Welfare · Parks · Gardens · Landscapes

## Introduction

Social horticulture research and theory often references the patchy but growing wealth of evidence that 'contact with nature' brings benefits to personal health and wellbeing (e.g. Kaplan 1992, Rohde and Kendle 1994, Ulrich 1983, Wilson 1984). This research usefully demonstrates an underlying need for human connection with the natural world but its focus on passive reception of nature-based experiences fails to encompass the benefits that come from a more active interaction between people and the natural world.

J. Stoneham (✉)
Sensory Trust, Watering Lane Nursery, Pentewan, St Austell,
PL26 6BE Cornwall, UK
e-mail: jstoneham@sensorytrust.org.uk

T. Kendle
Eden Project, Bodelva, St. Austell,
PL24 2SG Cornwall, UK
e-mail: tkendle@edenproject.com

G. R. Dixon, D. E. Aldous (eds.), *Horticulture: Plants for People and Places, Volume 3*,
DOI 10.1007/978-94-017-8560-0_27, © Springer Science+Business Media Dordrecht 2014

Horticulture centres on an active engagement. Cultivation is an act of human intervention and the role of the person in the process of engaging with nature is as critical to what emerges as the response of the natural components. More fundamentally it is grounded on a philosophy that accepts and encourages such an intervention. Horticulturists work with elements of nature, but those elements can be manipulated and shaped to suit human needs (Kendle and Forbes 1997).

The interplay between horticulture and society has evolved into a rich mix of activity and experience. In private gardens, public parks, flower festivals, allotments, school and hospital gardens people experience the social evidence of horticultural activity in their daily lives, whether or not they realise it and regardless of their level of participation.

In this chapter we will explore how the work of horticulturists contributes to social health and wellbeing. To do this we need to reflect on the two key words in the title of this chapter.

## The Essence of Horticulture

The root meaning of 'horticulture' is very different from 'nature'. It is about humans actively engaging with, and influencing, how plants grow. Horticulturists work with plants and other natural organisms. Horticulturists can facilitate contact with nature, and create environments where nature can flourish, where human contact with plants may be easier and safer to access, and more focused on what people need from a given place. But the art of horticulture is not passive and it is not about leaving nature alone (Kendle and Forbes 1997).

Much wellbeing research focuses on the innate values of nature as a healing or beneficial setting e.g. the theories of biophilia (Kellert 1993; Wilson 1984). Some researchers have attempted to identify which natural elements appear to trigger the strongest responses and there is evidence that we sometimes need, or at least value, experiences of being completely away from the human world, experiencing an environment where we are lost and absorbed in a truly 'natural' or wilderness setting (Kaplan and Kaplan 1989; Özgüner and Kendle 2006). Horticulturists cannot provide that wilderness experience for people, or at best they can fake it. Any environment created by a horticulturist has the handprint and intent of humankind woven through it (Fig. 27.1).

There is also a body of research exploring the value of horticulture as a therapeutic intervention (Brown and Jameton 2000; Lewis 1996; Relf 1992; Sempik et al. 2003). In these cases even when evidence of benefit is clear it is hard, inevitably, to know to what degree the benefits are a result of the contact with nature/plants or a result of the work of the human mediators and therapists. Is horticulture therapy innately different from occupational therapy, art therapy or music therapy because of the nature component? It is hard to believe that the moment of seeing the fruition from seeds you germinated isn't in some way special and inspiring and significant for mental health, but we have little evidence.

**Fig. 27.1** Horticulturists can create the feel of getting away from day to day life but not losing the sense of being in a man made place

This is an under-researched area. Certainly the research that has been carried out by environmental psychologists on the importance of human-nature contact rarely addresses how the effects of contact with 'gardened' nature and 'wild' nature are different. We know anecdotally that some individuals have strong affinities to wilderness and don't like gardens, and vice versa, but we can't say much more. Research is confounded by many problems, such as scale of focus—a single daisy growing in a crack in the pavement is arguably as natural in itself as one growing wild on a mountain scree.

If therapeutic benefits from contact with nature differ in essence from ones that are firmly grounded in the human world, then it may be because of the value of contact with something 'other than us', something greater and more ineffable. Many people describe their moments of high creativity as being fuelled by contact with something greater than themselves. It may be that something similar holds true for other creative therapies such as art therapy. But horticulture also allows that contact on a certain level and provides us also with agency and responsibility—the opportunity to nurture something living (Matsuo 1995). Again this is an aspect requiring further study.

If we do not have a clear understanding of these issues, one thing is for sure—we keep returning to the key point that horticulture is an active intervention not a passive one. It is not just about a human-nature contact, it is about human-nature manipulation—in some ways maybe a partnership, in many ways an attempt at control. There is a deep philosophical point here that is worthy of exploration. The very concept of 'natural' often refers to situations where human presence or human action is not present, or at least not dominant. The most important conservation areas in the world are often identified and mapped by the absence of human impact.

The reality of the world is complex though, and this definition of 'natural' is rife with paradox. Many animals dominate and impact on their environment in profound ways—e.g. huge herds of grazing animals perpetuate the grasslands they depend upon. These areas can still be seen as natural, so why is human impact different? Many areas traditionally seen as wilderness in reality have strong human footprints where indigenous human groups have lived in some sort of relatively balanced existence within the landscape and when they disappear the landscape begins to change. Conservationists have found themselves having to intervene to maintain ecosystems in a 'natural state' e.g. through fire management programmes to control invasive species.

Making sense of this picture requires a broader perspective and we need to question why such definitions are important. The idea that any one area is more or less 'natural' is actually a human cultural construct—the ultimate paradox—so why is this idea valuable to us? There are different reasons that are often conflated but it is worth teasing them apart.

First is that landscapes not impacted by people are rare and shrinking and many species do not cope well with human contact. Second, as mentioned above, leaving the human world, even if it is only through film or photography, is clearly important to our wellbeing and we need to protect wild spaces even if only through our own self interest.

However much more difficult is the principle embodied in much conservation policy, nature writing etc., that human impact is *de facto* and in its essence some form of degradation that makes the world worse than it was without human intervention (Kendle and Rose 2000; Hlubik and Betros 1994). This is based on doctrine more than objective analysis and is a perception that ultimately leads to justifying actions such as the clearance of indigenous people from national parks so that the wilderness can be enhanced. It is also a self-fulfilling position that reinforces and maintains the distinction between people and nature that many people argue is the root of our negative influence on the world. It is a position that defines the greatest good as having minimal impact, not having agency.

The reason for this analysis is that it gives us a clearer perception of why, by some definitions, nothing that horticulturists do can be 'natural' and gardens are not 'nature'. Horticulturists act on the natural world and change it.

But is horticulture thereby another form of human degradation of nature, another example of a human footprint that should be minimised? Obviously there are many times when this is true—when we do the wrong things in the wrong places, where the energy, water, pollution, waste and even social impacts of horticultural activities are detrimental and need to change. But horticulture also crystallises a deeply important set of questions. Is it possible to have an active relationship with nature which is more like partnership than dominance? Is it possible for us to act with agency, and be part of the world rather than apart from it? Is it within the human gift to sometimes leave the world better than we found it?

These are arguably the greatest human and nature wellbeing issues of all. We will return to this theme later.

## The Essence of Society

The second keyword is 'society'. We want to make a distinction here from the idea of the human population i.e. a mass of people who obviously rely on horticulture for food and other services. 'Society' refers to how people come together and determine the shared rules for living, shared endeavour and shared culture. Citizens in a society do not see themselves solely as individuals but recognise that we exist in a set of relationships to each other and with a shared identity at some level.

## *Society as Relationships*

A healthy society also does not come about passively, it itself needs to be cultivated. This active process is reflected in the root meaning of associated words such as community. A community is not just a group of people who live near each other or share a common interest. The word community derives from the Latin roots Communos, together in gift (Esposito 2009). A community is not created by living near each other in housing estates, or even by sharing common interests such as bird watching, but by a set of relationships and interactions between us.

Why should the essence of those relationships be somehow embodied in the notion of gift giving? It is a deeply profound idea that encapsulates the very particular nature of relationships that strengthen human bonds (Hyde 2007). Gift economies are recognised as very important in many traditional societies. Gift giving is not the same as transaction (or shouldn't be). A gift that is given without expectation of an immediate return is an investment in the wellbeing of the other person or at least an acknowledgement that they are important to us. Groups of people that operate gift economies have a strong and explicit sense that they need each other, that their lives are lessened or made less secure if they don't have others around them

Gift economies are intimately linked with food harvest. Prior to the development of sophisticated storage or long distance distribution, gluts in production were often shared in the expectation that this investment in the well being of others in society would be reciprocated at a later date. Abundance was not only a time of physical security, but also a time when social bonds could be strengthened. It is interesting by comparison how much modern capitalism prizes scarcity over abundance. Sharing food is a fundamental part of human bonding, which is hardly surprising. Again the roots of words we use daily can reveal a great deal—'companion' is derived from words meaning 'people who share bread together'.

Today we live in such an individualistic world that we have lost a lot of the conscious or implicit understanding of what makes society function, and what conditions and experiences we need to foster in order to make sure that people learn to live together in mutual respect. For example we are much more likely to be referred to as a consumer than as a companion in daily discourse. When we live as collections of individual people rather than real communities undesirable things happen. Robert Putnam analysed many of these in his seminal book Bowling Alone demonstrating how civic life declines when people fail to form associations (Putnam 2000), The point is very clear—society can disappear even when the people are all still there.

But even at a more mundane level a huge set of problems arise when people become isolated—they have no one to help with the shopping or may even find they have to call ambulances when they have minor ailments such as twisted ankles because they know no one who can help, consumption rises as everyone buys duplicates of everything rather than sharing, ungrounded fear of crime spirals and people become afraid to go out or talk to others.

Of course we have access to various new forms of community today such as through social networking etc and people may appear isolated but have huge networks of friends across the globe. But these are communities of interest or friendship that function because of the commonalities between the members. Crucially maintaining a healthy society also needs us to maintain relationships with the people we don't necessarily like, whose interests we don't necessarily share, whose opinions we don't necessarily agree with—and with the people who we may one day depend upon, or already do, but who we just don't really know.

Again as our culture has become more individualistic we have lost a sense of what element in our lives, our environments, and our experiences bring us together in a way that cultivates society (Leyden 2003). Ray Oldenburg identified one critical strand when he devised the concept of 'third places' (Oldenburg 1999). These are the physical locations where people come together to mix that are neither home nor work. They can include the civic spaces such as parks and playgrounds, but also cafes and pubs and some types of shops—anywhere where people bump into each other, chat, get to hear other worldviews. They are where we get to be familiar with the fact that nearly all strangers are well meaning. They are where new friends may be made, but they are also the places where mutual trust and empathy for others who are 'not like you' are fostered. They are ultimately places that help keep our sense of social justice and democracy alive.

This relates to horticulture of course because we are often the custodians and facilitators of much of that 'third place' infrastructure. Oldenburg analyses the decline in third place provision in the USA, but it is not hard to see the same thing happening in the UK. We build housing estates not living estates, places without civic spaces, community centres, pubs, corner shops or anywhere to get together. People are expected to meet all of their needs within their own homes, or by driving to shopping malls or flying somewhere else on holiday.

To defend against this trend, horticulturists need to become more literate and articulate about why society needs the things we do. We need to remember that there were powerful reasons why public parks were first created that were far more fundamental than 'leisure provision'. Many of them were created to promote a healthier and more coherent and civilised society. The health benefits associated with parks have a growing research base (Maller et al. 2002). Parks are not simply things we can substitute with a home exercise machine and a movie—they are part of the machinery of a healthy society that we dismantle at our jeopardy.

But in order to take that position with credibility we also need to challenge ourselves and question whether, and how, we really create spaces that function as they should. Not all gardens could or should be primarily social spaces—and that is certainly not the reason why they are all created. But we need to be very clear when this is a key role for a particular place, and understand the approaches, tools and measures that ensure that role is delivered.

We could look on this as a form of horticulture therapy that is focused not so much on the wellbeing of people as individuals, but on their wellbeing as a community and a society. But as with any therapy we need a framework that allows us to judge whether an intervention is successful.

## *Society as Identity*

Another critical element of society is that there needs to be some common sense of identity and culture that allows the participants to understand that they live within a bounded set that is distinguishable in some ways from other manifestations of human life and human possibility.

Of course this is a hugely complex set of issues that determine cultural identity that are never fully understood, stable or uncontested. A society will be made up of many different interpretations of this identity, and contain many people who don't feel they belong, challenge and evolve the boundaries. Identity changes also through time so the society inhabited by previous generations may be as much a 'foreign country' as a distant land is.

It is impossible to fully understand the culture we live in as we are so close to it that we will never get a clear view. But inevitably the gardens we create manifest and reflect our culture.

The timeframes that they develop in, however, and the importance of the heritage values they represent, do tend to mean that we protect the cultural visions of the past with more focused effort than we sometimes allow for new cultural manifestations.

Inevitably interwoven into this is the question of who gets to decide what happens with land. For the most part recognised 'heritage' gardens are survivors, remnants not just of the ravages of time but also of the cultural and political filters of what we find interesting, worthy of note and worthy of protection and on-going care. They are nearly always manifestations of grand investment, if not necessarily great design, but they also reflect the cultural elements of the past that we see as important.

Contemporary gardens are much more eclectic places, where people do very different things, where different possibilities play out, where people explore some of the dimensions of what society means to them. Very few of these will become heritage properties. Some of them don't have strong designs, they are places where different activities take place and would be meaningless to preserve if the activity stopped. But they are places that reflect our culture in myriad form and, more subtly, are where we maintain an exploration about what that culture is and could be.

The critical point here is that this exploration cannot happen unless access to land and the right to create gardens in different forms and manifestations is democratised. It is really important that different groups have a chance to participate, have agency and be creative in the creation of public landscapes.

## Practical Implications

Emerging from the discussions above we have a chance to identify what elements and functions we would expect to see happening in public gardens, to deliver the best contribution to society.

Firstly if society is all about relationships and common action then the core issue is that the environments that we create need to be inclusive and to encourage coming together, and also inspiring enough to ensure that people do come. Green space is not public space unless there are members of the public in it.

There are obvious design issues here that include the need to address the barriers that different people may encounter in accessing public gardens.

Can people get to and around these spaces and are they available to all members to society? The Making Connections survey (Price and Stoneham 2001, Stoneham and Thoday 1994) and the Diversity Review (Countryside Agency 2005) highlighted the extent to which our public greenspaces are only available for some segments of the population, and some groups (disabled people, older people, people with chronic health issues) are impacted by a range of barriers to access.

Inclusive design puts a focus on designing to include the widest range of people. The Access Chain is a methodology created by the Sensory Trust to identify and overcome these barriers by seeing a visit through the perspective of a visitor. The Social Sustainability Toolkit: inclusive design was developed by the Sensory Trust and Eden Project to integrate an inclusive approach within the wider framework of sustainable design (Stoneham and Thompson 2011).

But access alone does not make a successful public space. The feel of the place, and the range of experiences on offer, is critical in determining how successful an outdoor space will be in attracting people, how long they are likely to stay and how they will choose to spend their time when they are there (Stoneham 2004).

Jane Jacobs of the most influential writers to put the focus on a sense of place, and how the makeup of a space impacts on the way it feels and how people behave (Jacobs 1992). Along with Oscar Newman (Newman 1973) and William Whyte (Whyte 1980) she highlighted that one of the simplest ways of improving the safety of a public space was to have people in it—a place that is well used feels safe.

One of the key ingredients is seating (Marcus and Francis 1997). This is one of the least expensive and easiest to provide, and yet one of the most consistently overlooked aspects of park design. The presence of seats is important on two levels. One is to provide resting points for people who need them, particularly older people and those with limited stamina. The second is to provide the opportunity for dwell time—moments when people are not on the move, when they have the chance to stop and absorb the place, time when they can be private or can socialise with others. This use of seating is given less attention, but it has a crucial influence on the overall behaviour of people in a public space.

The design and arrangement of seats is important—park benches work well for couples but they do not lend themselves to groups who want to socialise, people need different seats and seating arrangements for that. The more inventive parks

**Fig. 27.2** A sunny day in Bryant Park is an excellent opportunity to see how effectively a park can provide for the diverse ways that people want to use the space

(e.g. Bryant Park in New York) have understood this well and provide a diversity of seats—moveable ones that people can arrange as they wish, benches, picnic style tables with chess tables, circular ones… the list is endless (Fig. 27.2).

In more natural settings there is often a tension between a desire to have people in the space, but without visual signs of them being there. The case for giving more attention to furnishings that enhance the visitor experience is not helped by examples that have littered their sites with furniture that is out of keeping with the landscape character and largely destroys the feel of the place. Furniture that is sensitively designed and located is surely the answer to ensuring these spaces can be used by the full range of people and offering experiences that include the opportunity to pause and absorb

Sharing food is one of the most simple and effective ways of inspiring people to interact with others and some research has recognised the value of food in this way (Morrison 1996). There is a trend for more urban parks to provide cafes and restaurants, recognising the attraction of drinking and eating in green surrounds. However, it is rare to find opportunities for more active participation and sharing of food.

The opportunity to engage in events, performance, exhibitions, celebrations is another important ingredient of a successful social space. It is one that connects to the earliest use of flowers, plants and green spaces.

Provision for sport and play has been headlined in recent years as part of the leisure and recreational role of green spaces, but in the main our green spaces provide for a relatively limited range of pursuits. Football pitches are important for the people who engage, and biking adds another popular dimension but these do not come near to providing for the full range of interests across the ages, genders and abilities. There are positive trends emerging, like the installation of outdoor table tennis tables, chess tables and other recreational facilities.

Ultimately though some of the real social values of horticulture can only be realised by giving members of that society an opportunity to participate in growing and garden making. Private gardens are of course important in this regard, but many people don't have them and in new developments the gardens that are provided are becoming tiny. But more fundamentally gardens that create society are gardens that inspire shared conversations. They need to be places where people meet on some sort of shared endeavour, and ideally visible so that other people stop and ask questions and talk.

The growth in community gardens and food growing projects has helped enormously in this regard (Draper and Freedman 2010; Hynes 1996; Lawson 2005; Lewis 1996). Not all food growing spaces are social. Allotments in the UK are typically very rule bound and defined very clearly as parcels of private space. The most interesting examples of social horticulture come from gardens where people grow collectively, producing food for shared use or sale, as this inspires conversation, collaboration and sharing. The value of such approaches has been recognised in the development of civic and urban agriculture (DeLind 2002).

Community gardens are not always about food growing, some are play spaces, ornamental gardens, recreational spaces (Payne and Fryman 2001). In many cases though they have inherited habits of design from public space that don't always encourage social and community strengthening—e.g. benches are isolated and there are no shared eating spaces. The essence of what makes a community garden is the social interaction and, ideally, mutual support between the gardeners that helps to create relationships.

## Conclusions

There is no doubt that public horticulture is in crisis in much of the world. Global economic recession and austerity drives by government have turned what has been a chronic pressure to reduce funds into an acute one (Harding 1999). To counter this trend and to save what is left we need stronger arguments that in turn can only come from a much stronger insights into the value that green spaces and parks really do provide.

This isn't a call for more research studies on environmental wellbeing so much as a call for a more robust narrative of the political, social and cultural importance of what we do and the use of clearer language to express it to others. No matter how much evidence there is that contact with nature does us good, and no matter how much evidence there is of the economic value that nature provides, our society is clearly ambivalent about public greenspace.

Many of us value holidays in natural spaces, many of us put enormous time, and resources, into growing our own gardens. And yet the commitment to funding parks and other community facilities flickers in and out (Harding 1999). Despite attempts to introduce and enforce minimum standards we still build houses and workplaces without meaningful access to green spaces—mainly because no one wants to pick up the maintenance bill.

Humans are social and cultural beings. We sometimes appreciate escaping from the trappings and business of our culture, to leave our common world behind in wilder places, but we return to our own societies as these are the fundamental setting for our lives. It is when horticulture aligns strongly with a social or cultural need, that people will respond and value what is created the most.

But more fundamental than that is the role of gardens as places where society is created and re-affirmed. Without shared land that we can all access together and meet each other we risk an even greater drift to a becoming a population of consumers rather than a society of citizens and companions.

Gardens also represent an interface between humans and nature. They are not, by most definitions, natural spaces they only exist because of human presence and agency. But that need not mean that they are in some way poorer for that. We have developed a strong sense and philosophy in our current culture that nature is something other than us, and it is best served by leaving it alone. If we could develop a parallel understanding that nature and people can sometimes work in beneficial partnership, and that humans can sometimes leave the world better than we found it, then the implications are profound. Gardens are where that partnership can be forged.

## References

Brown KH, Jameton AL (2000) Public health implications of urban agriculture. J Pub Health Policy 21:20–39

Countryside Agency (2005) Diversity review. Countryside Agency, Cheltenham

DeLind LB (2002) Place, work, and civic agriculture: common fields for cultivation. Agric Hum Values 19(3):217–224

Draper C, Freedman D (2010) Review and analysis of the benefits, purposes, and motivations associated with community gardening in the United States. J Community Pract 18(4):458–492

Esposito R (2009) Communitas: the origin and destiny of community. Stanford University Press, Stanford

Harding S (1999) Towards a renaissance in urban parks. Cult Trends 9(35):1–20

Hlubik WT, Betros H (1994) Nurturing people-plant relationships in order to foster environmental and community stewardship: the Rutgers Environmental and Community Stewardship (R.E.A.C.S.) program. In: Flagler J, Poincelot RP (eds) People-plant relationships: setting research priorities. Food Products Press, New York, pp 373–381

Hyde L (2007) The gift: how the creative spirit transforms the world. Canongate Books, Edinburgh

Hynes HP (1996) A patch of Eden, America's inner-city gardeners. Chelsea Green Publishing Company, White River Junction

Jacobs J (1992) The death and life of great American cities. Knopf Doubleday Publishing Group, New York

Kaplan S (1992) The restorative environment: nature and human experience. In: Relf D (ed) The role of horticulture in human well-being and social development: a national symposium. Timber Press, Portland

Kaplan R, Kaplan S (1989) The experience of nature: a psychological perspective. Cambridge University, Cambridge

Kellert SR (ed) (1993) The biophilia hypothesis. Island, Washington DC

Kendle AD, Forbes S (1997) Urban nature conservation: landscape management in the urban countryside. E&FN Spon, London

Kendle AD, Rose JE (2000) The aliens have landed! What are the justifications for 'native only' policies in landscape plantings? Landscape and urban planning, 47(1):19–31

Lawson LJ (2005) City bountiful, a century of community gardening in America. University of California Press, Berkeley
Leyden KM (2003) Social capital and the built environment: the importance of walkable neighborhoods. J Inf 93(9):1546–1551
Lewis CA (1996) Green nature/human nature: the meaning of plants in our lives. University of Illinois, Chicago
Maller C, Townsend M, Brown P, St. Leger L (2002) Healthy parks, healthy people: the health benefits of contact with nature in a park context. Deakin University, Melbourne
Marcus CC, Francis C (eds) (1997) People places: design guidlines for urban open space. Wiley, Hoboken
Matsuo E (1995) Horticulture helps us to live as human beings: providing balance and harmony in our behaviour and though and life worth living. ISHS Acta Hortic. 391:19–30. (ISHS)
Morrison M (1996) Sharing food at home and school: perspectives on commensality. Sociol Rev 44(4):648–674
Newman O (1973) Defensible space; crime prevention through urban design. Macmillan Publishing Company, New York
Oldenburg R (1999) The great good place. cafes, coffee shops, bookstores, bars, hair salons, and other hangouts at the heart of a community. Marlowe & Company, Emeryville
Özgüner H, Kendle AD (2006) Public attitudes towards naturalistic versus designed landscapes in the city of Sheffield (UK). Landsc Urban Plan, 74(2):139–157
Payne K, Fryman D (2001) Cultivating community: principles and practices for community gardening as a community-building tool, American Community Gardening Association, Columbus, Ohio
Price R, Stoneham J (2001) Making connections: a guide to accessible greenspace. Sensory Trust, St Austell
Putnam RD (2000) Bowling alone. Simon and Schuster Paperbacks, New York
Relf D (ed) (1992) The role of horticulture inhuman well-being and social development: a national symposium. Timber Press, Portland
Rohde CLE, Kendle AD (1994) Report to english nature-human well-being, natural landscapes and wildlife in urban areas. A review. english nature, Peterborough
Sempik J, Aldridge J, Becker S (2003) Social and therapeutic horticulture: evidence and messages from research. Thrive and Centre for Child and Family Research, Loughborough University
Stoneham JA (2004) Eden Project, a living theatre of people and plants-inclusive approaches to public communication and involvement. Acta Hortic (ISHS) 643:189–194
Stoneham JA, Thoday PR (1994) Landscape design for elderly and disabled people. Garden Art Press, Woodbridge
Stoneham JA, Thompson C (2011) Social sustainability toolkit: inclusive design. Sensory Trust, Cornwall
Ulrich RS (1983) Aesthetic and affective response to natural environment. In: Altman I, Wohlwill JF (eds) Behavior and the natural environment, vol 6. Plenum, New York. pp 85–127
Whyte W (1980) The social life of small urban spaces. Project for Public Spaces, New York
Wilson EO (1984) Biophilia. Harvard University, Cambridge

# Chapter 28
# Fruit and Vegetables and Health: An Overview

**Yves Desjardins**

**Abstract** A growing body of evidences suggests that the regular consumption of a diet rich in fruit and vegetables (FAV) reduces the risk of chronic human illnesses and increase lifespan and quality of life. FAV are considered energy poor, are rich sources of minerals, fibers, vitamins and most of all of many phytochemicals belonging to four main classes: polyphenols, terpenoids, sulphur compounds and alkaloids. Polyphenols, and to a certain extent carotenoids and sulphur containing compounds have been shown through epidemiological cohort studies or through mechanistic *in vitro* or animal studies, to prevent coronary heart diseases, chronic inflammatory diseases, obesity, diabetes, neurodegenerative diseases, cancer, macular degeneration, and many others. Owing to their particular chemical structure, theses phytochemicals display strong antioxidant capacity in vitro. Yet due to their poor bioavailability and their short residence time in the organism, it is more and more admitted that these molecules trigger detoxification mechanisms in the body and induce genes associated with energy metabolism, anti-inflammation and endogenous-antioxidant network at the cellular level.

This chapter describes the different phytochemicals found in FAV with emphasis on polyphenols, the most important class of compounds in relation to health benefits and amounts ingested on a daily basis in our diet. The contribution of these chemicals to the prevention of chronic diseases is covered and new insights on their possible mode of action are discussed. The scope of this chapter is broad and intends to brush an overview of this very complex and dynamic field of research, at the interface between plant and human physiology. The reader is guided and often referred to bibliographic reviews on topics as diverse and eclectic as phytochemicals biosynthesis, bioavailability, inflammatory responses, cancer etiology, appetite control, insulin resistance, and cognition.

**Keywords** Fruits · Vegetables · Health · Polyphenols · Carotenoids · Sulphur compounds · Cancer · Coronary heart disease · Obesity · Diabetes · Neurodegenerative disease · Bioavailability

---

Y. Desjardins (✉)
Institute of Nutrition and Functional Foods/Horticulture Research Center,
Laval University, Québec City, QC G1V 0A6, Canada
e-mail: yves.desjardins@fsaa.ulaval.ca

G. R. Dixon, D. E. Aldous (eds.), *Horticulture: Plants for People and Places, Volume 3*,
DOI 10.1007/978-94-017-8560-0_28, © Springer Science+Business Media Dordrecht 2014

## Introduction

It is implicitly accepted that fruit and vegetables (FAV) are good for you. Actually, many nutritionists and clinicians now consider fruit and vegetables consumption as a solution to many "diseases of civilization". These horticultural products bring diversity and stimulate our senses by having organoleptic properties like color, flavor, and texture and contribute to our appetite. FAVs have long been recognized for their nutritive value. They are excellent sources of minerals, essential fatty acids and fibers, but are also unique sources of vitamins (C, E, B, and folic acid). Most of all, they are rich sources of bioactive phytochemicals. They are considered energy poor and contribute, through their high content in non-digestible fibers, to the feeling of satiety. For these reasons, the consumption of FAV and plants in general are at the base of most for food pyramid (Anon 2011). For some, FAV are the most important component of the diet and contribute to a healthy living (Hung et al. 2004). For example, the Mediterranean diet, which is reputed for its quality and is largely composed of fish, alpha-linoleic acid and FAV has been associated with the low incidence of cardiovascular disease of population living in the Mediterranean bassin (de Lorgeril et al. 1994). Other diets, like the Portfolio diet, a reconstitution of the diet of our simian ancestors, is relying on the consumption of high levels of fibers, phytosterols, vegetables and nuts and has been show repetitively to confer significant cholesterol-lowering capacity and reduce the incidence of atherosclerosis (Kendall and Jenkins 2004). Indeed, a growing body of evidence suggests that the regular consumption of a phytochemical-rich diet reduces the risk of many chronic human illnesses and increases life span and quality in humans (Anon 2007; He et al. 2006).

New evidences support the fact that FAV are important in the prevention of cardiovascular (Van't Veer et al. 2000), vision (Snodderly 1995), bone (Baile et al. 2011), and pulmonary health (Trichopoulou et al. 2003). The World Health Organization has recognized this fact and is actively promoting the consumption of FAV to reduce the incidence of chronic disease (Anon 2007) and public health and growers organization in different countries, regrouped under the umbrella of IFAVA (Anon 2006) are actively promoting the consumption of FAV through the different 5 to 10 a day programs worldwide. There are many epidemiological studies linking the consumption of FAV and/or their constituents to beneficial health effects. For instance, many cohort and case-control epidemiological studies and even intervention studies show the beneficial effects of FAV. In general, an inverse association is found between FAV consumption and cardiovascular disease (SUVIMAX cohort study) (Bazzano 2006; Hercberg et al. 2004), chronic inflammatory diseases (Hermsdorff et al. 2010), diabetes (Bazzano 2005; Hamer and Chida 2007), obesity (Carlton Tohill 2005), neurodegenerative diseases (Cherniack 2012), and many more. However, the evidence for this effect is not as solid for cancer (World Cancer Research Fund 2007) and recently some doubts have been expressed on the link between FAV and coronary hearth disease prevention (Dauchet et al. 2009) since FAV consumption is often confounded with other general healthier life habits like non-smoking, reduced alcohol consumption, just to list a few.

The majority of the studies published over the last 20 years have focused on the identification and demonstration of the activity of bioactive compounds of FAV *in vitro*. They are mostly observational and results are often conflicting. The validity of *in vitro* studies is contested because they provide an incomplete and often biased image of the benefits of FAV to health. Other parameters must thus be considered since the responses of humans to the food they consume are complex and influenced by many confounding factors. Too many studies have not taken into account the poor bioavailability, the interactions between the phytonutrients and have often used supra-optimal non-physiological doses of bioactive compounds to demonstrate their effect and have thus lead to incorrect conclusions on their potential effects. Moreover, these effects have proved to be difficult to reproduce in human clinical trials. Taking into account these caveats, new hypothesis on the mode of action of phytonutrients lean toward a general anti-inflammatory and cell-signaling action.

This chapter is thus intended to briefly review the most pertinent scientific literature on the topic of health effects of FAV and highlights which components are responsible for disease protection and the most probable mechanisms by which they confer these effects. Many excellent reviews on the topic are also available (Crozier et al. 2006, 2009).

## FAV are Rich Sources of Nutrients

FAV are rich sources of minerals and vitamins in the diet. They provide large amount of phosphorus, potassium, calcium, magnesium, iron and zinc. They also contain unsaturated lipids and are a very rich source of vitamins and in particular vitamin-C (Table 28.1). Interestingly, they contain a high proportion of water, and have a high content in non-digestible fibers, which have been shown to reduce their energy density (Carlton Tohill 2005). Adding FAV to the diet reduces the overall energy density, increasing the amount of food that can be consumed for a given level of calories. Many comprehensive reviews have evaluated the effect of dietary fibre content on satiety, overall energy intake and body weight (Kim and Park 2011).

## Phytochemicals Found in FAV

FAV accumulate several hundred of thousands of so-called "secondary metabolites" to protect themselves from biotic stress like bacteria, fungi and insects (Kliebenstien 2004) and abiotic stress (Dixon and Paiva 1995). These chemicals are essential components of the adaptive arsenal of the plant to the environment and are involved in biotic and abiotic stress protection, cell signaling, plant development, pollinator attraction, plant-microorganism interaction, plant defense, herbivore repulsion and seed dispersion. These phytochemicals are regrouped into four broad classes according to their chemical structure: polyphenols, terpenoids, sulphur compounds, and alkaloids (Fig. 28.1 and Table 28.2).

**Table 28.1** Composition of typical fruit and vegetables. Values are per 100 g F.W. (Adapted from the USDA national nutrient database (Anon 2004, 2006)

| | Water % | Energy (Kcal) | Protein | Lipid | Fibers | Sugar | Minerals (mg) | | | | | | | Vitamins | | | | |
|---|---|---|---|---|---|---|---|---|---|---|---|---|---|---|---|---|---|---|
| | | | (mg) | | | | Ca | Fe | Mg | P | K | Na | Zn | C (mg) | B (mg) | E (mg) | A (µg) | Folate (µg) |
| *Fruit* | | | | | | | | | | | | | | | | | | |
| Apples | 85 | 52 | 0.3 | 0.2 | 2.4 | 13.8 | 6 | 0.1 | 5 | 11 | 107 | 1 | 0.05 | 4.6 | 0.2 | 0.2 | 3 | 3 |
| Avocado | 73 | 160 | 2 | 14.7 | 6.7 | 8.5 | 12 | 0.6 | 29 | 52 | 485 | 7 | 0.6 | 10 | 2.1 | 2 | 7 | 81 |
| Banana | 74 | 89 | 1 | 0.3 | 2.6 | 12.2 | 5 | 0.3 | 27 | 22 | 358 | 1 | 0.15 | 8.7 | 1 | 0.1 | 3 | 20 |
| Blackberry | 88 | 43 | 1.4 | 0.5 | 5.3 | 9.1 | 29 | 0.6 | 20 | 22 | 162 | 1 | 0.5 | 21 | 0.7 | 1.2 | 11 | 25 |
| Blueberry | 84 | 57 | 0.7 | 0.3 | 2.4 | 14.5 | 6 | 0.3 | 6 | 12 | 77 | 1 | 0.2 | 9.7 | 0.5 | 0.6 | 3 | 6 |
| Cranberry | 87 | 46 | 0.4 | 0.1 | 4.6 | 12.2 | 8 | 0.3 | 6 | 13 | 85 | 2 | 0.1 | 13 | 0.2 | 1.2 | 3 | 1 |
| Figs | 79 | 74 | 0.8 | 0.3 | 2.9 | 16.3 | 35 | 0.4 | 17 | 14 | 232 | 1 | 0.2 | 2 | 0.6 | 0.1 | 7 | 6 |
| Grapefruit | 88 | 42 | 0.8 | 0.1 | 1.6 | 10.6 | 22 | 0.1 | 9 | 18 | 135 | | 0.1 | 31 | 0.3 | 0.1 | 58 | 13 |
| Grapes | 81 | 67 | 0.6 | 0.4 | 0.9 | 17.2 | 14 | 0.3 | 5 | 10 | 191 | 2 | 0.04 | 4 | 0.5 | 0.2 | 5 | 4 |
| Melon | 90 | 34 | 0.8 | 0.2 | 0.9 | 8.6 | 9 | 0.2 | 12 | 15 | 267 | 16 | 0.2 | 36 | 0.8 | 0.1 | 169 | 21 |
| Oranges | 86 | 47 | 0.9 | 0.1 | 2.4 | 11.7 | 40 | 0.1 | 10 | 14 | 181 | | 0.1 | 53 | 0.5 | 0.2 | 11 | 30 |
| Pears | 84 | 57 | 0.4 | 0.1 | 3.1 | 9.8 | 9 | 0.2 | 7 | 12 | 116 | 1 | 0.1 | 4.3 | 0.2 | 0.1 | 1 | 7 |
| Raspberry | 85 | 52 | 1.2 | 0.7 | 6.5 | 11.9 | 25 | 0.7 | 22 | 29 | 151 | 1 | 0.4 | 26 | 0.7 | 0.9 | 2 | 21 |
| Strawberry | 91 | 32 | 0.7 | 0.3 | 2 | 7.7 | 15 | 0.4 | 13 | 24 | 153 | 1 | 0.1 | 59 | 0.5 | 0.3 | 1 | 24 |
| Vegetables | | | | | | | | | | | | | | | | | | |
| Artichoke | 85 | 47 | 3.3 | 0.2 | 5.4 | 1 | 44 | 1.3 | 60 | 90 | 370 | 94 | 0.5 | 12 | 1.2 | 0.2 | 14 | 68 |
| Asparagus | 93 | 20 | 2.9 | 0.2 | 2.8 | 2.5 | 32 | 2.9 | 19 | 70 | 271 | 3 | 0.7 | 7.5 | 1.7 | 1.5 | 51 | 70 |
| Beans green | 90 | 31 | 1.8 | 0.2 | 2.7 | 3.3 | 37 | 1.0 | 25 | 38 | 211 | 6 | 0.2 | 122 | 0.9 | 0.4 | 35 | 33 |
| Beets | 88 | 34 | 2.8 | 0.4 | 2.8 | 6.8 | 16 | 0.8 | 23 | 40 | 325 | 78 | 0.4 | 4.9 | 0.5 | 0.04 | 2 | 109 |
| Broccoli | 89 | 34 | 2.8 | 0.4 | 2.6 | 1.7 | 47 | 0.7 | 21 | 66 | 216 | 33 | 0.4 | 89 | 0.8 | 0.8 | 31 | 57 |
| Brussels sp. | 86 | 43 | 3.4 | 0.3 | 3.8 | 2.2 | 42 | 1.4 | 23 | 69 | 389 | 25 | 0.4 | 85 | 0.9 | 0.9 | 38 | 61 |
| Cabbage | 92 | 25 | 1.3 | 0.1 | 2.5 | 3.2 | 40 | 0.5 | 12 | 26 | 170 | 18 | 0.2 | 37 | 0.3 | 0.2 | 5 | 43 |
| Carrot | 88 | 41 | 0.9 | 0.2 | 2.8 | 4.7 | 33 | 0.3 | 12 | 35 | 320 | 69 | 0.2 | 6 | 1.1 | 0.7 | 835 | 19 |
| Cauliflower | 92 | 25 | 1.9 | 0.3 | 1.9 | 2.0 | 22 | 0.4 | 15 | 44 | 299 | 30 | 0.3 | 48 | 0.6 | 0.1 | 0 | 57 |
| Celery | 95 | 16 | 0.7 | 0.2 | 1.6 | 1.8 | 40 | 0.2 | 11 | 24 | 260 | 80 | 1.3 | 3.1 | 0.4 | 0.3 | 22 | 36 |
| Garlic | 59 | 149 | 6.4 | 0.5 | 2.1 | 1.0 | 181 | 1.7 | 25 | 153 | 401 | 17 | 1.2 | 31 | 1.0 | 0.1 | 0 | 3 |
| Lettuce | 95 | 17 | 1.2 | 0.3 | 2.1 | 1.2 | 33 | 1.0 | 14 | 30 | 247 | 8 | 0.2 | 4 | 0.4 | 0.1 | 436 | 136 |
| Onion | 89 | 40 | 1.1 | 0.1 | 1.7 | 4.2 | 23 | 0.2 | 10 | 29 | 146 | 4 | 0.2 | 7.4 | 0.2 | 0.02 | 0 | 19 |
| Pepper | 94 | 20 | 0.9 | 0.2 | 1.7 | 2.4 | 10 | 0.3 | 10 | 20 | 175 | 3 | 0.1 | 80 | 0.5 | 0.4 | 18 | 10 |
| Spinach | 91 | 23 | 2.9 | 0.4 | 2.2 | 0.4 | 99 | 2.7 | 79 | 49 | 558 | 79 | 0.5 | 28 | 1.0 | 2.0 | 469 | 194 |
| Tomato | 95 | 18 | 0.9 | 0.2 | 1.2 | 2.6 | 10 | 0.3 | 11 | 24 | 237 | 5 | 0.2 | 14 | 0.6 | 0.5 | 42 | 15 |

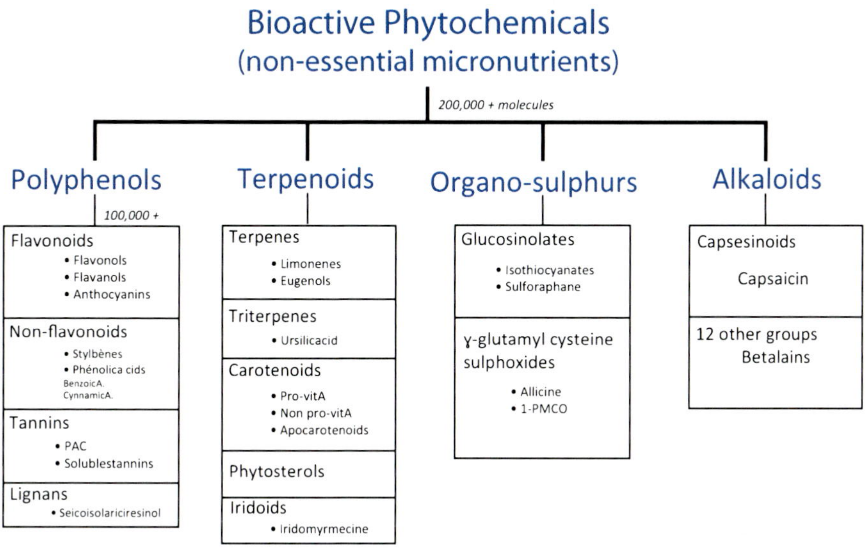


**Fig. 28.1** Different classes of bioactive phytochemicals found in FAV

**Table 28.2** Photochemicals found in fruit and vegetables

| Bioactive compound family | Primary source in fruits and vegetables | Database/source |
|---|---|---|
| *Terpenoids* | | |
| Carotenoids | Leafy vegetables, red and yellow fruits and vegetables | USDA nutrient database (Kimura and Rodriguez-Amaya 2003; Rodriguez-Amaya et al. 2008) |
| Monoterpenes | Citrus, cherries, mint and herbs | |
| Saponins | Alliaceae, asparagus | (Güçlü-Üstünda and Mazza 2007) |
| Apocarotenoids | Fruits | (Bouvier et al. 2005) |
| *Polyphenols* | | |
| Phenolic acids | Small fruits, apples, fruit and vegetables | (Rothwell et al. 2012) |
| Hydrolysable tannins | FAV, pomegrenate, raspberry | (Clifford and Scalbert 2000) |
| Stylbenes | Grapes, small fruits | (Waffo-Teguo et al. 2008) |
| Proanthocyanidins | FAV, cacao, small fruits, cranberry, blueberry | (Anon 2003) |
| Monophenolic alcohols tyrosol | Olive oil, wine | (Romero et al. 2002) |
| *Organo sulphur compounds* | | |
| Glucosinolates | Brassicaceae | (McNaughton and Marks 2003) |
| γ-Glutamyl cysteine sulphoxides | Alliaceae | (Griffiths et al. 2002) |
| *Alkaloids* | | |
| Capsaicin | Chili Pepper | (Surh and Sup Lee 1995) |
| Betalain | Red Beet, Prickly pear, Pittaya | (Stintzing and Carle 2007) |

## Polyphenols

The evolution of terrestrial plants has coincided and probably been rendered possible through the acquisition of the capacity for phenol biosynthesis from phenylalanine. Central to plant biology is the fact that these phenols polymerized to form lignin which provided mechanical support to plants, and through combined action of cutin and suberin, provided protection against desiccation and consequently the conquest of dry environments (Parr and Bolwell 2000). These molecules designated as polyphenols constitute a very heterogeneous group of molecules with almost 100,000 individual chemical species. This large number of known structures owe to the glycoside complexity of flavonoids, the variable stereochemistry of the molecules and their capacity to form polymers (Harborne 1977).

Polyphenols are characterized by the presence of one or more benzene ring bearing one or many hydroxyl groups. They can be very simple ring molecules of 6C, but can be much more complex structures with many functional groups or polymers (Table 28.3). Within this complex class, flavonoids are the most relevant to biology and food technology. The flavonoids are made of 15 C and are regrouped into ten classes based on the structure of the central heterocycle (Fig. 28.2) and their degree of oxidation. The most oxidized form corresponds to the anthocyanins, which confer color to fruits, while the reduced form correspond to flavan-3-ols, known for their astringency and health properties. The vast majority of polyphenols is water-soluble and is sequestered in vacuoles in glycosylated form, while some are lipophilic (flavone, flavonol methyl esters) and will thus dissolve in waxes and be encountered in the epidermis of plants. The glycosylated flavonoids need to be stripped of their sugar moiety before absorption by the gut epithelium. Polyphenols are generally nucleophilic on the basis of their oxygen atom in the heterocyclic pyrane C ring, also the presence of many double bounds in the aromatic rings and the presence of hydroxyl groups in ortho- and para- position on the A and B ring. They are thus strong antioxidants. They will often complex with metal ions and contribute to the vivid blue color of some flowers. The presence of metal ions also multiplies the antioxidant capacity. Due to the presence of many hydroxyl groups, most flavonoids will interact strongly with proteins (Dangles and Dufour 2006) and with enzymes (McDougall and Stewart 2005). They will contribute to the sensation of astringency by binding proline and tyrosine found in saliva and mouth epithelial proteins (Haslam and Lilley 1988).

## Food Sources of Polyphenols

FAV and beverage like wine are especially rich sources of polyphenols and in particular of flavonoids (Table 28.4). Their specific content depends on the species, the degree of maturity of the crop, the cultural management, the processing, the way they are cooked and stored. It is generally considered that the total flavonoid

**Table 28.3** Main classes of phenolic compounds found in fruit and vegetables. (Macheix et al. 1990)

| Carbon skeleton | Class | Type | Source |
|---|---|---|---|
| C6 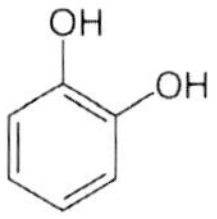 | Simple phenol | Catechol | Many species, degradation products |
| C6-C1 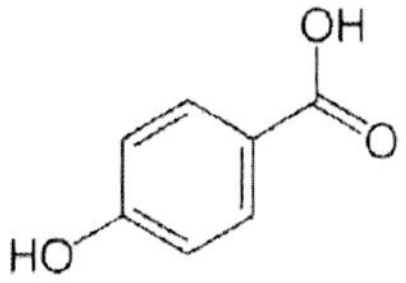 | Hydroxybenzoic acid | P-benzoic acid, gallic acid, protocatecuic acid | Spices, strawberry, raspberry, blackberry |
| C6-C3 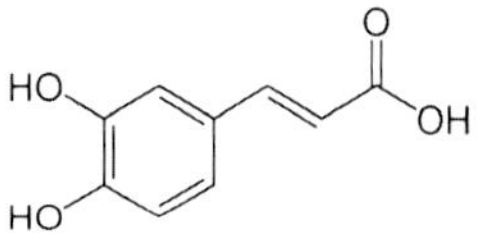 | Hydroxycinnamic acid, coumaric acid | Caffeic acid | Apples, citrus, potatoes, coffee (green), blueberry, spinach, |
| C6-C4 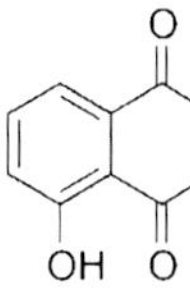 | Naphtoquinones | Juglone | Nuts |
| C6-C2-C6 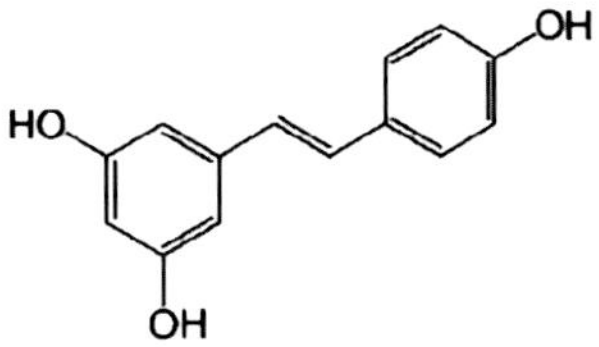 | Stylbenes | Resveratrol, viniferine | Grapes, wine |
| C6-C3-C6 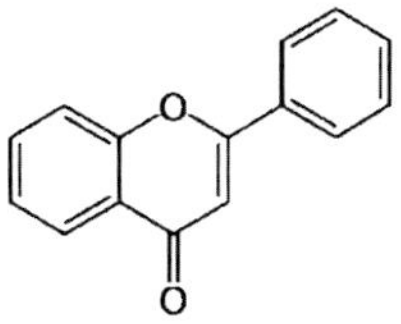 | Flavonoids | Quercetine, anthocyanin | Fruits, onion |
| $(C6\text{-}C3)_2$ 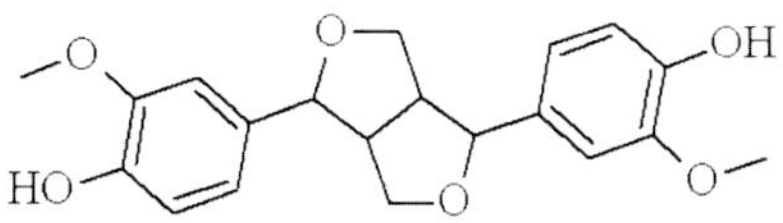 | Lignans | Pinoresinol, seicoisolariciresinol | Pine, kale, broccoli, apricot, strawberry |

**Table 28.3** (continued)

| Carbon skeleton | Class | Type | Source |
|---|---|---|---|
| $(C6\text{-}C3)_n$ | Lignin | | Stone fruits |
| $(C6\text{-}C3\text{-}C6)_n$ 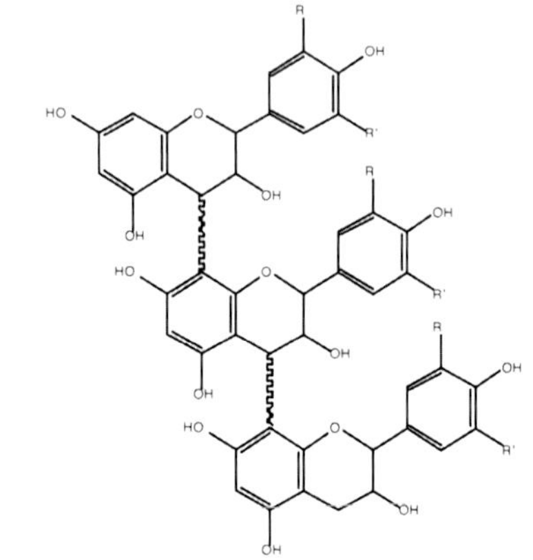 | Condensed tannins | Proanthocyanidins | Most fruits, cranberry, persimmon, nuts, chocolate |
| 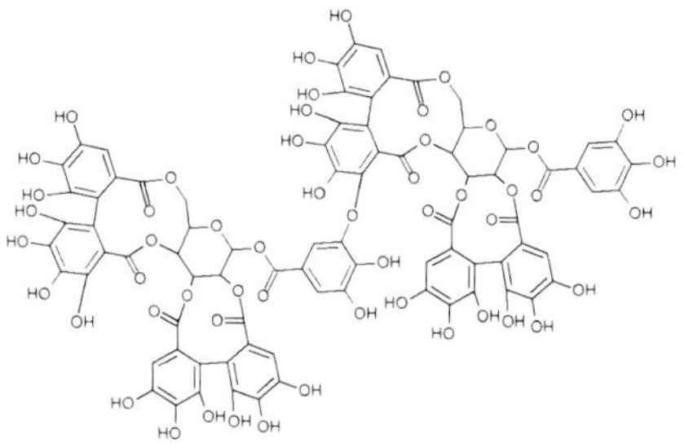 | Hydrosoluble tannins | Ellagitannins, Sanguiin H10 | Strawberry, rubus, pomegranate, nuts |

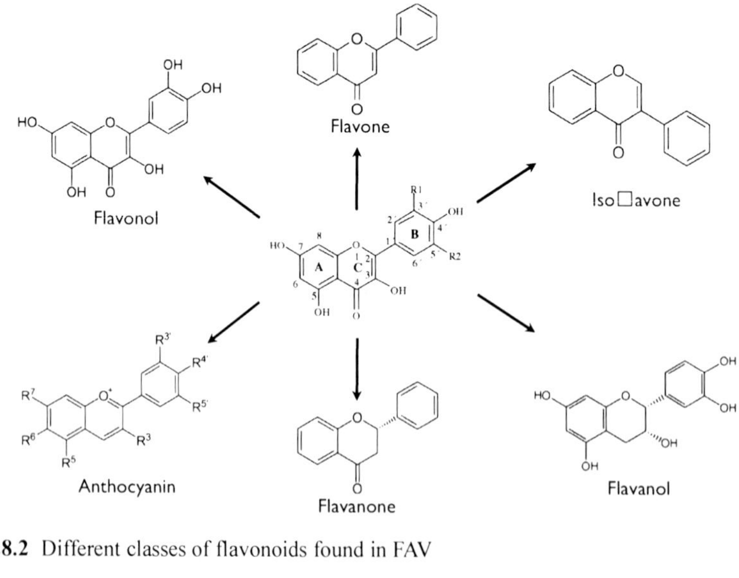


**Fig. 28.2** Different classes of flavonoids found in FAV

**Table 28.4** Sources of flavonoids in fruit and vegetables

| Subclass | Compounds | Primary source |
|---|---|---|
| Flavonols | Quercetin, myricetin, kaempferol, rutin, isorhamnetin | Onion, apples, cranberry, broccoli, berries, olives, bananas, lettuce, plums, grapes, wine |
| Favanones | Hesperin, hesperidin, naringin, naringenin, eriodictol | Citrus |
| Flavan-3-ols | Catechin, epicatechin, galloylated derivatives | Tea, plums, apple, cranberries, berries, chocolate |
| Flavones | Luteolin, apigenin | Apples, Apiaceae, celery, sweet red pepper, parsley, oregano, lettuce, beet |
| Anthocyanins | Cyanidin, delphinidin, pelargonidin, malvidin, peonidin, petunidin (mostly as glycosides) | Berries, red fruits, red cabbage, eggplant |

intake of occidentals is about 1 g/d (Kuhnau 1976; Scalbert and Williamson 2000). According to Brat et al. (2006), FAV consumption in France accounts for about 30 % of the overall daily polyphenol intake, which reach about 287 mg GAE/d. Other sources of polyphenols in the diet come from beverages like tea, coffee and wine but also from cereals. Moreover, humans consume a significant proportion of their polyphenols in a polymeric form as proanthocyanidins (PAC), which are often neglected (Saura-Calixto 2012). It is believed that oligomeric and polymeric forms of PAC are not absorbed by the enterocyte (Deprez et al. 2001) and are thus broken down by the gut bacteria where they provide prebiotic benefits (Williamson and Clifford 2010). The PAC are found in large quantities in small fruits like blueberry, cranberry and strawberry and are also abundant in nuts and especially hazelnuts, pecan, pistachio and almonds where their concentration can reach 500, 494, 237 and 184 mg/100 g F.W. respectively (Anon 2003).

## Sulphur Compounds

### *Glucosinolates*

Glucosinolates are amino-derived secondary plant metabolites containing a β-thioglucosyl moiety linked to an α-carbon forming a sulphated ketoxime (Fig. 28.3). They are found in the family of *Brassicacea* and are involved in plant/insects-pathogens interactions, and in plant development. The glucosinolate molecule is not involved as such in the biotic interactions but requires an hydrolysis catalyzed by a β-thioglucosidase, also called myrosinase to release the toxic isothiocyanate molecules. More than 120 glucosinolates have been identified in different species (Rosa et al. 1997) (Table 28.5).

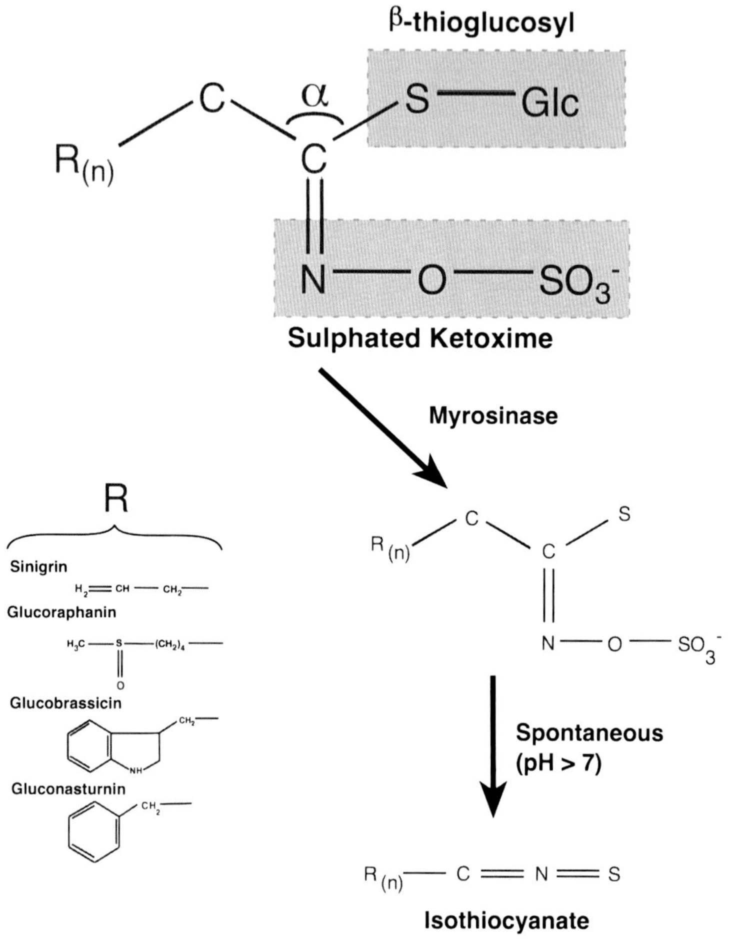


**Fig. 28.3** Chemical structure of different common glucosinolates and ensuing thiocyanate

Glucosinolates levels in plants are largely determined by their genetic make up (Rosa et al. 1997) but they are also influenced by abiotic factors like nitrogen, sulphur, or potassium supply (Verkerk et al. 2009). However, genes involved in glucosinolate biosynthesis are also induced by herbivore and pathogen attacks (Agrawal and Kurashige 2003; Brader et al. 2001). Jasmonate and salicylate involved in wounding and herbivory signal transduction increases glucosinolate concentration (Doughty et al. 1995; Kiddle et al. 1994).

Plants have developed efficient defenses against herbivores and pathogens whereby glucosinolate are transformed into isothiocyanate when placed in presence of a thioglucosidase also known as myrosinase (Fig. 28.4). Under normal conditions, the precursor molecule and enzyme are compartmentalized in different tissues; glucosinolates are scattered in vacuoles of most organs while the glucosidase occur only in specific cells called myrosin cells, scattered throughout the plant

**Table 28.5** Type of glucosinolate accumulated in different Brassica species, and their related thiocyanate. (Adapted from Verkerk et al. (2009)

| Species | Glucosinolates | Average range (mg/100 g F.W.) |
|---|---|---|
| Cabbage | Sinigrin | 21 |
| *Brassica oleracea var. capitata* | Indole glucosinolate | 30 |
| *F. alba* | Glucoiberin | |
| Radish | Dihydroerucin | 168 |
| *Raphanus sativus var. sativus* | Gluraphenin | 7 |
| | Indole glucosinolate | 6 |
| Mustard | Sinigrin | 330 |
| *Brassica juncea* | Gluconapin | 96 |
| | Indole glucosinolate | 44 |
| Rocket (Arugula) | 4-mercaptobutyl | 51 |
| *Eruca sativa* | Gluraphanin | 3 |
| | Glucoerucin | 3 |
| | Indole glucosinolate | 0.5 |
| Broccoli | Glucoraphanin | 20 |
| *Brassica oleracea var italic* | Indole glucosinolate | 6 |
| Cauliflower | Glucoraphanin | 11 |
| *Brassica oleracea var botrytis* | Sinigrin | 8 |
| | Glucoiberin | 3 |
| Turnip | Pro-goitrin | 10 |
| *Brassica rapa ssp. Rapa* | Indole glucosinolate | 15 |
| | Gluconasturnin | 10 |
| Red cabbage | Glucoraphanin | 11 |
| *Brassica oleracea var. capitata* | Sinigrin | 10 |
| *F. alba* | Glucoiberin | 9 |
| | Indole glucosinolate | 22 |
| Chinese broccoli | Gluconapin | 76 |
| *Brassica rapa var. alboglabra* | Glucoraphanin | 39 |
| | Pro-goitrin | 19 |
| | Indole glucosinolate | 127 |
| Brussels sprouts | Sinigrin | 23 |
| *Brassica oleracea var gemmifera* | Glucoiberin | 9 |
| | Indole glucosinolate | 36 |
| Chinese cabbage | Glucobrassiconapin | 4 |
| *Brassica campestris spp.* | Progroitrin | 3 |
| *pekinensis* | Indole glucosinolate | 6 |
| | Aromatic glucosinolate | 5 |

along the vascular system. Following physical damages, crushing or biting, glucosinolates will be placed in physical contact with the hydrolyzing enzyme and will release the thiocyanate. The type of hydrolysis product generated depends on the chemical nature of the parent glucosinolate side-chain and is also modulated by the presence of proteins associated to the enzyme myrosinase (epithiospecifier protein) responsible for cleaving glucose from its bond to the sulphur atom of the molecule (Kliebenstein et al. 2005) (Table 28.6).

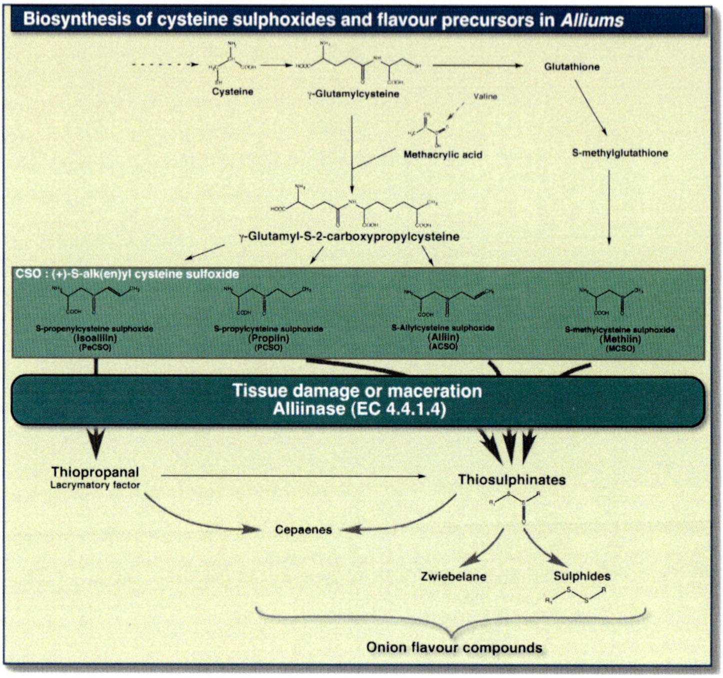


**Fig. 28.4** Biosynthetic pathway of S-Alk(en)yll cysteine sulphoxides in Alliaceae

Glucosinolate *per se* have no beneficial biological activity once ingested. Yet, once the sugar moiety is cleaved by myrosinase or by digestive enzymes, the resulting unstable aglycone form iso-thiocyanate or thiocyanate, which display beneficial health properties. One of these, sulphoraphane, is the active hydrolysis compound of the glucosinolate glucoraphanin found especially in broccoli florets, stems and sprouts (Fahey et al. 1997). Keck and Finley (2004) have shown that sulphoraphane is a strong inducer of phase II enzymes and can thus conjugate xenobiotics and transform them is such a way that they can be excreted through urine of the digestive tract (Verkerk et al. 2009).

## S-Alk(en)yl-Cysteine Suphoxides

Allium species are important agronomic crops worldwide. They possess characteristic flavor, conferred by specific sulphur compounds and by numerous volatiles and are largely utilized by different societies around the world as a staple food. The volatile sulphur compounds are generated through enzymatic reactions of non-volatile precursors (S-alk(en)yl L-cysteine sulphoxides). Different alliums will accumulate different amounts of these precursors; onion for instance will majorly accumulate 1-propenyl(vinyl-methyl), while garlic accumulate the allyl (methyl-vinyl 2-prope-

**Table 28.6** Polyphenol content of typical fruit and vegetables. (Adapted from Phenol-Explorer 2 (Rothwell et al. 2012; Brat et al. 2006))

| Fruit and vegetable types | Polyphenol content (mg/100 g F.W.) |
|---|---|
| Apple | 328 |
| Asparagus | 23 |
| Banana | 52 |
| Blueberry | 630 |
| Blackcurrent | 621 |
| Broccoli | 98 |
| Cabbage | |
| Cashew nuts | 295 |
| Carrots | 16 |
| Cauliflower | 13 |
| Cranberry | 17 |
| Cherry | 94 |
| Onion | 76 |
| French bean | 10 |
| Grapes | 195 |
| Guava | 186 |
| Leek | 33 |
| Papaya | 27 |
| Lettuce | |
| Pear | 69 |
| Pineapple | 103 |
| Mango | 68 |
| Melon | 8 |
| Oranges | 31 |
| Starfruit | 66 |
| Watermelon | 12 |
| Tomato | 14 |

nyl) derivatives and chives accumulate 5-propyl cysteine sulphoxides (Fig. 28.5). Sulphur compounds are integral part of allium metabolism and cysteine sulphoxide in some alliums represent up to 1 % of their fresh weight (Kubec et al. 2000). According to Lancaster and Kelly (1983), non-protein cysteine and glutathione and its derivatives account for almost 5 % of the plant dry weight. As for glucosinolates in brassicas, sulphur compounds in alliums are believed to participate to defense protection against pathogens and herbivores (Brewster 1994). They present both antifungal and antibacterial properties. The enzyme is also stored in a different cellular compartment (vacuoles) than its substrate (cytoplasm) and will generate the highly reactive sulfenic acids upon disruption of the cellular integrity after either slicing, mashing or bruising the bulbs. Sulfenic acids will spontaneously condense and inter-react to form many different thiosulfinates, a class of highly volatile and strong smelling compounds characteristics to most allium species. More than 80 volatile compounds of this class have been identified in the head-space of fresh or cooked alliums (Brewster 1994). The kinetics of cysteine sulphoxide hydrolysis and the reactivity of the initial sulfenic acid generated influence the type of thiosulfinates formed, hence the difference in flavor of fresh, boiled or fried onions.

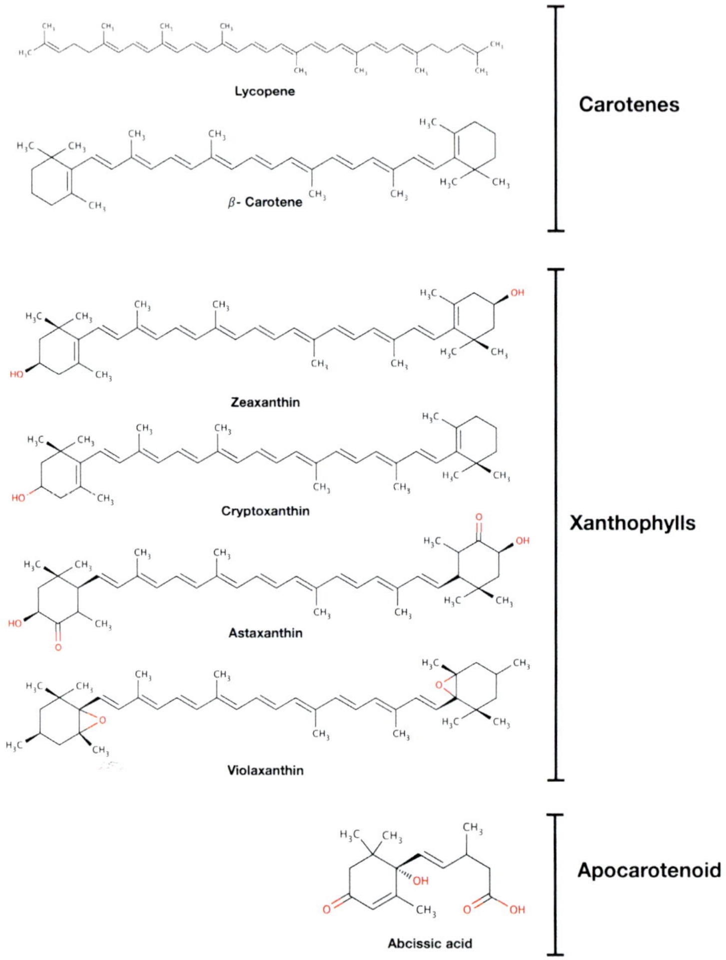


**Fig. 28.5** Typical structure of carotenoids and apocarotenoids found in fruit and vegetables

Both volatile and non-volatile compounds from alliums are reported to be effective in the prevention of many diseases (Griffiths et al. 2002). Traditional wisdom, and scientific literature to date, which represent more the 3000 publications, have confirmed the health benefits of onion and garlic in particular (Corzo-MartÌnez et al. 2007). These benefits include reduction of risk factors for cardiovascular diseases (Ali et al. 2000; Milner 2001), reduction in cancer incidence (Fleischauer and Arab 2001), reduction of inflammatory response (Srivastava 1986), enhanced xenobiotic detoxification (Munday et al. 2003), antioxidant properties (Prasad et al. 1995), antibiotic and antifungal properties (Lancaster and Kelly 1983; Rose et al. 2005).

**Table 28.7** Total carotenoid content of common FAV. (Adapted from van den berg et al. (2000) and Almeida-Melo et al. (2006)

| | Total carotenoids (mg/100 g F.W.) | Type of carotenoids |
|---|---|---|
| Fruits | | |
| Banana | 1 | β -carotenoid, lutein |
| Guava | 4.3 | Lycopene, β-carotene |
| Mango | 0.3–7.6 | β -carotene, β -cryptoxanthin |
| Melon | | β -carotene, β -cryptoxanthin |
| Orange | 2.8 | β -cryptoxanthin, β -carotene, α-carotene |
| Papaya | 2.4–7.3 | Lycopene, β -carotene, β -cryptoxanthin, |
| Peach | 0.5–2.1 | β -carotene, β -cryptoxanthin |
| Watermelon | 2.4–7.3 | Lycopene, β -cryptoxanthin, β -carotene |
| Vegetables | | |
| Broccoli | 1–44 | Lutein, β -carotene |
| Carrots | 1–64 | β -carotene, α -carotene |
| Green bean | 3 | β -carotene, lutein |
| Kale | | |
| Lettuce | 7.5 | Lutein, β -carotene |
| Pepper red | 5 | β -carotene, β -cryptoxanthin, lutein |
| Spinach | 2.8–44 | β -carotene, lutein |
| Sweet potato | 0.3–8 | β -carotene, β -cryptoxanthin |
| Tomato | 1–63 | Lycopene, lutein, β -carotene |

# Terpenoids

## *Carotenoids*

Carotenoids belong to a widespread group of plant pigments, represented by more than 600 structurally different molecules (Fraser and Bramley 2004). According to Bendich (1993), more than 60 sources of carotenoids are found in the human diet and they provide a number of beneficial effects on health. FAV constitute the major source of carotenoids in the human diet (Table 28.7). Carotenoids are isoprenoid polymers (C40) made up of a long symmetric aliphatic chain with many double bounds (Fig. 28.5). This molecule can undergo many conformational changes and is found mostly as *trans*- stereoisomers in FAV. Most non-oxygenated forms of carotenoids display pro-vitamin A activity (Fraser and Bramley 2004), act as strong antioxidants (Gramann and Gerald 2005; Palozza and Krinsky 1992), enhance immune function (Rao and Rao 2007), can protect the skin from UV radiation (Mathews-Roth 1993), and can prevent macular degeneration (Snodderly 1995). Moreover and more generally, carotenoids have been shown to reduce the incidence of certain types of cancer (Knekt et al. 1999; Limpens et al. 2006), reduce the incidence of cardiovascular diseases (Klipstein-Goldberg et al. 2000; Voutilainen et al. 2006), reduce diabetes (Ford et al. 1999), and strengthen the immune system (Hughes 1999;

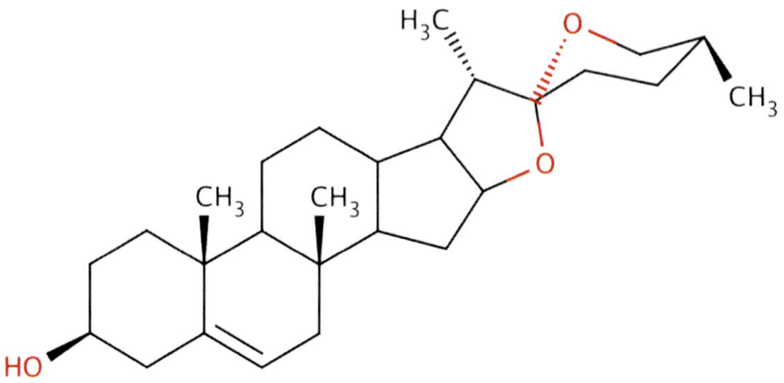


**Fig. 28.6** Chemical structure of Diosgenin a common saponin found in FAV

Riso et al. 2006). These beneficial effects are believed to derive from the presence of the many conjugated double bonds (up to 13). This unique arrangement of double bonds on the aliphatic chain, imparts the characteristic yellow, orange and red colour of carotenoid pigments. It also explains the strong antioxidant capacity of these molecules in lipophilic environments, owing to their singlet oxygen quenching capacity and electron delocalisation. Yet carotenoids as most antioxidant molecules can become pro-oxidants in certain conditions. As a matter of fact, two studies, the Alpha-Tocopherol Beta-Carotene (ATBC) (Goodman et al. 2004) and the β-Carotene and Retinol Efficiency Trial (CARET) (Omenn et al. 1996) respectively showed that β-carotene increased the incidence of cardiovascular diseases and increased mortality in groups of smokers probably through a pro-oxidant activity.

## Apocarotenoids

Greater attention is recently being placed on the health effects of carotenoids degradation products also known as apocarotenoids (Bouvier et al. 2005). Among this group of molecules, ABA and ABA metabolites are standing out for their bioactivity. Indeed, recent report from Guri's laboratory demonstrate that administration of pure ABA is involved in the etiology of diabetes (Bassaganya-Riera et al. 2010; Guri et al. 2010a), atherosclerosis (Guri et al. 2010b), and inflammatory bowel disease (Guri et al. 2007). Work by Bruzzone et al. (2008) show that picomolar concentration of ABA can influence insulin release from human pancreatic islets via cyclic adenosine signaling cascade. These authors claim that ABA is endogenously produced by human granulocytes and act as pro-inflammatory cytokines (Bruzzone et al. 2007). Since ABA was identified in mammalian brain (Lepage-Degivry et al. 1986), it is believed that it could also act as a neuromodulator (Bodrato et al. 2009). Berries are rich sources of ABA and ABA-glucose esters so are other seed sources (Jia et al. 2011; Zifkin et al. 2012).

## Triterpene Saponins

Saponins are constituted of a triterpene backbone to which are attached different glycosides (Fig. 28.6). They are present in oats, allium species (leek, garlic), asparagus, tea, spinach, sugarbeet, and yam (Price et al. 1987). Saponins have been reported to possess a wide range of biological activities and in particular to have analgesic, anti-inflammatory, antimicrobial, antimutagenic and antiobesity) properties (Güçlü-Üstünda and Mazza 2007).

## Alkaloids

### *Capsaicins*

Capsaicins are amide derivatives of vanillylamine and branched fatty acids chain. Typically, chilli pepper (*Capsicum sp.*) accumulates capsaicinoids in their fruits. For example, this molecule class displays a very strong pungency due to its ability to interact with non-selective cation channel protein receptors and thus trigger a general sensation of pain and heat in mammalians (Jordt et al. 2003). Healthwise, Surh and Lee (1995) have shown that capsaisins have anticarcinogenic properties and can induce apoptosis and thus display antitumoral activity. Capsaicinoids are also potent antioxidants (Kogure et al. 2002) and have been associated with increased energy expenditure in human and a decreased in long-term excess energy intake (Doucet and Tremblay 1997) and may thus be used therapeutically to control weight (Reinbach et al. 2009).

## Health Effects of FAV Phytochemicals

### *FAVs and the Burden of Chronic Diseases*

A huge body of evidence indicates that our current lifestyle, which includes smoking, low physical activity and poor diet, has a major influence on our health. The WHO global report entitled 'Preventing Chronic Diseases: A Vital Investment' (Anon 2007), informs us that 60 % of all death on the globe are due to chronic disease and that 80 % of these deaths occur in low and low-middle income countries. Shockingly, a large proportion of these casualties could be avoided by simply adopting a healthier lifestyle. Indeed, a recent paper by Khaw et al. (2007) evaluated the impact of behavioral factors to health and clearly demonstrated that adopting a healthy lifestyle could improve life expectancy of a population by 14 years. It can also reduce the incidence of diabetes by 80 % and cancers by 40 % (Anon 2004). Nutrition is probably the single most important factor affecting the health status of the population; diet has long been linked to the development of chronic diseases and dietary modifications are one of the cornerstones of chronic disease prevention.

There is a substantial and growing body of evidence linking increase in FAVs consumption to a reduction of the risk of chronic diseases, increases lifespan and quality of life, while decreasing medical costs (Tomas-Barberan and Gil 2008). However, the components of FAVs responsible for these beneficial health effects are not entirely identified and the manner by which they exert this effect is still open to debate. FAVs accumulate several thousands phytochemicals with shown biological activity against a number of illnesses *in vitro*. For example, glucosinolate and isothiocyanate found in Brassicas have been linked to cancer prevention (Talalay and Fahey 2001), polyphenols have been linked to cancer prevention, anti-inflammatory responses and prevention of coronary heart diseases (CHD) (Habauzit and Morand 2012), carotenoids found in leafy vegetables and carrots have been associated with a reduced incidence certain types of cancer (Gallicchio et al. 2008), prevention of cardiovascular diseases (Riccioni 2009), of macular degeneration (Sabour-Pickett et al. 2012) and with the strengthening of the immune system (Riso et al. 2006), while sulphur compounds found in onions have been related to CHD and cancer prevention (Corzo-Martínez et al. 2007).

## Antioxidant Capacity

Much work has been conducted to determine the antioxidant capacity of FAV (Proteggente et al. 2002). Antioxidants are protective agents that inactivate reactive oxygen species and therefore significantly prevent oxidative damages. Most organisms living on earth have developed strong defenses against oxidation and rely in particular on superoxide dismutases, catalases, and glutathione peroxidases to attenuate the reactive oxygen species generated by the metabolism. In addition, antioxidants such as vitamin E, vitamin C, carotenoids and to a certain extent polyphenols are available from foods (Anon 2004, 2006). The role of dietary antioxidants in disease prevention has received much attention over the years. It is nowadays less prominent since recent evidences from research are shedding new light on the role of antioxidants in the etiology of diseases; the tenet that dietary antioxidants are responsible for the prevention of diseases is now much contested (Hollman et al. 2011a; Scalbert et al. 2005). As a matter of fact, very few studies have shown that antioxidants were active *in vivo* (Frankel and German 2006; Halliwell et al. 2005) and some report that antioxidants can even have harmful effects *in vivo* (Perera and Bardeesy 2011; Ristow et al. 2009). Halliwell et al. (2005) caution that reports of increase in plasma total antioxidant activity after flavonoid intake, must be interpreted with care since these may be caused by many confounding factors. For instance, the antioxidant effect of polyphenol on plasma has recently been challenged since any change in plasma antioxidant capacity after fruit consumption may be caused by fructose mediated increases in uric acid rather than fruit-derived antioxidants (Godycki-Cwirko et al. 2010; Lotito and Frei 2006). According Halliwell et al.

(2005), results obtained from *in vitro* experiments demonstrating positive response of antioxidants on disease end-points may simply be measurement artifacts and biases of experimentation (Long et al. 2000). Yet, it appears clear that many classes of phytochemicals found in FAV can stimulate the natural antioxidant capacity of the organism and prime the defense system against a number of diseases (Hollman et al. 2011b; Traka and Mithen 2009).

## Bioavailability

For any phytochemicals to exert a systemic activity in the organism it has first to be bioavailable, in other words it has absorbed, enter the systemic circulation and reach the target tissues or organs at adequate levels and in an active form. Most bioactive phytochemicals derived from plants are recognized as xenobiotics and are thus poorly absorbed, or are aggressively conjugated in order to make them more water-soluble and ease their excretion. For instance, many researchers have reported that the bioavailability of plant polyphenols is very low (Del Rio et al. 2010; Kroon et al. 2004; Manach et al. 2005; McGhie et al. 2003; Milbury et al. 2010; Prior and Wu 2006; Scalbert and Williamson 2000; Spencer 2008). It is also accepted that the polyphenol glycoside moiety has to be removed before absorption. Yet, depending on the position of the glycoside attached to the flavonoid, the degree of polymerization or galloylation of flavan-3-ols, different quantity of the compound will be absorbed. Considering the complex and varying nature of the flavonoid glycosides found in FAV, one can understand that the bioavailability of the different polyphenols will fluctuate markedly in line with the different benefits stemming from their consumption. It is believed that interaction between different polyphenolic species can result in synergism or antagonisms (Scheepens et al. 2010). Knowing that specific polyphenols transporters are found at the surface of the intestine brush border, in particular the sodium-dependent glucose transporter (SGLT1), and that multidrug resistance proteins also participate in the elimination of xenobiotics and polyphenols, which are incidentally recognized as extraneous chemicals, it is assumed that different polyphenols can competitively interact at their site of absorption or elimination. In this context, the consumption of whole FAV, bringing a variety of polyphenols, is assume to confer more benefits than isolated molecules owing to the potential synergies that can develop between the many polyphenols and an improved bioavailability at the target sites of action.

On the contrary, carotenoids are more bioavailable and can be found in relatively higher concentration unchanged in the plasma and in certain tissues and organs (Verkerk et al. 2009). Being liposoluble, they will be absorbed by passive diffusion after being incorporated into micelles formed by dietary fats and bile acids. The micellar carotenoids are then incorporated into chylomicrons and eventually into lipoproteins to be released into the blood stream (van den Berg et al. 2000). The carotenoids are mostly included into plasma membrane or stored in adipose tissues. Interestingly, cooking and heat liberate the carotenoids from the food matrix and augment their bioavailability. It has been well demonstrated that absorption of lycopene from processed tomato products for instance is higher than from raw tomatoes (Rao and Agarwal 1998). The half life of the carotenoids in the system is about 2 to 3 days (Stahl and Sies 2005). β-carotene have been shown to accumulate in fat tissues but lycopene accumulate in human adrenal glands, prostate, breasts, testes, and liver. Intake is about 4–25 mg/d in North America (Rao and Rao 2007).

A large proportion of the glucosinolates and metabolites reach the intestine where they undergo a massive attack and degradation by the colonic microflora. The remaining glucosinolates and thiocyanates are absorbed passively through the gut epithelium. Once in the enterocyte, they will be rapidly conjugated with glutathione and transported in the systemic circulation to be metabolized via the mercapturic acid pathway for their subsequent urinary excretion. As for polyphenols, glucosinolates and their bioactive products are recognized as xenobiotics and are rapidly eliminated from the body after extensive metabolism in the liver, and enterocytes. Yet, by the action of sulfotransferases, isothiocyanate conjugates can release the free isothiocyanate at specific target tissue to display their biological activity (Traka and Mithen 2009). These molecules strongly bind to proteins like albumin and other glycoproteins. Sulforaphane is considered a poorer substrate that other glucosinolate to phase I enzymes which explains its longer residency in the body and thus higher bioactivity. The metabolites are most likely the molecules responsible for the beneficial effects.

## Cardiovascular Diseases

Nutritional epidemiology has provided convincing evidence for the cardio-protective effects of the frequent consumption of FAV prompting health authorities to promote their consumption (Ness and Powles 1997). For instance, in the Kuopio Ischemic Heart Disease Risk Factor Study (KIHD), Rissanen et al. (2003) observed a clear association between FAV consumption and cardiovascular health. However, recent reassessment of the data reveals that FAV consumption is weakly associated with reduced risk of coronary heart disease in cohort studies; evidences for FAV consumption preventing cardiovascular disease remains scarce (Dauchet et al. 2009). Yet, flavonoid and carotenoid intake have been linked to decreased morbidity and mortality from coronary heart diseases (CHD) (Hertog et al. 1995; Riccioni 2009; Voutilainen et al. 2006). The reported beneficial effects of these molecules on CHD risk are more than likely the result of a reduction in inflammation (Loke et al. 2008b), and a reduced inhibition of LDL oxidation, which has been demonstrated both *in vitro* and *in vivo* (Fuhrman and Aviram 2001). The underlying mechanisms for these beneficial effects are believed to include improved endothelial function through improved nitric oxide balance (Schewe et al. 2008), decrease in cellular oxidative stress (Steffen et al. 2008), and inhibition of inflammation (Loke et al. 2008b).

Oxidative stress and inflammation play a pivotal role in the initiation and progression of atherosclerosis and CHD. Atherosclerosis is a condition affecting the coronary arteries in which gradual uptake of oxidized lipoproteins by the endothelium and the resulting inflammatory response leads to deposition of plaques in the arterial walls and eventual restriction of blood flow which can aggravate or produce hypertension and eventually cause irreparable damage to the heart. The accumulation of oxidized LDLs in the intima and their uptake by macrophages are early events in atherosclerosis that could be lessened by the presence of polyphenols. Flavonoids contained in FAV may decrease the risk of developing atherosclerosis, due to their ability to inhibit low-density lipoprotein (LDL) oxidation (Arai et al. 2000; Basu et al. 2010), (Perez-Vizcaino et al. 2006), to up-regulate antioxidant enzyme expression (Wu et al. 2010), to reduce platelet aggregation and adhesion (Hubbard et al. 2006; Ostertag et al. 2010; Steffen et al. 2008), to reduce inflammatory response of the vascular tissues (Perez-Vizcaino et al. 2006; Xie et al. 2011; Youdim et al. 2000), while also inducing endothelium-dependent vasodilation (Andriambeloson et al. 1998; Kalea et al. 2009; Loke et al. 2008a) and reducing blood pressure (Edwards et al. 2007). Many researches show that polyphenols interact with the signaling pathways of immune and inflammatory cells (DeFuria et al. 2009; Rechner and Kroner 2005; Youdim et al. 2000). Such an action has been attributed to an improved vascular reactivity (Kalea et al. 2009) and in particular to an effect on iNOS activity and up-regulation of the endothelial nitric oxide synthase, both of which play a crucial role in maintaining cardiovascular homeostasis by favorably modulating blood pressure and reducing endothelial dysfunction, so as to maintain normal vascular function and blood pressure.

## Obesity

Obesity, is fast becoming a worldwide health problem and has dramatically increased in every continents in the last decade. The recent survey from the USA National Health and Nutrition Examination Survey (NHANES) show that approximately 35 % of men and women are obese (Flegal et al. 2012). In Europe, the prevalence of obesity (body mass index $\geq 30$ kg/m$^2$) has reached epidemic proportions, affecting more than 25 % of the population in countries like Spain, Poland, Czech Republic and Italy, in both men and women. A dramatic increase in overweight and obesity prevalence has also been observed in mainland China with 22.8 % and 7.1 %, respectively, which represent an increase of 41 % and 97 % of the respective incidence when compared to 1992 (Chen 2008). This disturbing reality is correlated with an the exponential rise in the prevalence of type 2 diabetes (T2D), which is estimated to reach the appalling rate of 439 million cases by 2030 (Shaw et al. 2010).

Obesity is a complex outcome influenced by a variety of interacting factors involving genetic, environmental, social and behavioral factors. It is ultimately the result of a disruption of the energy balance equation where energy intake surpasses its expenditure, resulting in the storage of excess energy into adipose tissue. Regulating energy intake is not as easy as it may sound, since eating is regulated by an intricate network of hormonal messages affecting the central nervous system at the level of the hypothalamus and regulating appetite (Woods and D'Alessio 2008).

Indeed, more than 20 peptides with hormone activity (insulin, GLP-1, CCK, PYY, ghrelins, leptin, adiponectin and others) are produced by peripheral tissues (adipose, muscular, pancreas, and the gut) linking adiposity levels and energy intake to the central nervous system (Kim and Park 2011).

A number of bioactive phytochemicals found in FAV have been shown to regulate energy balance and have an effect on weight gain and energy homeostasis (Kim and Park 2011). Specifically, many excellent literature reviews on the effect of polyphenols on energy metabolism and reduction type-2 diabetes (T2D) have recently been published (Basu and Lyons 2012; Cherniack 2011). T2D, a sequel of obesity and characterized by an hyperinsulinemia and insulin resistance belongs to a constellation of factors (hyperglycemia, hypertension, insulin resistance, glucose intolerance and dyslipidemia) leading to a diet sensitive condition called the metabolic syndrome (Bland 2011). Adipocyte dysfunction is at the origin of the syndrome and is associated with macrophage infiltration in adipose tissue leading to the release of pro-inflammatory cytokines and activation of inflammatory signalling pathways, which can interfere with insulin action in skeletal muscle, liver and adipose tissue thus the concept of insulin resistance.

Mechanistic studies using *in vitro* models have provided evidence of the beneficial effects of FAV polyphenols on hyperglycemia and hypertension, two significant CHD risk factors that coexist in metabolic syndrome (Hanhineva et al. 2010). For instance, polyphenols have been reported to exert beneficial effects on glucose homeostasis by (i) inhibiting sugar and lipid digestive enzyme (McDougall et al. 2008), (ii) inhibiting glucose absorption (Serrano et al. 2009) (iii) protecting from glucotoxicity through the reduction of advance glycation product formation (McIntyre et al. 2009) and from pancreas β-cells toxicity(Martineau et al. 2006) (iv) increasing insulin secretion (Adisakwattana et al. 2008), (v) improving glucose uptake in muscle and adipocytes (Grace et al. 2009), (vi) increasing hepatic glucokinase activity, (vii) suppressing gluconeogenesis (Burton-Freeman 2010) and (viii) controlling satiety (Molan et al. 2008). They can also protect against T2D through anti-inflammatory effect (González et al. 2011; Comalada et al. 2005). Indeed these authors showed that FAV polyphenols reduced a number of inflammatory biomarkers linked to insulin resistance and hyperglycemia such as TNF-a, IL-6, MCP-1, and iNOS (González et al. 2011). It is generally considered that polyphenols like quercetin (Dias et al. 2005), proanthocyanidins (Serrano et al. 2009), hydroxycinnamic acids (Barone et al. 2009), and stylbenes (Alberdi et al. 2011; Baile et al. 2011) prevent the occurrence of T2D by modify carbohydrate, lipid and energy metabolisms. Moreover, many polyphenols and in particular flavonols and stylbenes present in large quantities in onions and many other berries exert a myriad of anti-inflammatory, anti-obesity, anti-steatosis and hypoglycemic effects through an AMPK-SIRT1-PPARγ-dependent mechanisms (Alberdi et al. 2011; Baile et al. 2011), can lead to adipocyte differentiation (Moghe et al. 2012; Vuong et al. 2007) and increased the number of mitochondria and the energy expenditure (Pajuelo et al. 2011). In this context, resveratrol has been shown to mimic caloric restriction, increase lifespan and reduce inflammatory response leading to reduce T2D and energy metabolism dysfunction (Aires et al. 2012; Brasnyó et al. 2011; Szkudelska and Szkudelski 2010).

## Cancer

It is well accepted that consumption of FAV is associated with decreased risk of developing cancer (Cooke et al. 2005; Potter 2005; World Cancer Research Fund 2007). Among all FAV, berries probably possess the best documented anti-tumoral properties (Duthie 2007; Neto 2007; Neto et al. 2008; Seeram 2008), but there are also strong epidemiological evidences showing that Alliaceae and Brassicaceae consumption are linked to reduced incidence of cancers (Griffiths et al. 2002; Verkerk et al. 2009). Cancer development is complex and is commonly recognized as a multi-factor process that requires the cumulative action of three main events: initiation, promotion and progression. At the base, the incipient causes leading to the initiation of cancers are DNA damages and the succeeding accumulation of mutations. Phytochemicals found in FAV have been shown to interfere at all stages of the etiology of cancer.

Polyphenols found in FAV have been shown to prevent the growth and progression of cancers in many *in vitro* and in animal models; (e.g. mice endothelial neoplasms (Gordillo et al. 2009), colorectal mucosal cells (Håkansson et al. 2012), prostate cancer cells (Matchett et al. 2006), colon cancer cells (Seeram et al. 2006; Suh et al. 2007), breast, cervix cancer cells (Wedge et al. 2001), HepG2 liver cancer cells (Kraft et al. 2006), HL-60 leukemia cells (Murphy et al. 2003), HCT-116 and HT-29 human colon cells (Murphy et al. 2003; Ono et al. 2002), mammary carcinoma 4T1 cell lines (Mantena et al. 2006), breast cancer cells (Adams et al. 2011; Adams et al. 2010) (Aiyer et al. 2012; Faria et al. 2010), prostate cancer cells (Matchett et al. 2006; Schmidt et al. 2006), meduloblastoma cell lines (Labbe et al. 2009), and lung cancer cell lines (Kausar et al. 2012)). Actually, Aiyers et al. (Aiyer et al. 2012) has thoroughly reviewed the effect of polyphenols found in Vacciniums on receptor signaling and induction of cell death pathway. Moreover, polyphenols can mitigate the initial formation of tumors by blocking the action of carcinogens responsible for mutations. For instance, a blueberry extract was shown to induce phase II detoxification xenobiotic enzymes (quinone reductase) (Bomser et al. 1996) and inhibit the initiation stage of chemically induced carcinogenesis in liver cancer cells (Smith et al. 2000).

In a similar manner, many prospective and epidemiological studies have also shown that the regular consumption of *Alliums* could have protective effects against cancer (Griffiths et al. 2002; Lampe 1999). For instance, there appears to be a strong link between the consumption of onions and the reduced incidence of stomach and intestine cancers (You et al. 2005). The Epic Prospective Study, conducted on more than half a million subjects, showed clear correlation between onion consumption and reduction in intestinal and stomach cancers (Gonzalez and Riboli 2006). A synthesis of case-control studies carried in Italy and Switzerland revealed that consumption of one to seven portions of onions per week reduce the risks of colon, ovary, larynx and mouth cancers (Galeone et al. 2006). Similar correlations are also observed for brain and stomach cancers in a case-control study in China (Hu et al. 1999) and breast cancer in France (Challier et al. 1998). Mortality due to prostate cancer also appears to be reduced by a diet making a large place to onions (Grant 2004). Onion is probably the most important source of polyphenols in the diet (Hollman and Arts 2000) and it has been showed in many studies to have anticarcinogenic properties *in vitro* (Wilms et al. 2005).

There is also an inverse relationship between the consumption of dark green vegetables, and in particular of brassica vegetables and the risk of colorectal cancer (Voorrips et al. 2000). Glucosinolates and isothiocyanates found in Brassicas appear to explain this reduced risk. These molecules are triggering phase I and II enzymes involved in carcinogen metabolism and detoxification (Talalay and Fahey 2001). They are also priming the natural antioxidant defenses as evidence by the decreased in DNA damages and inhibition of aberrant crypt foci formation in animal studies. This protection appears to be mediated by antioxidant response elements in the promoter region of phase II detoxification enzymes and antioxidant enzymes and specifically through the activation of the Nfr2 transcription cascade (Jeong et al. 2006).

## Neurodegenerative Diseases and Cognition

Normal aging is accompanied by a decline in motor and cognitive performance (Lau et al. 2005). The mechanisms responsible for behavioral and neuronal changes seen during aging are not fully understood, but it appears that dietary FAV supplementation can slow or even reverse various age-related neuronal declines (Gu et al. 2010). The molecular mechanisms involved in the beneficial action of FAV on the brain remain unclear but likely relate to the modulation of processes, such as neuronal plasticity affected during aging. Alzheimer's disease is multifactorial, with a complex combination of genetic and non-genetic components but share a common biochemical pathway, that is, the altered production of the amyloid β peptide, which leads to neuronal death and dementia.

*In vitro* mechanistic investigations have begun to elucidate the molecular mechanisms involved in the beneficial effect of dietary polyphenols found in FAV on cognition and neurodegenerative diseases. These studies suggest that flavonoids can i) reduce the pro-inflammatory state (Frisardi et al. 2010), characteristic of the metabolic syndrome; ii) modulate intracellular signaling pathways controlling neuronal cell apoptosis (Choi et al. 2012; Spencer 2008); iii) have a neuroprotective effect on neurons and glial cells (Galli et al. 2006; Vuong et al. 2010); decrease cerebral inflammation through a retardation of the systemic vascular inflammation (Williams and Spencer 2011); and improve cerebral blood flow leading to new hippocampal cells formation and enhanced memory (Ghosh and Scheepens 2009).

In particular, there are many recent reviews on the beneficial effects of berry on neurodegeneration and improvement of cognition (Giacalone et al. 2011; Ramassamy 2006; Shukitt-Hale et al. 2008; Williams and Spencer 2011). These reviews specifically show that polyphenols convey beneficial effects on memory and learning in both animals and humans. However, there are few epidemiological data correlating the consumption of berry per say to neurodegenerative diseases and cognitive decline. One of these, the PAQUID study following a population over 10 years, showed that and average intake of about 14 mg/day of flavonoids was associated with a reduction in cognitive decline (Letenneur et al. 2007). There is also a paucity of human intervention study on berry polyphenols and cognition (Lamport et al. 2012). Among those published, Krikorian et al. (2010) were able to demonstrate, in a single blind clinical study trial with only 9 adults displaying mild cognitive impairment, that the consumption of a blueberry juice, providing 1.8 mg polyphenol/d for 12 weeks, had a significantly better verbal paired associate learning, but there was no difference in the California Visual Learning Test. In this study, there was a trend for the berry group to have a better mood while have a normalized glycemia and insulin level.

If there are only a few human studies on the effect of polyphenols on cognition, there are many animal studies showing that polyphenol supplementation and in particular blueberry polyphenols can prevent cognitive decline and memory (Cherniack 2011). For instance, Andres-Lacueva and Shukitt-Hale (2005) showed that blueberry anthocyanins could reach the cerebral cortex, the hippocampus, the striatus and the cerebellum and that they correlated with the performance of mice in a maze test. Similarly, Galli et al. (2006) showed that a blueberry diet fed for 10 weeks to aged rats improved had a better capacity to generate heat shock protein (HSP-70) a reflection of their ability to support neurodegenerative process in the brain. Blueberry supplementation also improved the ability of elderly rats to recognize objects through a preservation of neurogenesis in the hippocampus. This beneficial effect was not linked to a loss of amyloid plaque. Young rats fed with a blueberry polyphenol extract had a better performance in the water maze test (Joseph et al. 2008), while old rats fed with a blueberry diet maintained a better balance while walking across a wire and had a better performance in the water maze test (Joseph et al. 1998). Williams et al. (2008) were able to show that rats fed for 12 weeks with a 2 % blueberry diet had an improved special working memory and an improved cognitive performance. This effect was attributed to an improved phosphorylation of C-AMP Responsive Element-Binding Protein, involved in signal transduction and associated with long-term memory. Papandreou et al. (2009) also showed that mice fed with a blueberry extract had improved cognitive performance, had higher brain antioxidant capacity and had an improved acetylcholinesterase activity. Blueberry supplemented diet improved the hippocampal neurogenesis and improved special memory through an activation of the Insulin-Growth Factor-1, a key protein in the learning process and the modulation of neurogenesis (Casadesus et al. 2004). Vuong et al. (2010) showed that a fermented blueberry juice had a neuroprotective effect by activating cell survival pathways associated with p38 and JNK pathways. Recently, Rendeiro et al. (2012) showed that elderly rats feed with a 2% blueberry diet had a better special memory, a faster rate of learning than the control group. This effect was linked to the activation of the ERK-CREB-BDNF pathway.

## Conclusions

In conclusions, research conducted over the last 15 years demonstrates that FAV consumption definitely provides positive effects against a number of chronic diseases. The exact mode of action of the different bioactive compounds on health is slowly being unraveled. It is becoming clear that their preventative influence is not only mediated directly through their antioxidant capacity but chiefly through their effects on specific gene expression. In particular, signaling cascades associated with anti-inflammatory responses and control of energy metabolism are clearly affected. Polyphenols, but also carotenoids and sulphur compounds may act at different levels. Due to their low bioavailability, much emphasis is now placed on the activity of the circulating metabolites on target organs and cells.

We are definitely at a turning point with respect to the demonstration of health effects of FAV phytochemicals. A consensus is slowly emerging that the beneficial properties of its phytochemicals must be studied by conducting human clinical trials or animal studies. The demonstration of positive effects in human will undoubtedly stimulate FAV consumption all over the world.

## References


Adams LS, Phung S, Yee N et al (2010) Blueberry phytochemicals inhibit growth and metastatic potential of MDA-MB-231 breast cancer cells through modulation of the phosphatidylinositol 3-kinase pathway. Cancer Res 70:3594–3605. doi:10.1158/0008-5472.CAN-09-3565

Adams LS, Kanaya N, Phung S et al (2011) Whole blueberry powder modulates the growth and metastasis of MDA-MB-231 triple negative breast tumors in nude mice. J Nutr 141:1805–1812. doi:10.3945/jn.111.140178

Adisakwattana S, Moonsan P, Yibchok-anun S (2008) Insulin-releasing properties of a series of cinnamic acid derivatives in vitro and in vivo. J Agric Food Chem 56:7838–7844. doi:10.1021/jf801208t

Agrawal AA, Kurashige NS (2003) A role for isothiocyanates in plant resistance against the specialist herbivore *Pieris rapae*. J Chem Ecol 29:1403–1415. doi:10.1023/A:1024265420375

Aires DJ, Rockwell G, Wang T et al (2012) Potentiation of dietary restriction-induced lifespan extension by polyphenols. Biochim Biophys Acta 1822:522–526. doi:10.1016/j.bbadis.2012.01.005

Aiyer HS, Warri AM, Woode DR (2012) Influence of berry polyphenols on receptor signaling and cell-death pathways: implications for breast cancer prevention. J Agric Food Chem 60(23): 5693–5708

Alberdi G, Rodríguez VM, Miranda J et al (2011) Changes in white adipose tissue metabolism induced by resveratrol in rats. Nutr Metabol 8:29. doi:10.1186/1743-7075-8-29

Ali M, Thomson M, Afzal M (2000) Garlic and onions: their effect on eicosanoid metabolism and its clinical relevance. PLEAFA 62:55–73

Almeida Mélo E, Galvao de Lima VLA, Maciel MIS (2006) Polyphenol, ascorbic acid and total carotenoid contents in common fruits and vegetables. Braz J Food Technol 9:89–94

Andres-Lacueva C, Shukitt-Hale B (2005) Anthocyanins in aged blueberry-fed rats are found centrally and may enhance memory. Nutr Neurosci 8:111–120. doi:10.1080/10284150500078117

Andriambeloson E, Magnier C, Haan-Archipoff G et al (1998) Natural dietary polyphenolic compounds cause endothelium-dependent vasorelaxation in rat thoracic aorta. J Nutr 128:2324–2333

Anon (2003) USDA database for the proanthocyanidin content of selected foods, 2nd ed. USDA-ARS, Beltsville
Anon (2004) Fruit and vegetables for health. Report of a joint FAO/WHO workshop 46
Anon (2006) Fruits, vegetables and health: a scientific overview, Canadian Produce Marketing Association. IFAVA, Ottawa
Anon (2007) Preventing chronic diseases: a vital investment. WHO, Geneva
Anon (2011) USDA's My Plate. In: fnic.nal.usda.gov. http://fnic.nal.usda.gov/dietary-guidance/myplatefood-pyramid-resources/usda-myplate-food-pyramid-resources. Accessed 27 Mar 2011
Arai Y, Watanabe S, Kimira M et al (2000) Dietary intakes of flavonols, flavones and isoflavones by Japanese women and the inverse correlation between quercetin intake and plasma LDL cholesterol concentration. J Nutr 130:2243–2250
Baile CA, Yang J-Y, Rayalam S et al (2011) Effect of resveratrol on fat mobilization. Ann New York Acad Sci 1215:40–47. doi:10.1111/j.1749-6632.2010.05845.x
Barone E, Calabrese V, Mancuso C (2009) Ferulic acid and its therapeutic potential as a hormetin for age-related diseases. Biogerontology 10:97–108. doi:10.1007/s10522-008-9160-8
Bassaganya-Riera J, Skoneczka J, Kingston DGJ et al (2010) Mechanisms of action and medicinal applications of abscisic acid. Curr Med Chem 17:467–478
Basu A, Lyons TJ (2012) Strawberries, blueberries, and cranberries in the metabolic syndrome: clinical perspectives. J Agric Food Chem 60:5687–5692. doi:10.1021/jf203488k
Basu A, Du M, Leyva MJ et al (2010) Blueberries decrease cardiovascular risk factors in obese men and women with metabolic syndrome. J Nutri 140:1582–1587. doi:10.3945/jn.110.124701
Bazzano LA (2005) Dietary intake of fruit and vegetables and risk of diabetis mellitus and cardiovascular diseases. Joint FAO/WHO Workshop fruit and vegetables for health 1–66
Bazzano LA (2006) The high cost of not consuming fruits and vegetables. J Amer Diet Asso 106:1364–1368
Bendich A (1993) Biological functions of carotenoids. In: Canfield LM, Krinsky NI, Olson JA (eds) Carotenoids in human health, 1st ed. New York Academy of Science, New York, pp 61–67
Bland JS (2011) Metabolic syndrome: the complex relationship of diet to conditions of disturbed metabolism. Func Foods Health Dis 1:1–12
Bodrato N, Franco L, Fresia C et al (2009) Abscisic acid activates the murine microglial cell line N9 through the second messenger cyclic ADP-ribose. J Biol Chem 284:14777–14787. doi:10.1074/jbc.M802604200
Bomser J, Madhavi DL, Singletary K, Smith MLA (1996) In vitro anticancer activity of fruit extract from vaccinium species. Planta Med 62:212–216
Bouvier F, Rahier A, Camara B (2005) Biogenesis, molecular regulation and function of plant isoprenoids. Progr Lipid Res 44:357–429
Brader G, Tas E, Palva ET (2001) Jasmonate-dependent induction of indole glucosinolates in Arabidopsis by culture filtrates of the nonspecific pathogen Erwinia carotovora. Plant Physiol 126:849–860. doi:10.1104/pp.126.2.849
Brasnyó P, Molnár GA, Mohás M et al (2011) Resveratrol improves insulin sensitivity, reduces oxidative stress and activates the Akt pathway in type 2 diabetic patients. Brit J Nutr 106:383–389
Brat P, George S, Bellamy A et al (2006) Daily polyphenol intake in France from fruit and vegetables. J Nutr 136:2368–2373
Brewster JL (1994) Onions and other vegetable alliums, 1st ed. CAB International, Wallingford
Bruzzone S, Moreschi I, Usai C et al (2007) Abscisic acid is an endogenous cytokine in human granulocytes with cyclic ADP-ribose as second messenger. PNAS 104:5759–5764. doi:10.1073/pnas.0609379104
Bruzzone S, Bodrato N, Usai C et al (2008) Abscisic acid is an endogenous stimulator of insulin release from human pancreatic islets with cyclic ADP Ribose as second messenger. J Biol Chem 283:32188–32197. doi:10.1074/jbc.M802603200
Burton-Freeman B (2010) Postprandial metabolic events and fruit-derived phenolics: a review of the science. Brit J Nutr 104:S1–S14. doi:10.1017/S0007114510003909
Carlton Tohill B (2005) Dietary intake of fruit and vegetables and management of body weight. In: WHO, Kobe, pp 1–52

Casadesus G, Shukitt-Hale B, Stellwagen HM et al (2004) Modulation of hippocampal plasticity and cognitive behavior by short-term blueberry supplementation in aged rats. Nutr Neurosci 7:309–316. doi:10.1080/10284150400020482
Challier B, Perarnau JM, Viel JF (1998) Garlic, onion and cereal fibre as protective factors for breast cancer: a French case-control study. Eur J Epidemiol 14:737–747
Chen CM (2008) Overview of obesity in Mainland China. Obes Rev 9 Suppl 1:14–21. doi:10.1111/j.1467-789X.2007.00433.x
Cherniack EP (2011) Polyphenols: planting the seeds of treatment for the metabolic syndrome. Nutrition 27:617–623. doi:10.1016/j.nut.2010.10.013
Cherniack EP (2012) A berry thought-provoking idea: the potential role of plant polyphenols in the treatment of age-related cognitive disorders. Br J Nutr 108:794–800. doi:10.1017/S0007114512000669
Choi SS, Lee DH, Lee SH (2012) Blueberry protects LPS-stimulated BV-2 microglia through inhibiting activities of p38 MAPK and ERK1/2. Food Sci Biotechnol 21:1195–1201. doi:10.1007/s10068-012-0156-4
Clifford MN, Scalbert A (2000) Ellagitannins—nature, occurrence and dietary burden. J Sci Food Agri 80:1118–1125. doi:10.1002/(SICI)1097-0010(20000515)80:7 <1118::AID-JSFA570> 3.0.CO;2-9
Comalada MN, Camuesco DE, Sierra S et al (2005) In vivo quercitrin anti-inflammatory effect involves release of quercetin, which inhibits inflammation through down-regulation of the NF-kB pathway. Eur J Immunol 35:584–592. doi:10.1002/eji.200425778
Cooke D, Steward WP, Gescher AJ, Marczylo T (2005) Anthocyans from fruits and vegetables—does bright colour signal cancer chemopreventive activity? Eur J Cancer 41:1931–1940
Corzo-Martínez M, Corzo N, Villamiel M (2007) Biological properties of onions and garlic. Trends Food Sci Technol 18:609–625
Crozier A, Yokota T, Jaganath IB et al (2006) Secondary metabolites in fruits, vegetables, beverages and other plant-based dietary components. In: Crozier A, Clifford MN, Ashihara H (eds) Plant secondary metabolites: occurence, structure and role in the human diet. Blackwell, Oxford, pp 208–302
Crozier A, Jaganath IB, Clifford MN (2009) Dietary phenolics: chemistry, bioavailability and effects on health. Nat Prod Rep 26:1001–1043
Dangles O, Dufour C (2006) Flavonoids-protein interactions. In: Andersen OM, Markham KR (eds) Flavans and proanthocyanidins. Taylor & Francis, Boca Raton, pp 443–469
Dauchet L, Amouyel P, Dallongeville J (2009) Fruits, vegetables and coronary heart disease. Nat Rev Cardiol 6:599–608. doi:10.1038/nrcardio.2009.131
de Lorgeril M, Renaud, Mamelle N et al (1994) Mediterranean alpha-linolenic acid-rich diet in secondary prevention of coronary heart disease. Lancet 343:1454–1459
DeFuria J, Bennett G, Strissel KJ et al (2009) Dietary blueberry attenuates whole-body insulin resistance in high fat-fed mice by reducing adipocyte death and its inflammatory sequelae. J Nutr 139:1510–1516
Del Rio D, Borges G, Crozier A (2010) Berry flavonoids and phenolics: bioavailability and evidence of protective effects. Brit J Nutr 104:S67–S90. doi:10.1017/S0007114510003958
Deprez S, Mila I, Huneau J-F et al (2001) Transport of proanthocyanidin dimer, trimer, and polymer across monolayers of human intestinal epithelial Caco-2 cells. Antiox Redox Signal 3:957–967. doi:10.1089/152308601317203503
Dias AS, Porawski M, Alonso M et al (2005) Quercetin decreases oxidative stress, NF-kappaB activation, and iNOS overexpression in liver of streptozotocin-induced diabetic rats. J Nutr 135:2299–2304
Dixon RA, Paiva NL (1995) Stress-induced phenylpropanoid metabolism. Plant Cell 7:1085–1097
Doucet E, Tremblay A (1997) Food intake, energy balance and body weight control. Eur J Clin Nutr 51:846–855
Doughty KJ, Kiddle GA, Pye BJ et al (1995) Selective induction of glucosinolates in oilseed rape leaves by methyl jasmonate. Phytochemistry 38:347350

Duthie SJ (2007) Berry phytochemicals, genomic stability and cancer: evidence for chemoprotection at several stages in the carcinogenic process. Mol Nutr Food Res 51:665–674

Edwards RL, Lyon T, Litwin SE et al (2007) Quercetin reduces blood pressure in hypertensive subjects. J Nutr 137:2405–2411

Fahey JW, Zhang Y, Talalay P (1997) Broccoli sprouts: an exceptionally rich source of inducers of enzymes that protect against chemical carcinogens. PNAS 94:10367–10372

Faria A, Pestana D, Teixeira D et al (2010) Blueberry anthocyanins and pyruvic acid adducts: anticancer properties in breast cancer cell lines. Phytother Res 24:1862–1869. doi:10.1002/ptr.3213

Flegal KM, Carroll MD, Kit BK (2012) Prevalence of obesity and trends in the distribution of body mass index among US adults, 1999-2010. JAMA 307:491–498. doi:10.1001/jama.2012.40

Fleischauer AT, Arab L (2001) Garlic and cancer: a critical review of the epidemiologic literature. J Nutr 131:1032S–1040S

Ford ES, Will JC, Bowman BA, Venkat Narayan KM (1999) Diabetes mellitus and serum carotenoids: findings from the Third National health and Nutrition Examination Survey. Amer J Epidemiol 149:168–176

Frankel EN, German JB (2006) Antioxidants in foods and health: problems and fallacies in the field. J Sci Food Agri 86:1999–2001. doi:10.1002/jsfa.2616

Fraser PD, Bramley PM (2004) The biosynthesis and nutritional uses of carotenoids. Prog lipid Res 43:228–265

Frisardi V, Solfrizzi V, Seripa D et al (2010) Metabolic-cognitive syndrome: a cross-talk between metabolic syndrome and Alzheimer's disease. Ageing Res Rev 9:399–417. doi:10.1016/j.arr.2010.04.007

Fuhrman B, Aviram M (2001) Flavonoids protect LDL from oxidation and attenuate atherosclerosis. Curr Op Lipid 12:41–48

Galeone C, Pelucchi C, Levi F et al (2006) Onion and garlic use and human cancer. Am J Clin Nutr 84:1027–1032

Galli RL, Bielinski DF, Szprengiel A et al (2006) Blueberry supplemented diet reverses age-related decline in hippocampal HSP70 neuroprotection. NBA 27:344–350. doi:10.1016/j.neurobiolaging.2005.01.017

Gallicchio L, Boyd K, Matanoski G et al (2008) Carotenoids and the risk of developing lung cancer: a systematic review. Amer J Clin Nutr 88:372–383

Ghosh D, Scheepens A (2009) Vascular action of polyphenols. Mol Nutri Food Res 53:322–331. doi:10.1002/mnfr.200800182

Giacalone M, Di Sacco F, Traupe I et al (2011) Antioxidant and neuroprotective properties of blueberry polyphenols: a critical review. Nutr Neurosci 14:119–125. doi:10.1179/1476830511Y.0000000007

Godycki-Cwirko M, Krol M, Krol B et al (2010) Uric acid but not apple polyphenols Is responsible for the rise of plasma antioxidant activity after apple juice consumption in healthy subjects. J Amer Coll Nutr 29:397–406

Gonzalez CA, Riboli E (2006) Diet and cancer prevention: where we are, where we are going. Nutr Cancer 56:225–231

González R, Ballester I, López-Posadas R et al (2011) Effects of flavonoids and other polyphenols on inflammation. Crit Rev Food Sci Nutr 51:331–362. doi:10.1080/10408390903584094

Goodman GE, Thornquist MD, Balmes J et al (2004) The Beta-Carotene and Retinol Efficacy Trial: incidence of lung cancer and cardiovascular disease mortality during 6-year follow-up after stopping β-carotene and retinol supplements. J Nat Cancer Inst 96:1743–1751

Gordillo G, Fang H, Khanna S et al (2009) Oral administration of blueberry inhibits angiogenic tumor growth and enhances survival of mice with endothelial cell neoplasm. Antiox Redox Signal 11:47–58. doi:10.1089/ars.2008.2150

Grace MH, Ribnicky DM, Kuhn P et al (2009) Hypoglycemic activity of a novel anthocyanin-rich formulation from lowbush blueberry, *Vaccinium angustifolium* Aiton. Phytomedicine 16:406–415. doi:10.1016/j.phymed.2009.02.018

Gramann J, Gerald L (2005) Terpenoids as plant antioxidants. In: Vitamins & Hormones. Academic Press, pp 505–535
Grant WB (2004) A multicountry ecologic study of risk and risk reduction factors for prostate cancer mortality. Eur Urol 45:271–279
Griffiths G, Trueman L, Crowther TE et al (2002) Onions—a global benefit to health. Phytother Res 16:603–615
Gu Y, Nieves JW, Stern Y et al (2010) Food combination and Alzheimer disease risk: a protective diet. Arch Neurol 67:699–706. doi:10.1001/archneurol.2010.84
Güçlü-Üstünda Ö, Mazza G (2007) Saponins: properties, applications and processing. Crit Rev Food Sci Nutr 47:231–258
Guri AJ, Hontecillas R, Bassaganya-Riera J (2007) Abscisic acid ameliorates experimental IBD by downregulating cellular adhesion molecule expression and suppressing immune cell infiltration. Clin Nutr 26:107–116
Guri AJ, Hontecillas R, Bassaganya-Riera J (2010a) Abscisic acid synergizes with rosiglitazone to improve glucose tolerance and down-modulate macrophage accumulation in adipose tissue: possible action of the cAMP/PKA/PPAR [gamma] axis. Clin Nutr 29:646–653
Guri AJ, Misyak SA, Hontecillas R et al (2010b) Abscisic acid ameliorates atherosclerosis by suppressing macrophage and CD4 + T cell recruitment into the aortic wall. J Nutr Biochem 21:1178–1185. doi:10.1016/j.jnutbio.2009.10.003
Habauzit V, Morand C (2012) Evidence for a protective effect of polyphenols-containing foods on cardiovascular health: an update for clinicians. Thera Adv Chron Dis 3(2):87–106
Håkansson A, Bränning C, Molin G et al (2012) Blueberry husks and probiotics attenuate colorectal inflammation and oncogenesis, and liver injuries in rats exposed to cycling DSS-treatment. PLoS ONE 7:e33510. doi:10.1371/journal.pone.0033510
Halliwell B, Rafter J, Jenner A (2005) Health promotion by flavonoids, tocopherols, tocotrienols, and other phenols: direct or indirect effects? Antioxidant or not? Am J Clin Nutr 81:268S–276S
Hamer M, Chida Y (2007) Intake of fruit, vegetables, and antioxidants and risk of type 2 diabetes: systematic review and meta-analysis. J Hypertension 25:2361–2369
Hanhineva K, Törrönen R, Bondia-Pons I et al (2010) Impact of dietary polyphenols on carbohydrate metabolism. Int J Mol Sci 11:1365–1402. doi:10.3390/ijms11041365
Harborne JB (1977) Phytochemistry of fruits and vegetables: an ecological overview. In: Tomas-Barberan FA, Robins RJ (eds) Proceedings of the phytochemical society of Europe—phytochemistry of fruit and vegetables. Clarendon Press, Oxford, pp 353–367
Haslam E, Lilley TH (1988) Natural astringency of foodstuffs—a molecular interpretation. CRC Rev Food Sci Nutr 27:1–40
He FJ, Nowson CA, MacGregor GA (2006) Fruit and vegetable consumption and stroke: meta-analysis of cohort studies. Lancet 367:320–326
Hercberg S, Galan P, Preziosi P et al (2004) The SU.VI.MAX Study: a randomized, placebo-controlled trial of the health effects of antioxidant vitamins and minerals. Arch Intern Med 164:2335–2342. doi:10.1001/archinte.164.21.2335
Hermsdorff HHM, Zulet MA, Puchau B, Martinez JA (2010) Fruit and vegetable consumption and proinflammatory gene expression from peripheral blood mononuclear cells in young adults: a translational study. Nutr Metabol 7:42. doi:10.1186/1743-7075-7-42
Hertog MG, Kromhout D, Aravanis C et al (1995) Flavonoid intake and long-term risk of coronary heart disease and cancer in the seven countries study. Arch Intern Med 155:381–386
Hollman PCH, Arts ICW (2000) Flavonols, flavones and flavanols—nature, occurrence and dietary burden. J Sci Food Agri 80:1081–1093. doi:10.1002/(SICI)1097-0010(20000515)80:7 <1081::AID-JSFA566> 3.0.CO;2–G
Hollman PCH, Cassidy A, Comte B et al (2011a) The biological relevance of direct antioxidant effects of polyphenols for cardiovascular health in humans is not established. J Nutr 141:989S–1009S. doi:10.3945/jn.110.131490
Hollman PCH, Cassidy A, Comte B et al (2011b) The biological relevance of direct antioxidant effects of polyphenols for cardiovascular health in humans is not established. The journal of Nutrition 141:989S–1009S. doi:10.3945/jn.110.131490

Hu J, La Vecchia C, Negri E et al (1999) Diet and brain cancer in adults: a case-control study in northeast China. Int J Cancer 81:20–23

Hubbard GP, Wolffram S, Lovegrove JA, Gibbins JM (2006) Ingestion of quercetin inhibits platelet aggregation and essential components of the collagen-stimulated platelet activation pathway in man: a pilot study. J Thromb Haem 2:2138–2145

Hughes DA (1999) Effects of carotenoids on human immune fonction. Proc Nutr Soc 58:713–718

Hung H-C, Joshipura KJ, Jiang R et al (2004) Fruit and vegetable intake and risk of major chronic disease. J Nat Cancer Inst 96:1577–1584. doi:10.1093/jnci/djh296

Jeong W-S, Jun M, Kong A-NT (2006) Nrf2: a potential molecular target for cancer chemoprevention by natural compounds. Antiox Redox Sign 8:99–106. doi:10.1089/ars.2006.8.99

Jia H-F, Chai Y-M, Li C-L et al (2011) Abscisic acid plays an important role in the regulation of strawberry fruit ripening. Plant Physiol 157:188–199. doi:10.1104/pp.111.177311

Jordt S-E, McKemy DD, Julius D (2003) Lessons from peppers and peppermint: the molecular logic of thermosensation. Curr Op Neurobiol 13:487–492

Joseph JA, Shukitt-Hale B, Denisova NA et al (1998) Long-term dietary strawberry, spinach, or vitamin E supplementation retards the onset of age-related neuronal signal-transduction and cognitive behavioral deficits. J Neurosci 18:8047–8055

Joseph JA, Fisher DR, Cheng V et al (2008) Cellular and behavioral effects of stilbene resveratrol analogues: implications for reducing the deleterious effects of aging. J Agric Food Chem 56:10544–10551. doi:10.1021/jf802279h

Kalea AZ, Clark K, Schuschke DA, Klimis-Zacas DJ (2009) Vascular reactivity is affected by dietary consumption of wild blueberries in the Sprague-Dawley rat. J Med Food 12:21–28. doi:10.1089/jmf.2008.0078

Kausar H, Jeyabalan J, Aqil F et al (2012) Berry anthocyanidins synergistically suppress growth and invasive potential of human non-small-cell lung cancer cells. Cancer Let 325:54–62. doi:10.1016/j.canlet.2012.05.029

Keck AS, Finley JW (2004) Cruciferous vegetables: cancer protective mechanisms of glucosinolate hydrolysis products and selenium. Integr Cancer Therap 3:5–12

Kendall CWC, Jenkins DJA (2004) A dietary portfolio: maximal reduction of low-density lipoprotein cholesterol with diet. Curr Atheroscler Rep 6:492–498

Khaw KT, Wareham NJ, Bingham SA et al (2007) Combined impact of health behaviours and mortality in men and women: The EPIC-Norfolk prospective population study. PLOS Med 5:e12

Kiddle GA, Doughty KJ, Wallsgrove RM (1994) Salicylic acid-induced accumulation of glucosinolate biosynthesis from *Brassica juncea* cell cultures. J Exp Bot 45:1343–1346

Kim KH, Park Y (2011) Food components with anti-obesity effect. Ann Rev Food Sci Technol 2:237–257

Kimura M, Rodriguez-Amaya DB (2003) Carotenoid composition of hydroponic leafy vegetables. J Agric Food Chem 51:2603–2607

Kliebenstein DJ, Kroymann J, Mitchell-Olds T (2005) The glucosinolate-myrosinase system in an ecological and evolutionary context. Curr Op Plant Biol 8:264–271

Kliebenstien DJ (2004) Secondary metabolites and plant/environment interactions: a view through Arabidopsis thaliana tinged glasses. Plant, Cell & Environment 27:675–684

Klipstein-Goldberg K, Launer LJ, Geleinjnse JM et al (2000) Serum carotenoids and atherosclerosis: the Rotterdam study. Atherosclerosis 148:49–56

Knekt P, Jarvinen R, Teppo L et al (1999) Role of various carotenoids in lung cancer prevention. J Nat Cancer Inst 91:182–184

Kogure K, Goto S, Nishimura M et al (2002) Mechanism of potent antiperodixidative effect of capsaicin. Biochim Biophys Acta 1573:84–92

Kraft TFB, Schmidt BM, Yousef GG et al (2006) Chemopreventive potential of wild lowbush blueberry fruits in multiple stages of carcinogenesis. J Food Sci 70:S159–S166. doi:10.1111/j.1365-2621.2005.tb07151.x

Krikorian R, Shidler MD, Nash TA et al (2010) Blueberry supplementation improves memory in older adults. J Agric Food Chem 58:3996–4000. doi:10.1021/jf9029332

Kroon PA, Clifford MN, Crozier A et al (2004) How should we assess the effects of exposure to dietary polyphenols in vitro? Am J Clin Nutr 80:15–21
Kubec R, Svobodova M, Velisek J (2000) Distribution of S-alk(en)ylcysteine sulfoxides in some Allium species. Identification of a new flavor precursor: S-ethylcysteine sulfoxide (Ethiin). J Agric Food Chem 48:428–433. doi:10.1021/jf990938f
Kuhnau J (1976) A class of semi-essential food components: their role in human nutrition. World Rev Nutr Diet 24:117–191
Labbe D, Provençal M, Lamy S et al (2009) The flavonols quercetin, kaempferol, and myricetin inhibit hepatocyte growth factor-induced medulloblastoma cell migration. J Nutr 139:646–652. doi:10.3945/jn.108.102616
Lampe JW (1999) Health effects of vegetables and fruit: assessing mechanisms of action in human experimental studies. Am J Clin Nutr 70:475S–490S
Lamport DJ, Dye L, Wightman J, Lawton CL (2012) The effects of flavonoid and other polyphenol consumption on cognitive performance: a systematic research review of human experimental and epidemiological studies Nutri Aging 1:5–25. doi:10.3233/NUA-2012-0002
Lancaster JE, Kelly KE (1983) Quantitative analysis of the S-alk(en)yl-L-cysteine sulphoxides in onion (*Allium cepa* L.). J Sci Food Agri 34:1229–1235. doi:10.1002/jsfa.2740341111
Lau FC, Shukitt-Hale B, Joseph JA (2005) The beneficial effects of fruit polyphenols on brain aging. NBA 26 Suppl 1:128–132. doi:10.1016/j.neurobiolaging.2005.08.007
Lepage-Degivry MT, Bidard NN, Rouvier E et al (1986) Presence of abscisic acid, a phytohormone, in mammalian brain. PNAS 83:1155–1158
Letenneur L, Proust-Lima C, Le Gouge A et al (2007) Flavonoid intake and cognitive decline over a 10-year period. Amer J Epidem 165:1364–1371. doi:10.1093/aje/kwm036
Limpens J, Schroder FH, de Ridder CMA et al (2006) Combined lycopene and vitamin E treatment suppresses the growth of PC-346C human prostate cancer cells in nude mice. J Nutr 136:1287–1293
Loke WM, Hodgson JM, Proudfoot JM et al (2008a) Pure dietary flavonoids quercetin and (−)-epicatechin augment nitric oxide products and reduce endothelin-1 acutely in healthy men. Am J Clin Nutr 88:1018–1025
Loke WM, Proudfoot JM, Stewart S et al (2008b) Metabolic transformation has a profound effect on anti-inflammatory activity of flavonoids such as quercetin: lack of association between antioxidant and lipoxygenase inhibitory activity. Biochem Pharmacol 75:1045–1053
Long LH, Clement MV, Halliwell B (2000) Artifacts in cell culture: rapid generation of hydrogen peroxide on addition of (−)-epigallocatechin, (−)-epigallocatechin gallate, (+)-catechin, and quercetin to commonly used cell culture media. Biochem Biophys Acta 273:50–53
Lotito SB, Frei B (2006) Consumption of flavonoid-rich foods and increased plasma antioxidant capacity in humans: cause, consequence, or epiphenomenon? Free Rad Biol Med 41:1727–1746
Macheix J-J, Fleuriet A, Billot J. (1990) Fruits phenolics. Boca Raton, Fla. CRC Press, p 378
Manach C, Williamson G, Morand C et al (2005) Bioavailability and bioefficacy of polyphenols in humans. I. Review of 97 bioavailability studies. Am J Clin Nutr 81:230S–242
Mantena SK, Baliga MS, Katiyar SK (2006) Grape seed proanthocyanidins induce apoptosis and inhibit metastasis of highly metastatic breast carcinoma cells. Carcinogenesis 27:1682–1691. doi:10.1093/carcin/bgl030
Martineau LC, Couture A, Spoor D et al (2006) Anti-diabetic properties of the Canadian lowbush blueberry *Vaccinium angustifolium* Ait. Phytomedicine 13:612–623
Matchett MD, MacKinnon SL, Sweeney MI et al (2006) Inhibition of matrix metalloproteinase activity in DU145 human prostate cancer cells by flavonoids from lowbush blueberry (*Vaccinium angustifolium*): possible roles for protein kinase C and mitogen-activated protein-kinase-mediated events. J Nutr Biochem 17:117–125. doi:10.1016/j.jnutbio.2005.05.014
Mathews-Roth MM (1993) Carotenoids in erythropoietic protoporphyria and other photosensitivity diseases. Ann New York Acad Sci 691:127–138
McDougall GJ, Kulkarni NN, Stewart D (2008) Current developments on the inhibitory effects of berry polyphenols on digestive enzymes. BioFactors 34:73–80
McDougall GJ, Stewart D (2005) The inhibitory effects of berry polyphenols on digestive enzymes. BioFactors 23:189–195

McGhie TK, Ainge GD, Barnett LE et al (2003) Anthocyanin glycosides from berry fruit are absorbed and excreted unmetabolized by both humans and rats. J Agric Food Chem 51:4539–4548

McIntyre KL, Harris CS, Saleem A et al (2009) Seasonal phytochemical variation of anti-glycation principles in lowbush blueberry (*Vaccinium angustifolium*). Planta Med 75:286–292. doi:10.1055/s-0028-1088394

McNaughton SA, Marks GC (2003) Development of a food composition database for the estimation of dietary intakes of glucosinolates, the biologically active constituents of cruciferous vegetables. Brit J Nutr 90:687–697

Milbury PE, Vita JA, Blumberg JB (2010) Anthocyanins are bioavailable in humans following an acute dose of cranberry juice. J Nutr 140:1099–1104. doi:10.3945/jn.109.117168

Milner JA (2001) Garlic: the mystical food in health promotion. In: Wildman REC (ed) Handbook of nutraceuticals and functional foods. CRC Press, Boca Raton, pp 193–207

Moghe SS, Juma S, Imrhan V (2012) Effect of blueberry polyphenols on 3T3-F442A preadipocyte differentiation. J Med Food 15:448–452

Molan AL, Lila MA, Mawson J (2008) Satiety in rats following blueberry extract consumption induced by appetite-suppressing mechanisms unrelated to in vitro or in vivo antioxidant capacity. Food Chem 107:1039–1044

Munday R, Munday JS, Munday CM (2003) Comparative effects of mono-, di-, tri-, and tetrasulfides derived from plants of the Allium family: redox cycling in vitro and hemolytic activity and Phase 2 enzyme induction in vivo. Free Rad Biol Med 34:1200–1211

Murphy BT, MacKinnon SL, Yan X et al (2003) Identification of triterpene hydroxycinnamates with in vitro antitumor activity from whole cranberry fruit (*Vaccinium macrocarpon*). J Agric Food Chem 51:3541–3545. doi:10.1021/jf034114g

Ness AR, Powles JW (1997) Fruit and vegetables, and cardiovascular disease: a review. Internat J Epidemiol 26:1–13

Neto CC (2007) Cranberry and blueberry: evidence for protective effects against cancer and vascular diseases. Mol Nutr Food Res 51:652–664. doi:10.1002/mnfr.200600279

Neto CC, Amoroso JW, Liberty AM (2008) Anticancer activities of cranberry phytochemicals: an update. Mol Nutr Food Res 52:S18–S27. doi:10.1002/mnfr.200700433

Omenn GS, Goodman GE, Thornquist MD et al (1996) Effects of a combination of beta carotene and vitamin A on lung cancer and cardiovascular disease. N Engl J Med 334:1150–1155. doi:10.1056/NEJM199605023341802

Ono M, Masuoka C, Koto M et al (2002) Antioxidant ortho-benzoyloxyphenyl acetic acid ester, vaccihein A, from the fruit of rabbiteye blueberry (Vaccinium ashei). Chem Pharm Bull 50:1416–1417

Ostertag LM, O'Kennedy N, Kroon PA et al (2010) Impact of dietary polyphenols on human platelet function–a critical review of controlled dietary intervention studies. Mol Nutri Food Res 54:60–81. doi:10.1002/mnfr.200900172

Pajuelo D, Díaz S, Quesada H et al (2011) Acute administration of grape seed proanthocyanidin extract modulates energetic metabolism in skeletal muscle and BAT mitochondria. J Agric Food Chem 59(8):4279–4287. doi:10.1021/jf200322x

Palozza P, Krinsky NI (1992) Antioxidant effects of carotenoids in vivo and in vitro: an overview. Meth Enzymol 213:403–420

Papandreou MA, Dimakopoulou A, Linardaki ZI et al (2009) Effect of a polyphenol-rich wild blueberry extract on cognitive performance of mice, brain antioxidant markers and acetylcholinesterase activity. Behaviour Brain Res 198:352–358. doi:10.1016/j.bbr.2008.11.013

Parr AJ, Bolwell GP (2000) Phenols in the plant and in man. The potential for possible nutritional enhancement of the diet by modifying the phenols content or profile. J Sci Food Agric 80:985–1012

Perera RM, Bardeesy N (2011) Cancer: when antioxidants are bad. Nature London 475:43–44. doi:10.1038/475043a

Perez-Vizcaino F, Duarte J, Andriantsitohaina R (2006) Endothelial function and cardiovascular disease: effects of quercetin and wine polyphenols. Free Radic Res 40:1054–1065. doi:10.1080/10715760600823128

Potter JD (2005) Vegetables, fruit, and cancer. Lancet 366:527–530

Prasad K, Laxdal VA, Yu M, Raney BL (1995) Antioxidant activity of allicin, an active principle in garlic. Mol Cell Biochem 148:183–189. doi:10.1007/BF00928155

Price KR, Johnson IT, Fenwick GR (1987) The chemistry and biological significance of saponins in foods and feeding stuffs. CRC Rev Food Sci Nutr 26:27–137

Prior RL, Wu X (2006) Anthocyanins: structural characteristics that result in unique metabolic patterns and biological activities. Free Radic Res 40:1014–1028. doi:10.1080/10715760600758522

Proteggente AR, Pannala AS, Paganga G et al (2002) The antioxidant activity of regularly consumed fruit and vegetables reflects their phenolic and vitamin C composition. Free Radic Res 36:217–233

Ramassamy C (2006) Emerging role of polyphenolic compounds in the treatment of neurodegenerative diseases: a review of their intracellular targets. Eur J Pharmacol 545:51–64. doi:10.1016/j.ejphar.2006.06.025

Rao A, Rao L (2007) Carotenoids and human health. Pharmacol Res 55:207–216. doi:10.1016/j.phrs.2007.01.012

Rao AV, Agarwal S (1998) Bioavailability and in vivo antioxidant properties of lycopene from tomato products and their possible role in the prevention of cancer. Nutr Cancer 31:199–203. doi:10.1080/01635589809514703

Rechner AR, Kroner C (2005) Anthocyanins and colonic metabolites of dietary polyphenols inhibit platelet function. Thromb Res 116:327–334. doi:10.1016/j.thromres.2005.01.002

Reinbach HC, Smeets A, Martinussen T et al (2009) Effects of capsaicin, green tea and CH-19 sweet pepper on appetite and energy intake in humans in negative and positive energy balance. Clin Nutr 28:260–265. doi:10.1016/j.clnu.2009.01.010

Rendeiro C, Vauzour D, Kean RJ et al (2012) Blueberry supplementation induces spatial memory improvements and region-specific regulation of hippocampal BDNF mRNA expression in young rats. Psychopharmacology 223:319–330. doi:10.1007/s00213-012-2719-8

Riccioni G (2009) Carotenoids and cardiovascular disease. Curr Atheroscler Rep 11:434–439. doi:10.1007/s11883-009-0065-z

Riso P, Visioli F, Grande S et al (2006) Effect of a tomato-based drink on markers of inflammation, immunomodulation, and oxidative stress. J Agric Food Chem 54:2563–2566. doi:10.1021/jf053033c

Rissanen TH, Voutilainen S, Virtanen JK et al (2003) Low intake of fruits, berries and vegetables is associated with excess mortality in men: the Kuopio Ischaemic Heart Disease Risk Factor (KIHD) study. J Nutr 133:199–204

Ristow M, Zarse K, Oberbach A et al (2009) Antioxidants prevent health-promoting effects of physical exercise in humans. PNAS 106:8665–8670. doi:10.1073/pnas.0903485106

Rodriguez-Amaya DB, Kimura M, Godoy HT, Amaya-Farfan J (2008) Updated Brazilian database on food carotenoids: factors affecting carotenoid composition. J Food Comp Anal 21:445–463

Romero C, Brenes M, García P, Garrido A (2002) Hydroxytyrosol 4-β- d-glucoside, an important phenolic compound in olive fruits and derived products. J Agric Food Chem 50:3835–3839. doi:10.1021/jf011485t

Rosa WA, Heaney RK, Fenwick GR, Portas CAM (1997) Glucosinolates in crop plants. Horticult Rev 19:99–215

Rose P, Whiteman M, Moore PK, Zhu YZ (2005) Bioactive S-alk(en)yl cysteine sulfoxide metabolites in the genus Allium: the chemistry of potential therapeutic agents. NPR 22:351–368

Rothwell JA, Urpi-Sarda M, Boto-Ordonez M et al (2012) Phenol-Explorer 2.0: a major update of the Phenol-Explorer database integrating data on polyphenol metabolism and pharmacokinetics in humans and experimental animals. Database 2012:1–8. doi:10.1093/database/bas031

Sabour-Pickett S, Nolan JM, Loughman J, Beatty S (2012) A review of the evidence germane to the putative protective role of the macular carotenoids for age-related macular degeneration. Mol Nutr Food Res 56:270–286. doi:10.1002/mnfr.201100219

Saura-Calixto F (2012) Concept and health-related properties of nonextractable polyphenols: the missing dietary polyphenols. J Agric Food Chem 60:11195–11200. doi:10.1021/jf303758j
Scalbert A, Williamson G (2000) Dietary intake and bioavailability of polyphenols. J Nutr 130:2073S–2085S
Scalbert A, Johnson IT, Saltmarsh M (2005) Polyphenols: antioxidants and beyond. Am J Clin Nutr 81:215S–217
Scheepens A, Tan K, Paxton JW (2010) Improving the oral bioavailability of beneficial polyphenols through designed synergies. Genes Nutr 5:75–87. doi:10.1007/s12263-009-0148-z
Schewe T, Steffen Y, Sies H (2008) How do dietary flavanols improve vascular function? A position paper. Arch Biochem Biophys 476:102–106. doi:10.1016/j.abb.2008.03.004
Schmidt BM, Erdman JW Jr, Lila MA (2006) Differential effects of blueberry proanthocyanidins on androgen sensitive and insensitive human prostate cancer cell lines. Cancer Let 231:240–246. doi:10.1016/j.canlet.2005.02.003
Seeram NP (2008) Berry fruits for cancer prevention: current status and future prospects. J Agric Food Chem 56:630–635
Seeram NP, Adams LS, Zhang Y et al (2006) Blackberry, black raspberry, blueberry, cranberry, red raspberry, and strawberry extracts inhibit growth and stimulate apoptosis of human cancer cellsin vitro. J Agric Food Chem 54:9329–9339. doi:10.1021/jf061750g
Serrano J, Puupponen-Pimiä R, Dauer A et al (2009) Tannins: current knowledge of food sources, intake, bioavailability and biological effects. Mol Nutr Food Res 53:S310–S329. doi:10.1002/mnfr.200900039
Shaw JE, Sicree RA, Zimmet PZ (2010) Global estimates of the prevalence of diabetes for 2010 and 2030. Diabetes Res Clin Pract 87:4–14
Shukitt-Hale B, Lau FC, Joseph JA (2008) Berry fruit supplementation and the aging brain. J Agric Food Chem 56:636–641. doi:10.1021/jf072505f
Smith MAL, Marley KA, Seigler D et al (2000) Bioactive properties of wild blueberry fruits. J Food Sci 65:352–356. doi:10.1111/j.1365-2621.2000.tb16006.x
Snodderly DM (1995) Evidence for protection against age-related macular degeneration by carotenoids and antioxidant vitamins. Am J Clin Nutr 62:1448S–1461
Spencer JPE (2008) Food for thought: the role of dietary flavonoids in enhancing human memory, learning and neuro-cognitive performance. Proc Nutr Soc 67:238–252
Srivastava KC (1986) Onion exerts antiaggregatory effects by altering arachidonic acid metabolism in platelets. Prostag Leukotr Med 24:43–50
Stahl W, Sies H (2005) Bioactivity and protective effects of natural carotenoids. Biochim Biophys Acta 1740:101–107
Steffen Y, Gruber C, Schewe T, Sies H (2008) Mono-O-methylated flavanols and other flavonoids as inhibitors of endothelial NADPH oxidase. Arch Biochem Biophys 469:209–219. doi:10.1016/j.abb.2007.10.012
Stintzing FC, Carle R (2007) Betalains—emerging prospects for food scientists. Trends Food Sci Technol 18:514–525. doi:10.1016/j.tifs.2007.04.012
Suh N, Paul S, Hao X et al (2007) Pterostilbene, an active constituent of blueberries, suppresses aberrant crypt foci formation in the azoxymethane-induced colon carcinogenesis model in rats. Clin Cancer Res 13:350–355. doi:10.1158/1078-0432.CCR-06-1528
Surh Y-J, Sup Lee S (1995) Capsaicin, a double-edged sword: toxicity, metabolism, and chemopreventive potential. Life Sci 56:1845–1855
Szkudelska K, Szkudelski T (2010) Resveratrol, obesity and diabetes. Eur J Pharmacol 635:1–8. doi:10.1016/j.ejphar.2010.02.054
Talalay P, Fahey JW (2001) Phytochemicals from cruciferous plants protect agains cancer by modulating carcinogen matabolism. Journal of Nutrition 131:3027–3033
Tomas-Barberan FA, Gil MI (2008) Improving the health-promoting properties of fruit and vegetable product. Woodhead Publishing Limited, CRC Press, Cambridge
Traka M, Mithen R (2009) Glucosinolates, isothiocyanates and human health. Phytochem Rev 8:269–282. doi:10.1007/s11101-008-9103-7

Trichopoulou A, Naska A, Antoniou A et al (2003) Vegetable and fruit: the evidence in their favour and the public health perspective. Int J Vitam Nutr Res 73:63–69

van den Berg H, Faulks R, Granado HF (2000) The potential for the improvement of carotenoid levels in foods and the likely systemic effects. J Sci Food Agri 80:880–912. doi:10.1002/(SICI)1097-0010(20000515)80:7 <880::AID-JSFA646> 3.0.CO;2–1

Van't Veer P, Jansen MCJF, Klerk M, Kok FJ (2000) Fruits and vegetables in the prevention of cancer and cardiovascular disease. Pub Health Nutr 3:103–107

Verkerk R, Schreiner M, Krumbein A et al (2009) Glucosinolates in Brassica vegetables: the influence of the food supply chain on intake, bioavailability and human health. Mol Nutr Food Res 53:S219–S256. doi:10.1002/mnfr.200800065

Voorrips LE, Goldbohm RA, van Poppel G et al (2000) Vegetable and fruit consumption and risks of colon and rectal cancer in a prospective cohort study: the Netherlands Cohort Study on Diet and Cancer. Amer J Epidem 152:1081–1092

Voutilainen S, Nurmi T, Mursu J, Rissanen TH (2006) Carotenoids and cardiovascular health. Amer J Clin Nutr 83:1265–1271

Vuong T, Martineau L, Ramassamy C et al (2007) Fermented Canadian lowbush blueberry juice stimulates glucose uptake and AMP-activated protein kinase in insulin-sensitive cultured muscle cells and adipocytes. Can J Physiol Pharm 85:956–965

Vuong T, Matar C, Ramassamy C (2010) Biotransformed blueberry juice protects neurons from hydrogen peroxide-induced oxidative stress and mitogen-activated protein kinase pathway alterations. Brit J Nutr 104:656–663

Waffo-Teguo P, Krisa S, Richard T, Mérillon J-M (2008) Grapevine stilbenes and their biological effects. In: Ramawat KG, Mérillon JM (eds) Bioactive molecules and medicinal plants. Springer, pp 25–54

Wedge DE, Meepagala KM, Magee JB et al (2001) Anticarcinogenic activity of strawberry, blueberry, and raspberry extracts to breast and cervical cancer cells. J Med Food 4:49–51. doi:10.1089/10966200152053703

Williams CM, El-Mohsen MA, Vauzour D et al (2008) Blueberry-induced changes in spatial working memory correlate with changes in hippocampal CREB phosphorylation and brain-derived neurotrophic factor (BDNF) levels. Free Rad Biol Med 45:295–305. doi:10.1016/j.freeradbiomed.2008.04.008

Williams RJ, Spencer J (2011) Flavonoids, cognition and dementia: actions, mechanisms and potential therapeutic utility for Alzheimer's disease. Free Rad Biol Med 52:35–45

Williamson G, Clifford MN (2010) Colonic metabolites of berry polyphenols: the missing link to biological activity? Brit J Nutr 104:S48–S66. doi:10.1017/S0007114510003946

Wilms LC, Hollman PCH, Boots AW, Kleinjans JCS (2005) Protection by quercetin and quercetin-rich fruit juice against induction of oxidative DNA damage and formation of BPDE-DNA adducts in human lymphocytes. Mut Res/Gen Toxicol Environ Mutagen 582:155–162. doi:10.1016/j.mrgentox.2005.01.006

Woods SC, D'Alessio DA (2008) Central control of body weight and appetite. J Clin Endocrinol Metab 93:s37–s50. doi:10.1210/jc.2008-1630

World Cancer Research Fund (2007) Food, nutrition, physical activity, and the prevention of cancer: a global perspective. World Cancer Research Fund, Washington DC: AICR

Wu X, Kang J, Xie C et al (2010) Dietary blueberries attenuate atherosclerosis in apolipoprotein E-deficient mice by upregulating antioxidant enzyme expression. J Nutr 140:1628–1632. doi:10.3945/jn.110.123927

Xie C, Kang J, Ferguson ME et al (2011) Blueberries reduce pro-inflammatory cytokine TNF-$\alpha$ and IL-6 production in mouse macrophages by inhibiting NF-$\kappa$B activation and the MAPK pathway. Mol Nut Food Res 55:1587–1591. doi:10.1002/mnfr.201100344

You WC, Li JY, Zhang L et al (2005) Etiology and prevention of gastric cancer: a population study in high risk area of China. Chin J Digest Dis 6:149–154

Youdim KA, Shukitt-Hale B, MacKinnon SL et al (2000) Polyphenolics enhance red blood cell resistance to oxidative stress: in vitro and in vivo. Biochem Biophys Acta 1523:117–122

Zifkin M, Jin A, Ozga JA et al (2012) Gene expression and metabolite profiling of developing highbush blueberry fruit indicates transcriptional regulation of flavonoid metabolism and activation of abscisic acid metabolism. Plant Physiol 158:200–224. doi:10.1104/pp.111.180950

# Chapter 29
# Health and Well-Being

**Ross W. F. Cameron**

**Abstract** Horticulture is considered to be one of a number of activities that promote green- or eco-therapy. Working with plants in a (semi-) natural landscape is thought to provide a range of health and social benefits. Benefits that would not necessarily accrue in an anthropogenic dominated landscape working with artificial materials. The benefit of horticultural activities relate to the promotion of relaxation in stressed individuals through physical activity and exposure to natural/living objects, the promotion of social skills in a socially non-demanding environment (e.g. outdoors, facilitating flexible and mobile group structures) or the ability to set goals, work and achieve results at one's own pace. Although most scientific data cites the stress avoidance/stress relief characteristics of horticulture and green spaces, there may also be an argument for stimulation in some user groups too, although this is less well-defined. The benefits of horticulture are most manifest through Horticultural Therapy and Social and Therapeutic Horticulture (although paradoxically these areas have least scientific evidence supporting them), but even exposure to green landscapes, for example, by viewing natural scenes from a hospital window can provide physiological and psychological benefits. This chapter highlights horticulture contribution to eco-therapy and alludes to how form, quantity and quality of landscape may impact of the extent of those benefits.

**Keywords** Health · Well being · Green-Infrastructure · Green-space · Natural environment · Created environment · Stress recovery · Attention recovery · Eco-therapy · Psychological health · Social benefit · Obesity · Cancer · Heart disease

## Introduction

The relationship between horticulture and health and well-being is a component of a wider spectrum of interactions between humans and natural/semi-natural landscapes. In this context, landscapes are often defined as 'green space' or 'green infrastructure' and many are intrinsic to horticultural activities and landscapes (landscape horticulture, allotments, gardening, park landscapes, arboreta, cemeter-

R. W. F. Cameron (✉)
Department of Landscape, University of Sheffield, Sheffield, United Kingdom
e-mail: r.w.cameron@sheffield.ac.uk

G. R. Dixon, D. E. Aldous (eds.), *Horticulture: Plants for People and Places, Volume 3*, 1001
DOI 10.1007/978-94-017-8560-0_29, © Springer Science+Business Media Dordrecht 2014

ies, interior landscapes, green roofs/walls and even 'guerrilla' gardening;—where volunteer groups cultivate and plant urban areas without any official authority to do so). Although not always referred to as 'horticulture'—horticultural-like activities (planting, pruning, thinning, mowing, weeding, soil cultivation, irrigation) are also responsible for the appearance of many other forms of green space—sports facilities, urban forests, nature reserves, wild flower meadows, roadside verges, green networks and corridors and even the maintenance of semi-naturalistic ecosystems such as chalk down-land or prairie grasslands. As such, this review will deal with horticulture in its widest definition and be used interchangeably with green space management or maintenance. Reference will be made however, to specific forms of green space or particular types of activity where these may influence the value or benefits associated with health and well-being.

The implicit relationship between green places and human health and well-being has been acknowledged for at least 200 years, although a formal scientific evidence base has only begun to form over the last 30 years or so. Indeed, it might be argued that community leaders in the 19th Century (e.g. the 'Victorians' in the UK) had a greater understanding of the relationship between green space and well-being, than subsequent generations of planners and designers (Parr 2007). Despite the industrialisation and rapid urban sprawl that was associated with the Victorian era, the Victorians were also champions of city parks, street trees and the requirement for the medical institutions and hospitals of the day to be placed in 'green and pleasant' surroundings. Many 'sanatoriums' were placed in spacious green field sites to help provide a peaceful and tranquil sanctuary for the patients. Interest in ornamental horticulture 'blossomed' during this era too—the collection and cultivation of garden plants became commonplace, particularly in Europe (Hadfield 1969). Horticultural societies were formed and provided a rest-bite to long hours of heavy manual or repetitive work for miners, weavers, mill-workers, potters and steel workers (Duthie 1988). Workers who moved to the cities, but still retained roots to a rural agricultural society, sought and were encouraged to engage in horticultural activities. The relationship between green space and health became much more ambiguous over the intervening years. The term 'environmental health' coined in the 1960's is symptomatic of the negative aspects of nature—e.g. diseases caused or transmitted by biological agents or by risks posed by the natural world (environmental extremes, features of the landscape or potentially injurious organisms, e.g. poisonous plants, fungi or venomous animals), rather than highlight the positive influences of the natural environment (Morris 2003). A key paradox has been that despite many people intrinsically engaging with nature and claiming it to have positive effects on wellbeing (denoted by everyday terms such as 'fresh-air' or 'natural goodness') only recently has the relationship between health and landscape influenced decisions about green space infrastructure. Even today, the arguments surrounding the development of green space tend to polarise between on the one hand the need for economic development, requirement for housing and the building of infrastructure to support industry and on the other hand, the impact on local biodiversity, reduction of aesthetic values or loss of a particular resource (e.g. place to 'walk the dog').

Only in the last 10 years of so (and in many places it is still absent) does the impact on human physical and psychological health enter the cost/benefit analysis of the decision-making process. It is largely because the costs of health care have increased so dramatically in recent years (due to enhanced population longevity, a more sedentary lifestyle and fat-rich diet—Eastwood 2011), that policy makers have begun to focus on the role that green space can play on lifestyle and health-related issues (Maller et al. 2006).

Defining the relationship between human well-being and green landscapes, however, is a complex process. Nevertheless, many organisations involved with managing the natural environment have connected with the principle that participation in the green landscape is beneficial in terms of well-being (e.g. The European Environment Agency; Natural England and the Royal Society for the Protection of Birds, UK: USDA Forest Service USA), a number of which have developed policies around health benefits to humans. In parts of the medical community too, the benefits of green space are now acknowledged (Maller et al. 2006).

> Some of 'exposure to nature' approaches appear to be just as effective in achieving health gains as traditional drug-oriented treatments. (St Leger 2003).

Medical practitioners in the UK are encouraged to recommend 'green' or 'eco-therapy' to their patients to improve physical fitness and aid psychological restoration (Bird 2007). Such activities include clearing scrub woodland, digging and maintaining ponds and planting trees and wildflowers. Many of the activities involve group work and the 'low key', unpressurised social dynamic is thought to help patients improve communication skills, regain confidence and improve self-esteem. For elderly or disabled people there is less stigma attached to these activities compared to competitive sports. There seems to be enhanced longevity of engagement too, with drop-out rates less than that of joining a conventional gym (Morris 2003).

Nevertheless, the evidence base for health and well-being related to green space remains controversial (Lee and Maheswaran 2011) or incomplete, particularly with respect to details on what sorts of activities or landscapes provide (most) benefit and can be used to treat or counteract specific illness or disorders.

## The Evidence for Health Benefits from Green Space/Horticulture

The health (and allied social) benefits associated with engagement with the green space have been increasingly researched in recent years (e.g. Cimprich and Ronis 2003; St Leger 2003; Hartig 2007; Berman et al. 2008; van den Berg et al. 2010; Mitchell 2012) and largely centre on the attention restoration theory (ART) (Kaplan 1995), or the psychophysiological stress recovery theory (PSRT) (Ulrich et al. 1991). ART states that nature or green space provides a restorative opportunity by acting at four levels; allowing patients to 'be away' (removed from stress inducing factors either physically or psychologically), promotes 'fascination' (effortless,

**Fig. 29.1** The natural world is thought to stimulate 'fascination' helping individuals to substitute 'directed attention' (often associated with pressures of work and lifestyle) for 'indirect attention' (effortless, thought processes on natural or abstract subject matter)

interest-driven engagement with natural objects, Fig. 29.1), provides 'extent' (scope for exploration and curiosity-driven discovery) and is 'compatible' (how closely the environment matches one's interests or needs at that time). Kellert and Wilson (1993) go further, arguing that engagement with nature is an essential component for human development and well-being—the 'Biophilia hypothesis'. Wilson (1984) stated that biophilia is determined by biological needs and is an emotional or spiritual relationship between humans and nature. It has been theorised that biophilia relates to our evolution, e.g. the fact that *Homo sapiens* developed in a 'parkland like' savannah habitat and has a particular affinity to certain landscape features even today (e.g. attraction to water—evolutionary important as a source of food and drink, a desire for vantage points and vistas to view the wider landscape and the need for refuge points to provide protection and shelter, Fig. 29.2).

The opportunities for enhanced physiological health through greater physical activity in green space are increasingly cited in relation to health policy documents. Research has identified key health benefits attributed to green space/activities as; faster recovery times from illness (Ulrich et al. 1991), short-term recovery from stress or mental fatigue (Kaplan 1995, 2001), longer term improvements in health and well-being (Verlarde et al. 2007), providing enhanced opportunity for physical activity (Björk et al. 2008) and not least, reduced mortality (Mitchell and Popham

**Fig. 29.2** Does today's garden patio reflect the evolutionary need for a sheltered location with a view? (Image: A. Clayden, University of Sheffield)

2008). In terms of psychological health, the benefits have been outlined by Maller et al. (2002, 2006) and Stone and Hanna (2003) as:

- Improved self-awareness, self-esteem, self-concept, and positive mood state (Barton et al. 2012).
- Reduction of negative feelings such as anger, fear, anxiety and frustration.
- Improved ability to recover from stressful episodes.
- Effective alleviation of the symptoms of anxiety, depression and psychosomatic illness (including irritability, insomnia, tension, headaches and indigestion).
- Improved emotional and cognitive aspects (including reduced symptoms of Attention Deficit Disorder—ADD).
- Restored capacity for concentration and attention.
- Enhanced spiritual, sensory and aesthetic awareness.
- Delays in the onset of dementia (Simons et al. 2006) and aid to those suffering from it (Jarrott et al. 2002; Lee and Kim 2008).

Physiological benefits include (Hartig et al. 2003; Tzoulas et al. 2007; Bowler et al. 2010):

- Cardiovascular-respiratory fitness and reduced heart/pulse rate (Ulrich et al. 1991; Hartig et al. 1991, 2003; Tsunetsugu et al. 2007).
- Pain relief (Ulrich 1984) including relief from headache (Moore 1981; Hansmann et al. 2007).
- Lower skin conductance, (Verlarde et al. 2007).
- Reduced muscle tension (Tzoulas et al. 2007).
- Enhanced physiological motor performance and maintenance of mobility (Cooper and Barnes 1999; Scholz and Krombholz 2007).
- Longer telomere length (less oxidative damage to DNA sections) (Woo et al. 2009).
- Cortisol (stress hormone) reduction (van den Berg and Custers 2011).
- Improved brain function (Park et al. 2007).

- Diminished risk of osteoporosis, falls and fractures (Cooper and Barnes 1999)
- Reduced digestive illness (Moore 1981).

Social benefits (Morris 2003) include:

- Enhanced personal and social communication skills
- Increased independence and an ability to assert personal control and increased sensitivity to one's own well-being.

In contrast, other studies suggest that generic links between health and green infrastructure are weak (Lee and Maheswaran 2011), or the response is dependent on factors such as sample population studied (Ottosson and Grahn 2005), the form, extent or quality of the green infrastructure (Milligan and Bingley 2007; Mitchell and Popham 2007), the nature of the 'green' activity undertaken (Barton and Pretty 2010) or ancillary factors such as ease of access, degree of motivation or perceptions of safety (Lee and Maheswaran 2011). Korpela et al. (2010) also suggest that although natural/semi-natural landscapes promote greater health benefits, there was no difference within a city context between green space (parks) and favourite placed within the built infrastructure. Therefore, familiar places selected on the basis of preference or emotional attachment (whether they are green or not) may have some restorative value.

Despite these observations Maller et al. (2006) stated that

> public health strategies have yet to maximise the untapped resource nature provides.

These authors commented on increased disengagement with the natural world and potential negative consequences for human health. A point echoed by Kellert and Wilson 1993; Pretty and Ward 2001; Pretty et al. 2003. Federighi (2008) cites examples of disconnection from nature in the USA.

- "In a single 6-year period (1997–2003), the proportion of American children aged 9–12 who spent time on outdoor activities such as hiking, gardening, and fishing fell by 50 %."
- "On average, American children aged 8–18 spend: almost 6.5 h per day with electronic media, including more than 3 h watching TV; and only 30 min per week on unstructured outdoor activities."
- "National undergraduate enrolment in natural resource science programmes has declined by 40 % since 1995."
- In essence, Pretty et al. (2003) suggest "that our whole physiology has been outpaced by the recent changes in society and lifestyle."

Increases in obesity and a sedentary lifestyle are increasing in industrialised societies. In England, UK, 22 % of men and 24 % of women aged $\geq 16$ were classified as obese in 2009 (Body mass index $\geq 30 \text{kg m}^{-2}$)(Eastwood 2011). When combined with information on where fat is laid down this correlates with 20 % of men and 23 % of women deemed to be in the 'very high health risk' category. In the same census 16 % of boys and 15 % of girls were classed as obese, an increase from 11 % and 12 % respectively from 1995. The costs of treating obesity related disease in the UK is thought to be £ 4.2 billion a year, but will rise rapidly to £ 6.6 billion by 2015.

**Fig. 29.3** Designed landscapes that mimic nature (in this case prairie or savannah grasslands) may be useful in terms of restoration from stress, although research on how different elements act or can be effectively be combined is still in its infancy. (Image: A. Clayden, University of Sheffield)

The impact of mental health disorders is even more stark. In the UK 18 % of the population suffer anxiety disorders or depression, exhaustion and fatigue syndromes and 1 % have a severe mental illness, such as bipolar disorder or schizophrenia. Mental health problems cost the UK an estimated £ 105.2 billion p.a. in healthcare, benefits and lost productivity (Anon 2010). Many of these forms of mental illness are thought to be activated by physiological stress. Short duration, stress reactions are a natural process, which in evolutionary terms helped humans escape threats and enhance sensory perceptions. Prolonged stress on the other hand is considered detrimental to human physiology and is linked to cardiovascular disease, neuro-hormonal imbalance, type II diabetes, susceptibility to pathogen infection and depression (Esch et al. 2002; Hartig et al. 2003).

## What Sort of Green Space?

People seem to prefer natural environments to help restore well-being as well as enjoyment. Kaplan et al. (1998) indicate that natural settings do not need to be remote wilderness, and emphasise the value of 'the everyday, often unspectacular natural environments' (Fig. 26.3), such as nearby parks and open spaces (Bedima-Rung et al. 2005), street trees (van Dillen et al. 2012), gardens (Cameron et al. 2012) and fields and forests (Nilsson et al. 2011; Park et al. 2011; Fig. 29.3).

## Physical Activity

The provision of green space can help encourage physical activity and decrease diseases associated with a sedentary lifestyle (Maas et al. 2006; Mitchell and Popham 2008; Macintyre et al. 2008; Coombes et al. 2010). Larger areas of green space may enhance health effects (Giles-Corti et al. 2005) but paradoxically be less prevalent in deprived areas where health problems are greatest (Mitchell et al. 2011). Proximity to green space, however, has been shown to have both a positive (Giles-Corti et al. 2005; Roemmich et al. 2006; Nielsen and Hansen 2007) or no (Hoehner et al. 2005; Hillsdon et al. 2006; Maas et al. 2008; Potwarka et al. 2008) influence on physical activity/health. Such inconsistent findings may reflect variations in user groups (e.g. children vs adults), dietary behaviour, degree, type or frequency of physical activity, or type and accessibility of green space. Coombes et al. (2010) demonstrated that proximity to location was important in their study; people living farther from a park or similar green space were less likely to use it, were less likely to meet the minimum guidelines for physical activity and were more likely to be overweight. Frequency of use was increased by closer proximity. Formal green spaces (parks) make them particularly suitable for physical activities claim Coombes et al. (2010) due to the network of pathways encouraging cycling, walking and jogging.

The proximity argument has been refuted to some extent by Schipperijin et al. (2010), who found that only 56 % of respondents used their nearest green space on a weekly basis. Nearby, accessible green space was seen as important for certain user groups, however; the immobile, elderly or those with young children. Dog walkers were also found to use local green space on a frequent basis. Interestingly, factors such as vegetation type, diversity of the area, presence of facilities, area shape, marital status, profession, level of stress, preference for different activities, preference for different green space elements, importance of maintenance and view on nature did not seem to change the frequency of use of local green space. Those without access to a domestic garden did not use local green spaces more frequently; people with gardens utilizing local green space as, or more, frequently, than those without. This may indicate that garden owners are more interested in spending time outside in general; in their own garden for certain activities and in urban green space for others (Grahn and Stigsdotter 2003; Maas et al. 2006, Maas et al. 2009). In contrast to those who selected local green space, many other groups would by-pass their local green space to travel further to larger green areas (larger areas being more popular e.g. >5 ha). Hence, Schipperijin et al. (2010) suggest it is important for city planners to find a balance between the provision of both areas of smaller and more extensive green space, within a reasonable distance of residents.

## Psychological Benefits

Psychological benefits largely relate to the restorative effect from stress and mental fatigue (Kaplan and Kaplan 1989; Bell et al. 2004). Korpela and Hartig (1996) suggest that restorative experiences figure in emotional and self-regulatory processes through which individuals develop an identity of place. They found people often went to their favourite places to relax, calm down and clear their mind of negative events. The favourite places identified, most often incorporated elements of greenery, water or scenic quality. Such places have coined the term 'therapeutic landscapes'.

According to Searles (1960) green space acts as a catalyst for creative processes that are important for restoration from stress. This relates to how humans process information. Information is coded and stored through 3 different processes: Subsymbolic (sensory, motor and somatic modes, as perceived by muscles, inner organs, skin), Symbolic imagery (visual pictures in a person's mind) and Symbolic verbal (verbal concepts and interpretations) (Grahn and Stigsdotter 2010). Simple relationships e.g. between an individual and inanimate objects, plants or animals do not overload these sensory processes, whereas more complex interactions (e.g. between humans during conversation) can tax them, especially in the symbolic imagery and verbal spheres. Those environments that allow engagement with the simple processes (natural landscapes, wilderness, parks, gardens) are deemed to have greater restoration potential than those that require more complex sensory processing (city centres, airports, shopping malls, business and work environments) (Hartig et al. 2003; Coon et al. 2011; Mitchell 2012). Scopelliti and Giuliani (2004) imply, however, that certain urban built environments can have restorative effects themselves.

Within this context are there differences in the restorative potential of distinct types of green space? Despite Kaplan's claims that restorative environments do not need to be pristine wilderness, there is evidence to suggest that scale, form and quality of landscape may impact on restoration potential. Korpela and Hartig (1996) and Korpela et al. (2001) found that restorative experiences were greatest in favourite exercise areas outdoors (Mitchell 2012), waterside locations and extensively managed natural settings (mainly urban woodlands) compared to either favourite places in built urban settings or green spaces in urban settings (mostly parks). Grahn and Stigsdotter (2010) argue that people looking for restoration may have slightly different preferences to those not suffering from stress or mental health problems. People suffering stress reported preference for landscapes that provide 'Refuge' and 'Nature' and that a combination of factors associated with 'Refuge', 'Nature' and 'Rich in Species', with low or no 'Social' components could be interpreted as the most restorative environments for stressed individuals.

Non-stressed individuals, however, ranked preference in landscapes in order of 'Serene', followed by 'Space', 'Nature' with 'Rich in Species' and 'Refuge' only as middle ranking factors. 'Culture' 'Prospect' and 'Social' were the least preferred factors. (See Table 29.1 for definition of terms).

Earlier work though by de Groot and van den Born (2003) suggested that seeking aspects such as serenity, solitude or refuge are not the only motivating factors. Their results showed that $\geq 50\%$ of respondents expressed preference for landscapes where one experienced the 'force of nature', with another 33% preferring wild, interactive landscapes. The authors state that preference for landscape should distinguish between visual or behavioural preferences; i.e. some people are attracted to certain landscapes for personal desires/interests, e.g. to partake in active outdoor sports. Such attractions may not be restricted to wilderness alone. In urban parks, visitors were expected to experience greater pleasure in those landscapes that possessed more trees and had less undergrowth present; as parks with these features were considered to be calming and relaxing (Hull and Harvey 1989). Preference for parks, however, increased in line with feelings of pleasure and arousal rather than relaxation. Indeed, preference related to traversing paths with thick undergrowth

**Table 29.1** Different components in the landscape that can influence stress restoration in humans as defined by Grahn and Stigsdotter (2010)

| Landscape component | Aspects |
|---|---|
| Nature: | Experiencing the inherent force and power of nature, designed and manifested on nature's own terms. |
| Culture: | Containing an essence of human culture and may relate to the inclusion of ornamental or historical components e.g. fountains, ponds and ornamental plants. |
| Prospect: | An open area with a view. Considered an evolutionary component where our ancestors tried to find vistas and vantage points in the landscape. |
| Social: | An environment that is equipped for social activities. Offers potential for people to meet, amuse themselves and watch one another. |
| Space: | A green environment with space (scope) and connecting components. Scope requires the environment to be large enough that one can move around in it to maximise an experience and connectedness is that various parts of the environment must be perceived as belonging to a larger whole. |
| Rich in species: | This comprises variables demonstrating the importance of finding a wide range of expressions of life: numerous birds, butterflies, flowers and other forms of wildlife. |
| Refuge: | An enclosed and safe environment, where people can play or watch other people being active. |
| Serene: | This factor is about being in an undisturbed, silent and calm environment, which can be interpreted as an environment for retreat—a virtually holy and safe place. |

which induced exhilaration and arousal as well as an element of fear into some visitors. The arousal-inducing characteristics were counter to the calming influence of parks expected by the researchers. This indicates that although feelings of calm and relaxation are major components of people's emotional reactions to nature, more animated response of stimulation or awe are also important (Rohde and Kendle 1994). Enjoyment of green areas may help people to relax or may actually give them fresh energy (Ulrich 1990).

Although some studies suggest larger green areas can enhance benefits, this should not be interpreted as smaller local areas having limited value. In a comparison between urban green spaces and streetscapes van Dillen et al. (2012) investigated how quality parameters (accessibility, maintenance, variation, naturalness, colourfulness, clear arrangement, shelter, absence of litter, safety and overall general impression) influenced three different health variates (perceived general health, acute health-related complaints and mental health scores). They found that the enhanced quality of streetscape greenery improved all three health indicators. With green spaces, however, enhanced quality was only linked to general health and acute health complaints. (A lack of correlation to mental health scores was attributed to shorter exposure times in the green spaces). These results suggest that immediate, everyday encounters with green space can influence health and well-being. Horticultural activities (design, plant selection, type and frequency and maintenance) that promote quality criteria, therefore may improve the health

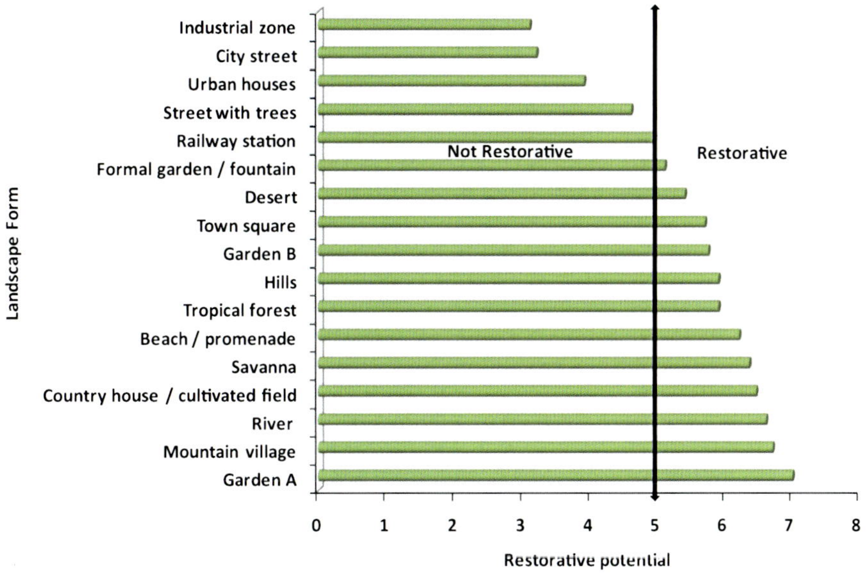


**Fig. 29.4** The relative restorative value of different landscape types. A value of ≥ 5 donating when landscapes were considered to be particularly positive in terms of restoration potential. Note the difference in score between 2 gardens with different styles of landscape. Modified from Ivarsson and Hagerhall (2008) who compared results from three different studies

potential of such landscapes. This reflects findings from Giles-Corti et al. (2005) and Ivarsson and Hagerhall (2008), the latter indicating restoration potential could vary radically based on the design of individual gardens—Fig. 29.4. A point echoed by Niepel (2004) who states that if gardens are therapeutic then designers need to understand the components that make them so. Aldous and Aldous (2008) suggests these components include intrinsic (people related) and extrinsic (plant/landscape related) variables and when combined effectively improve the design and development of the therapeutic landscape. In dealing with trade-offs between landscape quantity and quality, van Dillen et al. (2012) suggest that rather than quality and quantity being able to compensate for one another, the one becomes more critical if the other is high.

## Visualizing Green

The benefits of green landscapes have been cited not only from active activities within the landscape, but having natural scenes (or representations of them) within view. Views of green space, both real and artificial (pictures, photographs, videos) are reported to increase recovery from stress. Ulrich cites that viewing natural settings can produce significant restoration within less than five minutes as indicated by positive changes, for instance, in blood pressure, heart activity, muscle tension, and brain electrical activity (Ulrich 1981, 2001; Ulrich et al. 1991; Hartig et al. 2003). Others argue that the presence of greenery around domestic dwellings is particularly important in reduced stress loads (Kaplan 2001; Kuo and Sullivan 2001a, b; De Vries et al. 2003; Grahn and Stigsdotter 2003). Even the provision of roadside vegetation may help drivers recover from stress or being more tolerant to it (Parsons et al. 1998).

Views and sounds of the natural landscape can reduce pain and are used in 'distraction therapy' (Diette et al. 2003). Murals of natural scenes placed at patient's bedsides, and nature sounds provided before, during, and after flexible bronchoscopy have shown better pain control than conventional approaches after data was adjusted for other factors. Older patients and those with better health status, particularly, reported significantly less pain. There was no difference, however, in patients-reporting anxiety between the treatment and control groups.

Possibly the most striking example of the impact of green views is that of Kuo and Sullivan (2001a, b) which demonstrated that domestic violence (aggression against a partner?or children?) increased by 25–35 % in housing estates in Chicago where large landscape trees were removed from the view of some housing blocks but not others (for communities with similar housing stock, and social-economic background). The authors related the level of aggression encountered to enhanced stress and anxiety in those dwellings where there was no view of greenery. Similar modes of action may explain the reductions in both gun crimes and vandalism across Philadelphia when vacant city spaces were re-landscaped with vegetation (Branas 2011).

## Gardening

Horticultural activities such as allotment gardening (Armstrong 2000; Twiss et al. 2003; Milligan et al. 2004), school gardening (Lee 2002), recreational gardening (Brown and Jameton 2000; York and Wiseman 2012) and hospital gardens all aid restoration from stress (Cooper and Barnes 1995, 1999; Ulrich 2002; Barnicle and Midden 2003). In the latter case, the restorative value was not just limited to patients, but evident too in staff and visitors (Whitehouse et al. 2001). Gardening is effective in providing relief from depression and stress, but also can elicit benefits through personal fulfilment as a hobby, social contact and physical exercise to prevent or manage disease or disability (Elings 2006; Kingsley 2009).

Studies on allotment gardening suggest that those ≥62 years-old scored significantly better on certain measures of health and well-being compared to neighbours of the same age who did not garden, but measures for younger gardeners did not differ from their equivalent age group (van den Berg et al. 2010). Beneficial effects of allotment gardens include enhanced physical activity, reduced levels of stress and mental fatigue but better social and cultural integration is also considered an important component (Schmelzkopf 1995; Milligan et al. 2004; Groenewegen et al. 2006; Phelps et al. 2010). Especially for older people, allotment gardens can provide a supportive environment that combats social isolation and contributes to the development of their social networks (Milligan et al. 2004). Family cohesion was deemed to be improved when families came together in a community garden project (Carney 2011), as well as health benefits associated with increased ingestion of fresh fruit and vegetables. Eating fruit and vegetables in a 'several times a day category' increased from 18 to 85 % for adults during the duration of the project (and from 24 to 64 % for children).

Parents and teachers working with children on gardening projects relate that activities help improve children's self-esteem and reduce stress or anxiety levels (Waliczek et al. 2000). The form of garden activity itself may be influential with Yamane et al. (2004) claiming the presence of plants was important in determining any benefits during a potting based activity. Groups of individuals either potted on soil alone, a non-flowering plant or a flowering plant. Before and after potting, they were screened for brain waves (electromyogram), eye blinking and assessed via a questionnaire on mood state. Results showed that potting plants reduced electromyogram activity significantly, whereas potting soil alone did not. Eye blink rates also decreased significantly after potting of flowering plants. Negative characteristics such as anxiety, depression and fatigue were significantly lower after the potting of flowering plants compared to the other two treatments. Whether the flowering plants were more stimulating because they were plants per se, or that additional alterations in form and colour were contributing to the positive responses remains a moot point however (no artificial equivalents were used as a control).

Gardening is one of a number of pastimes that are linked with encouraging greater physical activity, including sustaining long-term engagement through opportunities for creativity, communication and self-expression (Blair et al. 1991; Fig. 26.5). Benefits associated with gardening and similar activities include reduced mortality, higher bone density, lower blood pressure and cholesterol levels (Pahor et al. 1994; Walsh et al. 2001; Milligan et al. 2004). Park et al. (2009) claim regular physical activity can reduce risk of coronary heart disease, ischemic stroke, Type II diabetes, hypertension, osteoporosis, anxiety, depression and certain types of cancer. The intensity of physical activity, however, will vary with type of gardening task, age and ability of participant (Dallosso et al. 1988). Although gardening has shown positive influences in reducing mortality rates, links to reduced cardio-vascular disease directly, however, remain unproven and more intensive activities (cycling, regular walking or stair climbing) may be required to help offset this particular risk (Stamatakis et al. 2009). Benefits tend to be maximised when moderate garden exercise is maintained throughout the year (Magnus et al. 1979). An additional benefit of gardening is that the activities may have longevity or may be more convenient for regular activity. In one study among elderly men in The Netherlands, participants spent a greater amount of time per week gardening than other activities such as walking or cycling (Caspersen et al. 1991). Gardening is not risk free though. In the USA 1 % of the population are thought to suffer some sort of injury in the garden every month (Powell et al. 1998). Gardening is blamed for injuries linked with the misuse of tools (especially lawn mowers, van Duijne et al. 2008), dermatitis (McMullen and Gawkrodger 2006) and certain pathogens (O'Connor et al. 2007; Fig. 29.5).

**Fig. 29.5** Horticulture provides opportunity for regular physical exercise, with the growing and other developmental aspects allowing people to retain their levels of engagement and interest. (Image: A. Clayden, University of Sheffield)

## Children and Greenspace

As with other groups green space is perceived to improve self-esteem and a sense of empowerment in children, as well as improve engagement with school (Maller 2009). Green space within the vicinity of the home may influence a child's development. A house surrounded by nature helps to boost a child's attention capabilities (Fig. 29.6). When children's cognitive functioning was compared before and after they moved from poor (low volume green space)—to better-quality housing (greater green space) differences emerged in attention capacity, even when the effects of the improved housing were taken into account (Wells 2000; Wells and Evans 2003).

The Forest School concept was initiated in Denmark and has spread across the globe (Maynard 2007). This embraces the notion that children's development and educational performance is enhanced when teaching and learning is conducted in a natural setting, rather than a 'bland' interior classroom. The impact of plants and greenery on attentional functioning may also explain why children with attention deficit disorder (ADD) appear to respond better when playing in a green or natural environment (Faber-Taylor et al. 2001). In this study, parents rated their child's attentional functioning after activities in several settings. Results indicated that children functioned better after activities in green settings, and increasing the component of greenery resulted in less severe attention deficit symptoms. Thus, contact with nature may support attentional functioning in those children who have special educational needs.

**Fig. 29.6** Access to green space from an early age can help cognitive development, as well as an appreciation of the natural world

Contemporary living environments and lifestyle, however, may affect perceptions of nature today. Secondary schoolchildren expressed fears about natural areas they encountered through school or recreational activities (Wohlwill 1983). Such negative perceptions have been linked to preferences for manicured paths, urban environments and indoor social recreation activities over rural activities (Bixler and Floyd 1997). Interestingly, counter to popular assumptions about urban attitudes to the natural world, it was mostly rural and suburban students that had these negative attitudes.

## Horticultural Therapy (HT), Social and Therapeutic Horticulture (S&TH) and Healing Gardens

Horticulture is frequently used as a therapeutic process for those suffering physical and mental disabilities. It is one of a number of 'green' or 'nature-based' care initiatives (Horticultural Therapy, Social and Therapeutic Horticulture, Healing or Therapy Gardens, Ecotherapy, Green Exercise, Wilderness Therapy, Care Farming and Animal Assisted Activity, Haubenhofer et al. 2010). Horticultural Therapy (HT) has predefined clinical goals and helps clients learn new skills or regain ones lost. It aims to help clients improve memory, initiate tasks, encourage attention to detail, improve responsibility, enhance problem solving skills or regain physical abilities such as coordination, balance and strength. A range of different client groups are thought to respond to HT; people with physical or mental disabilities or in rehabilitation from illness, injury addiction or abuse (Haubenhofer et al. 2010). In contrast to HT, Social & Therapeutic Horticulture (S&TH) has a more general focus on well-being improvements and is not necessarily set against clinical objectives. Both of these forms of green care, however, centre-around plants and horticultural activities. Healing or therapeutic gardens can be used for HT and S&TH activities, but are also associated with hospitals and care homes where much of the interactions may tend to be relatively passive (e.g. viewing and hearing nature, smelling and touching plants).

Anecdotal reports of HT and S&TH tend to be very positive and benefits reported for a variety of settings and a diverse range of special need groups. In the UK 21,000 clients are through to engage with S&HT every week (Haubenhofer et al. 2010) although empirical data proving the benefits remains elusive. Gigliotti and Jarrott (2005), however, working with dementia sufferers showed greater levels of client engagement with HT approaches (78 %) compared to conventional activities (28 %), although the latter type of activity varied between care homes/needs of individual clients (a drawback in this sort of study). As well as greater engagement, clients demonstrated more positive responses during HT compared to control activities. Despite horticulture indicating some advantages over conventional interventions for clients suffering mental health problems, structured comparisons between different forms of therapy, e.g. music or art therapy (and hence the relative advantages/drawbacks of HT) still remain to be determined. The nature of HT itself makes clinical analyses difficult. Due to the need to provide high quality care, client numbers tend to be limited at any one location. This combined with differences in approaches, methodologies and client groups make large randomly controlled experiments difficult to accomplish (Sempik 2007).

Although Gonzalez et al. (2010, 2011) were also frustrated by an inability to complete randomised controlled experiments, their data on changes on client's scores before and after a HT programme strengthens the evidence base for the value of such interventions. Clients in their study suffered from clinical depression and were assessed using five different mental health assessments (Gonzalez et al. 2011). Significant beneficial changes were observed in all the variables measured during the HT programme with large reductions in depression severity occurring over the first 4 weeks. Moreover, benefits were still significantly greater than original baseline values 3 months after the HT finished. The value of HT in relation to clinical depression is thought to activate the 'being away' and 'fascination' components of Kaplan's attention restoration theory (Gonzalez et al. 2010) as well as provide social interactions and cohesion (Stepney and Davies 2004; Gonzalez et al. 2011).

The majority of studies on HT and the associated evidence base tend to rely on case studies or interview/questionnaire survey methods– often on carers perceptions of the effectiveness of any interventions. In a review of HT in the UK, Sempik et al. (2005) contacted 24 garden projects and interviewed 137 clients, 88 project staff and carers, and 11 health professionals. Key findings included social as well as medical benefits; some clients extending their existing social networks while others made new and significant friendships. It was found that gardening acts as a common medium and offers clients the opportunity to engage with members of the wider community who share an interest in horticulture. As such, these projects help to develop closer relationships between vulnerable (socially excluded) and non-vulnerable (socially included) members of society and help to promote the abilities of vulnerable adults, as opposed to their disabilities. The clients often contribute to the entire spectrum of horticultural activities—cultivating produce through to selling it, and are involved in aspects of training, education and marketing. These processes help promote self-confidence and independence as well as contribute to physical and mental health and well-being (Sempik et al. 2005). With respect to the latter, HT and S&TH programmes were considered to offer clients opportunities for self-reflection, relaxation and restoration as well as increased access to 'growing things', nature, 'being outside' and the peace of the natural environment. Parr (2007) suggested the process of cultivating plants was a particularly effective part of healing, and that positive attitudes and experiences relayed by garden staff to clients was an additionally important component.

The provision of healing gardens allied to hospitals and hospices too is thought to justify the expenditure. Whitehouse et al. (2001) studying the impact of a healing garden at a children's hospital in California demonstrated it had positive effects on the users; 54% reported they were more relaxed and less stressed, 12% felt refreshed and rejuvenated, 18% more positive and able to cope, and only 10% reported no difference in mood. Even short visits appear to be beneficial, as nearly half of visitors spent less than five minutes in the garden at any one time.

## Conclusions

The evidence base for enhanced health and well-being benefits associated with horticulture and allied activities is increasing. Perhaps a more apt question now is to ask how effective are these forms of treatments and preventions compared to other activities or therapies. The human brain is complex and developing methodologies that clearly demonstrate benefits of horticultural activities and green landscapes in line with human psychological function is challenging, especially when meeting the needs for ethical criteria and robust, balanced experimental designs. Larger, more astute, epidemiologically-designed studies are required to address some of existing limitations to existing studies. One area still relatively unexplored is the concept of 'compatibility' cited by Kaplan (1995) i.e. how universal are the benefit of horticultural activities, do people need to have a certain affinity or understanding of the basic concepts of nature before they engage? If so, what are the implications for an increasingly urbanised and technology-driven society, where dissociation with nature is a concern? Perhaps on the other-hand horticulture will become one of a number of activities that provide a counterbalance to a hectic, stressful, sedentary lifestyle. If so, the provision of appropriate green space, including the domestic garden, the local park and the nearby soccer pitch need to remain high on the political agenda. This also alludes to the ideas being promoted by Grahn and Stigsdotter (2003), van Dillen et al. (2012) and others that form and quality of green space need to be given greater consideration to ensure benefits are maximised, or tailored to specific user groups (van den Berg and van Winsum-Westra 2010). Horticulture should have a central role in defining this agenda and ensuring its effective delivery.

## References

Aldous DE, Aldous MD (2008) Intrinsic and extrinsic parameters in designing therapeutic landscapes for special populations and uses. Acta Hort 775:99–105

Anon (2010) The economic and social costs of mental health problems in 2009/10. Centre for Mental Health, UK. http://wwwcentreformentalhealthorguk/news/2010_cost_of_mental_ill_healthaspx

Armstrong D (2000) A survey of community gardens in upstate New York: implications for health promotion and community development. Health Place 6:319–327

Barnicle T, Midden KS (2003) The effects of a horticulture activity program on the psychological well-being of older people in a long-term care facility. HortTechnology 13:81–85

Barton J, Pretty J (2010) What is the best dose of nature and green exercise for improving mental health? A multi-study analysis. Environ Sci Technol 44:3947–3955

Barton J, Griffin M, Pretty J (2012) Exercise-, nature- and socially interactive-based initiatives improve mood and self-esteem in the clinical population. Perspect Public Heal 132:89–96

Bedimo-Rung AL, Mowen, AJ, Cohen DA (2005) The significance of parks to physical activity and public health: a conceptual model. Am J Prev Med 28:159–168

Bell S, Morris N, Findlay C et al (2004) Nature for people: the importance of green spaces to East Midlands communities, (Research report 567). English Nature, Peterborough

Berman MG, Jonides J, Kaplan S (2008) The cognitive benefits of interacting with nature. Psychol Sci 19:1207–1212

Bird W (2007) Natural greenspace. Br J Gen Pract 57:69

Bixler RD, Floyd, MF (1997) Nature is scary, disgusting and uncomfortable. Environ Behav 29:443–456

Björk J, Albin M, Grahn P et al (2008) Recreational values of the natural environment in relation to neighbourhood satisfaction, physical activity, obesity and wellbeing. J Epidemiol Community Health 62:e2. doi:101136/jech2007062414

Blair D, Giesecke C, Sherman S (1991) A dietary, social and economic evaluation of the Philadelphia urban gardening project. J Nutr Educ 23:161–167

Bowler D, Buying-Ali L, Knight T et al (2010) The importance of nature for health: is there a specific benefit of contact with greenspace? Collaboration of Environmental Evidence Systematic review No. 40

Branas CC, Cheney RA, MacDonald JM et al (2011) A difference-in-differences analysis of health, safety, and greening vacant urban space. Am J Epidemiol 174:1296–1306

Brown KH, Jameton AL (2000) Public health implications of urban agriculture. J Public Health Pol 21:20–39

Cameron RWF, Blanuša T, Taylor JE. (2012) The domestic garden—its contribution to urban green infrastructure. Urban For Urban Gree 11:129–137

Carney M (2011) Compounding crises of economic recession and food insecurity: a comparative study of three low-income communities in Santa Barbara County. Agric Human Values:1–17. doi:101007/s10460-011-9333-y

Caspersen CJ, Bloemberg BP, Saris WH et al (1991) The prevalence of selected physical activities and their relation with coronary heart disease risk factors in elderly men: The Zutphen Study, 1985. Am J Epidemiol 133:1078–1092

Cimprich B, Ronis DL (2003) An environmental intervention to restore attention in women with newly diagnosed breast cancer. Cancer Nurs 26:284–292

Coombes E, Jones AP, Hillsdon M (2010) The relationship of physical activity and overweight to objectively measured green space accessibility and use. Soc Sci Med 70:816–822

Coon JT, Boddy K, Stein K et al (2011) Does participating in physical activity in outdoor natural environments have a greater effect on physical and mental wellbeing than physical activity indoors? A systematic review. Environ Sci Technol 45:1761–1772

Cooper MC, Barnes M (1995) Gardens in healthcare facilities: uses, therapeutic benefits, and design recommendations Martinez, CA. The Centre for Health Design, Concord, California

Cooper MC, Barnes M (1999) Healing gardens therapeutic benefits and design recommendations. Wiley, New York
Dallosso HM, Morgan K, Bassey EJ et al (1988) Levels of customary physical activity among the old and the very old living at home. J Epidemiol Community Health 42: 121–127
De Groot WT, van den Born RJG (2003) Visions of nature and landscape type preferences: An exploration in The Netherlands. Landscape Urban Plan 63:127–138
De Vries S, Verheij RA, Groenewegen PP et al (2003) Natural environments—healthy environments? An exploratory analysis of the relationship between greenspace and health. Environ Plann A 35:1717–1731
Diette GB, Lechtzin N, Haponik E et al (2003) Distraction therapy with nature sights and sounds reduces pain during flexible bronchoscopy: a complementary approach to routine analgesia. Chest 123:941–948
Duthie R (1988) Florists' flowers and societies Shire garden history. Shire Publications Ltd, Aylesbury
Eastwood P (2011) The NHS Information Centre, Lifestyles Statistics National Health Service Information Centre, UK. wwwicnhsuk
Elings M (2006) People–plant interaction: The physiological, psychological and sociological effects of plants on people. In: Hassink J, van Dijk M (eds) Farming for Health. Springer, Netherlands pp 43–55
Esch T, Stefano GB, Fricchione GL et al (2002) The role of stress in neurodegenerative diseases and mental disorders. Neuroendocrinol Lett 23:199–208
Faber-Taylor A, Kuo FE, Sullivan WC (2001) Coping with ADD: the surprising connection to green play settings. Environ and Behav 33:54–77
Federighi S (2008) Today's challenges and opportunities—kids in the woods. US Department of Agriculture, Forest Service, Office of Communication. http//wwwfsfedus/emphasis/products/kids-factspdf
Gigliotti CM, Jarrott SE (2005) Effects of horticulture therapy on engagement and affect. Can J Aging 24:367–377
Giles-Corti B, Broomhall MH, Knuiman M et al (2005) Increasing walking: how important is distance to, attractiveness, and size of public open space? Am J Prev Med 28:169–176
Gonzalez MT, Hartig T, Patil GG et al (2010) Therapeutic horticulture in clinical depression: a prospective study of active components. J Adv Nurs 66:2002–2013
Gonzalez MT, Hartig T, Patil GG et al (2011) A prospective study of existential issues in therapeutic horticulture for clinical depression. Iss Men Health Nurs 32:73–81
Grahn P, Stigsdotter U (2003) Landscape planning and stress. Urban For Urban Gree 2:1–18
Grahn P, Stigsdotter U (2010) The relation between perceived sensory dimensions of urban green space and stress restoration. Landscape Urban Plan 94:264–275
Groenewegen PP, van den Berg AE, de Vries S et al (2006) Vitamin G: effects of green space on health, well-being, and social safety. BMC Public Health 6:149
Hadfield M (1969) A history of British gardening, 3rd edn. Spring Books, New Orleans
Hansmann R, Hug S-M, Seeland K (2007) Restoration and stress relief through physical activities in forests and parks. Urban For Urban Gree 6:213–225
Hartig T (2007) Three steps to understanding restorative environments as health resources. In: Ward Thompson C, Travlou P (eds) Open space: people space. Taylor and Francis, London, pp 163–179
Hartig T, Mang M, Evans GW (1991) Restorative effects of natural environment experiences. Environ and Behav 23:3–26
Hartig T, Evans GW, Jamner et al (2003) Tracking restoration in natural and urban field settings. J Environ Psychol 23:109–123
Haubenhofer DK, Elings M, Hassink J et al (2010) The development of green care in western European countries. Explore 6:106–111
Hillsdon M, Panter J, Foster C et al (2006) The relationship between access and quality of urban green space with population physical activity. Public Health 120:1127–1132

Hoehner CM, Brennan Ramizrez LK, Elliott MB et al (2005) Perceived and objective environmental measures and physical activity among urban adults. Am J Prev Med 28:105–116
Hull RB, Harvey A (1989) Exploring the emotions people experience in suburban parks. Environ Behav 21:323–345
Ivarsson CT, Hagerhall CM (2008). The perceived restorativeness of gardens—assessing the restorativeness of a mixed built and natural scene type. Urban For Urban Gree 7:107–118
Jarrott SE, Kwack HR, Relf D (2002) An observational assessment of a dementia-specific horticultural therapy program. HortTechnology 12:403–410
Kaplan S (1995) The restorative benefits of nature: toward an integrative framework. J Environ Psychol 15:169–182
Kaplan R (2001) The nature of the view from home: psychological benefits. Environ Behav 33:507–542
Kaplan R, Kaplan S (1989) The experience of nature: a psychological perspective. Cambridge University Press, Cambridge
Kaplan R, Kaplan S, Ryan RL (1998) With people in mind. Design and the management of everyday nature. Island Press, Washington DC
Kellert SR, Wilson EO (1993) The Biophilia hypothesis. Island Press, Washington DC
Kingsley J, Townsend M, Henderson-Wilson C (2009) Cultivating health and wellbeing: Members' perceptions of the health benefits of a Port Melbourne community garden. Leisure Studies 28:207–219
Korpela K, Hartig T (1996) Restorative qualities of favorite places. J Environ Psychol 16:221–233
Korpela KM, Hartig T, Kaiser FG et al (2001) Restorative experience and self-regulation in favourite places. Environ Behav 33:572–589
Korpela KM, Ylén M, Tyrväinen L et al (2010) Favourite green, waterside and urban environments, restorative experiences and perceived health in Finland. Health Promot Int 25:200–209
Kuo FE, Sullivan WC (2001a) Environment and crime in the inner city: Does vegetation reduce crime? Environ Behav 33:343–367
Kuo FE, Sullivan WC (2001b) Aggression and violence in the inner city: effects of environment via mental fatigue. Environ Behav 33:543–571
Lee S (2002) Community gardening benefits as perceived among American-born and immigrant gardeners in San Jose, California. socratesberkeleyedu/~es196/projects/2002final/LeeSpdf
Lee Y, Kim S (2008) Effects of indoor gardening on sleep, agitation, and cognition in dementia patients—a pilot study. Int J Geriatr Psych 23:485–489
Lee ACK, Maheswaran R (2011) The health benefits of urban green spaces: a review of the evidence. J Public Health 33:212–222
Maas J, Verheij RA, Groenewegen PP et al (2006) Green space, urbanity and health: how strong is the relation? J Epidemiol Community Health 60:587–592
Maas J, Verheij RA, Spreeuwenberg P et al (2008) Physical activity as a possible mechanism behind the relationship between green space and health: a multilevel analysis. BMC Public Health 8:206 doi:101186/1471–2458-8-206
Maas J, van Dillen SME, Verheij RA et al (2009) Social contacts as a possible mechanism behind the relation between green space and health. Health Place 15:586–595
Macintyre S, Macdonald L, Ellaway A (2008) Lack of agreement between measured and self-reported distance from public green parks in Glasgow, Scotland. Int J Behav Nutr Phy 5:26. doi:101186/1479-5868-5-26
Magnus K, Matroos A, Strackee J (1979) Walking, cycling, or gardening, with or without seasonal interruption, in relation to coronary events. Am J Epidemiol 110: 724–733
Maller C (2009) Promoting children's mental, emotional and social health through contact with nature: a model. Health Education 109:522–543
Maller C, Townsend M, Brown P et al (2002) Healthy parks healthy people: the health benefits of contact with nature in a park context: a review of the literature. Deakin University and Parks Victoria, Melbourne
Maller C, Townsend M, Pryor A et al (2006) Healthy nature healthy people: 'Contact with nature' as an upstream health promotion intervention for populations. Health Promot Int 21:45–54

Maynard T (2007) Forest schools in Great Britain: an initial exploration. Cont Iss Early Child 8:320–330
McMullen E, Gawkrodger DJ (2006) Physical friction is under-recognized as an irritant that can cause or contribute to contact dermatitis. Brit J Dermatol 154:154–156
Milligan C, Bingley A (2007) Restorative places or scary spaces? The impact of woodland on the mental well-being of young adults. Health Place 13:799–811
Milligan C, Gatrell A, Bingley A (2004) 'Cultivating health': therapeutic landscapes and older people in northern England. Soc Sci Med 58:1781–1793
Mitchell R (2012) Is physical activity in natural environments better for mental health than physical activity in other environments. Soc Sci Med. doi:101016/jsocscimed201204012
Mitchell R, Popham F (2007) Greenspace, urbanity and health: relationships in England. J Epidemiol Community Health 61:681–683
Mitchell R, Popham F (2008) Effect of exposure to natural environment on health inequalities: an observational population study. Lancet 372:1655–1660
Mitchell R, Astell-Burt T, Richardson EA (2011) A comparison of green space indicators for epidemiological research. J Epidemiol Community Health 65:853–858
Moore EO (1981) A prison environment's effect on health care service demands. J Environ Systems 11:17–34
Morris N (2003) Health, Well-being and open space, literature review. http://wwwopenspaceecaacuk/healthwellbeinghtm
Nielsen TS, Hansen KB (2007) Do green areas affect health? Results from a Danish survey on the use of green areas and health indicators. Health Place 13:839–850
Niepel A (2004) Therapeutic gardens—deficiencies and potentials. ISHA. Acta Hort 643:205–207
Nilsson K, Sangster M, Konijnendijk CC (2011) In: Nilsson K, Sangster M, Gallis C et al (eds) Forests, trees and human health and well-being. Springer, Netherlands
O'Connor BA, Carman J, Eckert K et al (2007) Does using potting mix make you sick? Results from a Legionella longbeachae case-control study in South Australia. Epidemiol Infect 135:34–39
Ottosson J, Grahn P (2005) A comparison of leisure time spent in a garden with leisure time spent indoors: on measures of restoration in residents in geriatric care. Landscape Res 30:23–55
Pahor M, Guralnik JM, Salive ME et al (1994) Physical activity and risk of severe gastrointestinal hemorrhage in older persons. J Am Med Assoc 272:595–599
Park B-J, Tsunetsugu Y, Kasetani T et al (2007) Physiological effects of shinrin-yoku (taking in the atmosphere of the forest)—using salivary cortisol and cerebral activity as indicators. J Physiol Anthropol 26:123–128
Park B-J, Furuya K, Kasetani T (2011) Relationship between psychological responses and physical environments in forest settings. Landscape Urban Plan 102:24–32
Parr H (2007) Mental health, nature work, and social inclusion. Environ Plann D 25:537–561
Parsons R, Tassinary LG, Ulrich RS et al (1998) The view from the road: Implications for stress recovery and immunization. J Environ Psychol 18:113–140
Phelps J, Hermann JR, Parker SP et al (2010) Advantages of gardening as a form of physical activity in an after-school program. J Extension 48, 6RIB5: 1–7
Potwarka LR, Kaczynski AT, Flack AL (2008) Places to play: association of park space and facilities with healthy weight status among children. J Commun Health 33:344–350
Powell KE, Heath GW, Kresnow M-J et al (1998) Injury rates from walking, gardening, weightlifting, outdoor bicycling, and aerobics. Med Sci Sport Exer 30:1246–1249
Pretty JN, Ward H (2001) Social capital and the environment. World Dev 29:209–227
Pretty J, Griffin M, Sellens M et al (2003) Green exercise: complementary roles of nature, exercise and diet in physical and emotional well-being and implications for public health policy, Occasional Paper 2003-1. University of Essex, p 38. http://www2essexacuk/ces/ResearchProgrammes/CESOccasionalPapers/GreenExercisepdf
Roemmich JN, Epstein LH, Raja S et al (2006) Association of access to parks and recreational facilities with the physical activity of young children. Prev Med 43:437–441

Rohde CLE, Kendle AD (1994) Human well-being, natural landscapes and wildlife in urban areas. English Nature, Peterborough
Schipperijn J, Ekholm O, Stigsdotter UK et al (2010) Factors influencing the use of green space: results from a Danish national representative survey. Landscape Urban Plan 95:130–137
Schmelzkopf K (1995) Urban community gardens as contested space. Geogr Rev 85: 364–381
Scholz U, Krombholz H (2007) A study of the physical performance ability of children from wood kindergartens and from regular kindergartens. Motorik Mar 1:17–22
Scopelliti M, Giuliania MV (2004) Choosing restorative environments across the lifespan: a matter of place experience. J Environ Psychol 24:423–437
Searles HF (1960) The nonhuman environment in normal development and in schizophrenia. International Universities Press, New York
Sempik J (2007) Researching social and therapeutic horticulture: a study of methodology reading. Thrive and Loughborough University CCFR
Sempik J, Aldridge J, Becker S (2005) Health, well-being and social inclusion, therapeutic horticulture in the UK. The Policy Press, Bristol
Simons LA, Simons J, McCallum J et al (2006) Lifestyle factors and risk of dementia: Dubbo Study of the elderly. Med J Australia 184:68–70
Stamatakis E, Hamer M, Lawlor DA (2009) Physical activity, mortality, and cardiovascular disease: is domestic physical activity beneficial? The Scottish health survey—1995, 1998 and 2003. Am J Epidemiol 169:1191–1200
Stepney P, Davies P (2004) Mental health, social inclusion and the green agenda: an evaluation of a land based rehabilitation project designed to promote occupational access and inclusion of service users in North Somerset, UK. Soc Work Health Care 39:357–397
St Leger L (2003) Health and nature—new challenges for health promotion. Health Promot Int 18:173–175
Stone D, Hanna J (2003) Health and nature: the sustainable option for healthy cities. English Nature. http://wwwprojectevergreencom/pdf/Health%20&%20Naturepdf
Tsunetsugu Y, Park, B-J, Ishii H et al (2007) Physiological effects of "Shinrin-yoku" (taking in the atmosphere of the forest) in an old-growth broadleaf forest in Yamagata prefecture. Jpn J Physiol Anthropol 26:135–142
Twiss J, Dickinson J, Duma S et al (2003) Community gardens: lessons learned from California healthy cities and communities. Am J Public Health 93:1435–1438
Tzoulas K, Korpela K, Venn S et al (2007) Promoting ecosystem and human health in urban areas using green infrastructure: a literature review. Landscape Urban Plan 81:167–178
Ulrich RS (1981) Natural versus urban scenes: some psycho-physiological effects. Environ Behav 13:523–556
Ulrich RS (1984) View through a window may influence recovery from surgery. Science 224:420–421
Ulrich RS (1990) The role of trees in well-being and health. In: Rodbell PD (ed) Proc fourth urban forestry conference, St Louis, Missouri 15–19 October 1990
Ulrich, RS (2001) Effects of healthcare environmental design on medical outcomes. In: Dilani A (ed) Design and health: proceedings of the second international conference on health and design, Stockholm, Sweden, Svensk Byggtjanst, pp 49–59
Ulrich RS (2002) Health benefits of gardens in hospitals. In: Plants for people symposium, reducing health complaints at work, Amsterdam, Netherlands
Ulrich RS, Simons RF, Losito BD et al (1991) Stress recovery during exposure to natural and urban environments. J Environ Psychol 11:201–230
Van den Berg AE, Custers MHG (2011) Gardening promotes neuroendocrine and affective restoration from stress. J Health Psychol 16:3–11
Van den Berg AE, van Winsum-Westra M (2010) Manicured, romantic, or wild? The relation between need for structure and preferences for garden styles. Urban For Urban Gree 9:179–186
Van den Berg AE, van Winsum-Westra M, de Vries S et al (2010) Allotment gardening and health: a comparative survey among allotment gardeners and their neighbours without an allotment. Environ Health 9:74

Van Dillen SME, de Vries S, Groenewegen PP et al (2012) Greenspace in urban neighbourhoods and residents' health: adding quality to quantity. J Epidemiol Community Health. doi:10.1136/jech.2009.104695
Van Duijne FH, Kanis H, Hale AR et al (2008) Risk perception in the usage of electrically powered gardening tools. Safety Sci 46:104–118
Yamane K, Kawashima M, Fujishige N et al (2004) Effects of interior horticultural activities with potted plants on human physiological and emotional status. Acta Hort (ISHS) 639:37–43
York M, Wiseman T (2012) Gardening as an occupation: a critical review. Brit J Occup Ther 75:76–84
Velarde MD, Fry G, Tveit M (2007) Health effects of viewing landscapes—Landscape types in environmental psychology. Urban For Urban Gree 6:199–212
Waliczek TM, Bradley JC, Lineberger RD et al (2000) Using a web-based survey to research the benefits of children gardening. HortTechnology 10:71–76
Walsh JME, Pressman AR, Cauley JA et al (2001) Predictors of physical activity in community-dwelling elderly white women. J Gen Intern Med 16:721–727
Wells NM (2000) At home with nature: effects of "greenness" on children's cognitive functioning. Environ Behav 32:775–795
Wells NM, Evans GW (2003) Nearby nature, a buffer of life stress among rural children. Environ and Behav 35:311–330
Whitehouse S, Varni JW, Seid M et al (2001) Evaluating a children's hospital garden environment: utilisation and consumer satisfaction. J Environ Psychol 21:301–314
Wilson EO (1984) Biophilia. Harvard Press, Harvard, USA
Wohlwill JF (1983) Concept of nature: a psychologist's view. In: Altman J, Wohlwill JF (eds) Behaviour and the natural environment: human behaviour and environment: advances in theory and research. Plenum, New York, pp 5–35
Woo J, Tang N, Suen E et al (2009) Green space, psychological restoration and telomere length. Lancet 373:299

# Chapter 30
# Human Dimensions of Wildlife Gardening: Its Development, Controversies and Psychological Benefits

**Susanna Curtin and Dorothy Fox**

**Abstract** A prevalent social discourse concerning climate change, loss of biodiversity and the importance of nature to human health currently dominates news articles, television programmes and political comment. These anthropogenic impacts on the natural environment question humankind's predominant relationship with nature; particularly in western developed cultures where people are usually perceived as separate from nature rather than part of it. Whilst the world's declining iconic species catch media attention, it is often local and indigenous wildlife that become the focus of communities at a local level. As a result, conservation organisation membership has increased over the last 5 years alongside a strong retail sector which encourages people to purchase, for example, wild bird food, bird feeders and nest boxes. As interest in feeding the wild birds that visit gardens has increased, so too has an appreciation of the need to conserve the wider aspects of the ecosystem such as plants, insects and amphibians which attract and support the birds and mammals that have become more welcome visitors to our gardens. There is also increasing recognition of the health and psychological benefits that wildlife gardening can bring to individuals and communities. Many prominent garden attractions and horticultural shows in England and throughout the world have developed a wild theme into their garden design which has captured the imagination of garden visitors who wish to marry their love of horticulture with their interest in wildlife. Such naturalistic and wild flower planting has thus become a more common element of home garden design reflected in the retail sector, media programmes and garden magazines and books.

**Keywords** Wildlife · Macro landscapes · Micro landscapes · Human interaction · Psychological benefits · Ecology · Eco-therapy · Biodiversity · Anthropogenic influence

---

S. Curtin (✉) · D. Fox
School of Tourism, Bournemouth University, Dorset House, Talbot Campus,
Fern Barrow, Poole, Dorset BH12 5BB, UK
e-mail: scurtin@bournemouth.ac.uk

D. Fox
e-mail: dfox@bournemouth.ac.u

G. R. Dixon, D. E. Aldous (eds.), *Horticulture: Plants for People and Places, Volume 3*,
DOI 10.1007/978-94-017-8560-0_30, 

## Introduction

With a little imagination and understanding, wildlife gardening provides the opportunity to bring nature back into our lives not only for the aesthetic beauty and pleasure that flora and fauna brings but also for the sensual pleasures that can be derived from the sound of birdsong, the croaking of amphibians and the movement and spectacle of insects such as bees, dragonflies and butterflies that can be attracted into our gardens. With the help of wildlife conservation agencies, the growing media related to gardening and conservation, and the awareness of global environmental degradation, wildlife gardening has recently become more mainstream. Forty years ago, the majority of people would have scorned the idea of gardening for wildlife. Gardens were a place where the control of any wildlife which prevented or reduced high production of flowers or produce was the primary *modus operandi*; gardeners were encouraged to reach for insect sprays at the first signs of damage. Today, however, we are realising that every living thing is part of a complex chain; a web of life, with a myriad of symbiotic relationships and connections from one species to the next. We do not fully understand these connections but by now we have witnessed the loss of species which have been systematically destroyed by urbanisation, a growing human population and agricultural practices directed purely towards maximum yields: the resultant loss of biodiversity is a distressing harvest to reap. At a time when natural habitats are declining at an alarming rate, conservation organisations see gardens as essential corridors; highways and oases of modified habitat which can be exploited by wildlife.

It is a general misconception that a wildlife garden is an unkempt space where nature has been allowed to take over. Quite the opposite: Ryrie (2003) suggests that there is greater biodiversity in a well managed wildlife garden which has a wide variety of plants and habitats rather than one which has been allowed to become tangled undergrowth. The purpose of this chapter is to highlight and examine the changing attitudes and understanding of wildlife gardening and the sometimes complex and conflicting relationship between wildlife and horticulture. It does this amidst a discussion of the human attraction of wildlife, the psychological benefits that can be gained from creating a sanctuary where wildlife can be enjoyed and the psychological processes that are involved in the personal emersion and enjoyment of nature whether in one's own garden or in a horticultural visitor attraction.

## Human Dimensions of Wildlife

Human affiliation and affection for wildlife is a very complex and dynamic phenomenon. There is general agreement amongst commentators that public values towards wildlife have changed considerably in the developed world over the last 50 years (Manfredo et al. 2003) during which time there has been a gradual shift away from traditional wildlife values that emphasise the use and management of wildlife for utilitarian reasons towards a greater appreciation of the aesthetic, psychological and ecological importance of wildlife. Inglehart and Baker (2000) propose that during the industrialisation and urbanisation phases of a nation's development, nature is regarded as something to be conquered or controlled; materialist values are focused on human 'material' needs such as security, housing, economic development and jobs. Following materialism and urbanisation, people can experience a 'call of the wild' in which they exhibit an inherent, biological need to reconnect with the nature that is missing in their busy, urban lifestyles (Wilson 1984). This is evidenced by a greater number of people keeping pets, gardening, contributing to conservation organisations, wildlife watching and feeding garden birds. This ultimate return to the natural world is perhaps not that surprising given that its beauty and diversity have been a constant source of inspiration throughout human history, influencing traditions, the way societies have evolved and supplying the basic goods and services upon which trade and the economy is built (van den Duim and Caalders 2002).

Relations between humans and wildlife have deep evolutionary roots and are particularly complex. Animals are our companions, our food, our clothing, a source of spiritual enlightenment and a focus of stories, fables, poetry, sport and art. The boundaries between 'animality' and humanity are thus socially, culturally and scientifically bound, and blurred, as we position ourselves as part of the animal kingdom on the one hand yet distinctly separate from it on the other. However, any observation of the animal kingdom can immediately recognise the connections between animal and human behaviours; their curiosity, playfulness, foraging for food, rearing young and belonging to social groups are the building blocks of our own existence. As animals cannot reveal their thoughts to us, it is human nature to impose our own anthropomorphic interpretations of their world given our shared common life domains of survival, acquisition of territory and reproduction. We may see the theatre of our own lives similarly displayed in theirs. As Mabey explains:

> An honest experience of nature would find that the natural world is an arena of endurance, tragedy and sacrifice as much as joy and uplift. It is about the struggle against the weather, the perils of migration, the ceaseless vigilance against predators, the loss of whole families and the brevity of existence. The natural world is like a theatre, a stage beyond our own, in which the dramas that are an irreducible part of being alive are played out without hatred or envy or hypocrisy. No wonder they tell us so much about ourselves and our own frailties. (2006, p. 13)

This 'mutuality of behaviour' makes animals a source of fascination because they are more than mere objects. Wild creatures are subjects that provide 'a window into which we can look and from which someone looks out' (Rolston 1987, p. 26). This can be particularly true in the case of fellow mammals but also to some degree the amphibians, birds, butterflies and other insects that visit our gardens. As they inhabit our 'created' garden spaces, it is easy to become involved and watchful over their day to day existence. Gardeners themselves are also becoming more aware that their gardens are important for the conservation of wildlife, especially given the gradual encroachment of development into once thriving wildlife habitats and the proliferation of house-building and urban development in countries such as the United Kingdom where the competition for land is high. There is therefore an added enjoyment and dimension to be had from a garden that entices wildlife and persuades it to stay; a haven that is purposively and lovingly created by its owner. Modern conceptions of nature are informed by a combination of personal experience, scientific understanding and social construction. Clayton and Myers remind us that 'beliefs about what nature is, as well as the way in which nature is valued, are created within a historical and cultural context' (2009, p. 15). Media representations of wildlife, popular narratives and wildlife marketing communications play a pivotal role in socially constructing ideals of nature and what constitutes charismatic or desirable wildlife in the context of gardening.

It is suggested that 'birds are the most visible and charming of the garden's inhabitants and visitors' (Harper et al. 1994, p. 114). Whether this is the sight of them or hearing bird song varies between individual householders. Similarly bees are nurtured not only for the honey that they provide, but also for their sound as Middleton recorded, '…I have spent many happy hours …listening to a sound like the deep diapason note of a great organ—the music of a thousand bees in the lime-tree up above' (Middleton 1939, p. 240). Similarly butterflies add another dimension of colour and movement. Why some species are more encouraged than others is likely to be based on personal reasons as well as sub-cultural ones. For example urban and rural sentiments can be astoundingly different when it comes to charismatic wildlife. In urban areas of England, the red fox (*Vulpes vulpes*) is often encouraged into gardens through feeding. However in rural areas it is subject to hunting (albeit limited in its form by legislation). Harper also suggests that cultural biases are "in favour of 'nice' (large, attractive, cuddly, rare) organisms and against those that: seem dull or common; sting or are poisonous: these so-called pests, and creepy-crawlies; are associated with death, decay, and excrement" (Harper et al. 1994, p. 123). That said, some insects do not fit this description at all and are welcomed because of their beauty, usefulness or charismatic qualities, i.e. butterflies, dragonflies, ladybirds, fireflies and glow-worms.

## Wildlife in Gardens

The study of wildlife in domestic gardens has increased both at an amateur and academic level over recent decades. Although it is not a new phenomenon, as two early British books demonstrate. The Book of Garden Animals by Daglish (1928) 'gives accounts from the naturalist's point of view of most of the animals generally found in gardens, with their life-histories' (Hadfield 1936, p. 551). Similarly for birds 'Every Garden a Bird Sanctuary' by Turner (1935) is described as 'An up-to-date book on the practical aspects of birds in the garden, and a full account of garden sanctuaries, feeding, nest-boxes, baths etc.' (Hadfield 1936, p. 552).

The relationships between people and wildlife in their gardens have probably always been mixed, with both positive and negative elements. Sudell (1950) suggests that from a horticultural perspective, birds can be classified as harmful, beneficial, or neutral, based on what they eat. A bird is harmful if, he argues, it eats food grown for human consumption, such as fruit. It is beneficial, if it consumes something lower in the food chain that damages people's crops, and neutral, if the bird's food source does not affect people (such as wild grasses). In fact, all wildlife in gardens can be thought of in this way, although it must be acknowledged, that many species may change in the way that they are perceived throughout the year. A bird that is viewed as a pest whilst it is consuming fruit is an ally when its diet returns to one of damage-causing insects for the greater part of the year. Similarly views can change during the life cycle. A gardener might seek to eliminate a caterpillar but the butterfly of the same species might be welcome. Also they may view something as a pest in one part of their garden, but tolerate it in another. Finally of course, not all garden owners view a particular species in the same way. For example, whilst some residents in Nova Scotia, Canada, feed sunflower seeds to Eastern chipmunks (*Tamius striatus*) for the pleasure of watching them, others use a half-filled bucket of water with the sunflower seeds floating on the surface in order to entrap them. As Ellis (ca. 1935, p. 112) noted however regarding birds in England 'we are perhaps apt to notice the relative amount of harm more than the good done by these beautiful and cheery inhabitants and visitors of our gardens'.

## Destruction of Garden Wildlife

There is a long tradition of eliminating wildlife in gardens if they are harmful. Harm may not only be to food crops as Sudel (1950) suggests but also includes damage to flowers (e.g. the sulphur-crested cockatoos (*Cacatua galerita*) which destroy the flowers of ornamental tulip (*Tulipa*) cultivars in the Sydney region of New South Wales, Australia and damage to lawns (e.g. the chinch bug, (*Blissus leucopterus hirtus*) which are sucking insects that attack St Augustine grass (*Stenotaphrum secunatum*) in the southern states of the USA (Wyman 1971). Similarly harm may occur in ponds or other water features and may not only be caused by herbivores, although this is the most damaging, but also by carnivores and omnivores, which one might think would be a gardener's allies.

Efforts to destroy or control unwanted wildlife using chemical and other methods are described in earlier chapters of these volumes but it is worth noting here that modern techniques especially those espoused by organic gardeners are less damaging to wildlife overall. Techniques such as using barriers, for example, frames to protect brassicas against pests such as the caterpillars of the large white butterfly (*Pierris brassicae*) and small white butterfly (*Pierris rapae*) in England, and similarly the cabbage looper (*Trichoplusia ni*) in the USA (Wyman 1971) have proved extremely effective in protecting the crops without destroying the wildlife.

## The Introduction of Non-native Flora and Fauna

Throughout the world, gardens are a combination of native and non-native species of both flora and fauna. Arrivals have been both natural through for example, seed dispersion and migration but also anthropogenic. There has been a long tradition of deliberate introduction of exogenous species for consumption, aesthetics, or for reasons of grandeur. The tomato (*Lycopersicon esculentum*), by way of example, originated in Mexico, was commonly eaten in Europe for centuries but regarded as poisonous in the USA and only grown in gardens there as an ornamental, known as the 'love apple' (Anon 2012a). The common Indian myna bird (*Acridotheres tristis*) was first introduced to Australia to control insect pests, and is often fed by unsuspecting householders. However, in 2000, the International Union for the Conservation of Nature (IUCN) declared it to be one of the world's most invasive species, as it is extremely aggressive, chasing out native birds, almost to extinction in Polynesia, Hawaii, and Mauritius (Thomas 2012).

In many countries the goldfish (*Carassius auratus*) sometimes known as the Golden Carp, has been introduced as a prized ornament in garden ponds. It is omnivorous and requires additional special fish meat meals, but it can come into conflict with coarse fish from natural ponds and streams as it often attacks other fish when breeding (Sudell 1950). However, many introduced species, particularly plants, bring great pleasure to gardeners, without adverse impact on the native biodiversity. A survey of 61 domestic gardens in the old industrial city of Sheffield, in England, showed a total of 1,166 species of flora, of which only 30% were native, although the gardens, irrespective of size, contained on average 45% natives. Seventy-nine per cent of the species that were recorded only once were alien, demonstrating the extent of plant introductions in British domestic gardens. However, the flora included 72% of the plant families recorded in the wild in Britain and Ireland (although the latter include native and naturalised species), suggesting that many of the aliens could be important sources of fruit, pollen or nectar for wildlife (Smith et al. 2006).

Movement of wildlife into additional gardens without direct human intervention has also occurred, for example the Northern cardinal (*Cardinalis cardinalis*) and American goldfinch (*Carduelis tristis*) have increased their northward expansion from the USA (Robb et al. 2008) into the gardens of Nova Scotia, Canada. Similarly the European goldfinch (*Carduelis carduelis*) was introduced at numerous places in south-eastern Australia in the nineteenth century, and their populations quickly increased and their range expanded greatly to where they now range from Brisbane, Queensland south to the Eyre Peninsula in South Australia.

The creation of gardens requires the destruction of other habitats, often natural habitats, which support extensive biodiversity. It must be acknowledged that much wildlife in a garden may not be seen or perceived, for example there are microscopic mites, including the parasitic mite (*Varroa*) destructor of honey bees, along with protozoa, bacteria and viruses (Harper et al. 1994). These are all part of the ecology of a garden with important roles to play. Nonetheless, as noted below, many organisations encourage householders to use their gardens to support wildlife although this may not be easy initially, as Harper et al. (1994, p. 58) note: 'creating a space for the benefit of wildlife involves unlearning many old patterns, a relaxation of control, and finding out what can be persuaded to live in your garden'.

## Means of Supporting Wildlife in a Garden

Wildlife can be encouraged to enter into a garden and then remain there, through two principal means—provision of an appropriate habitat and supplementary feeding. For example, leaving leaf litter and mulching in New South Wales, Australia provides the common or eastern blue-tongued lizard (*Tiliqua scincoides scincoides*) with shelter and a habitat for its diet of snails and other garden pests. Similarly providing a pile of fallen logs for beetles, or leaving seed heads and dead stems over winter for ladybirds (ladybugs) (*Coccinellidae*) is also effective. Wildlife in temperate zones may also need suitable habitats for hibernation and piles of old leaves or straw provide appropriate materials for the European or common hedgehog (*Erinaceus europaeus*) to hibernate as well as nest. Other simple actions are also suggested by conservation organisations such as advising householders to leave small gaps at the base of walls and fences to afford movement of hedgehogs between gardens.

Nesting materials, such as grasses and moss are sought by many bird species and in the USA, Wyman (1971, p. 137) suggests 'providing thickets of shrubbery for nesting purposes'. In European countries the house martin (*Delichon urbica*), a summer migrant from Africa, can be encouraged by providing water from the edge of streams, ponds or even puddles to mix the mud needed to build nests under the eaves of buildings. Other water sources, such as ponds are also beneficial. There are over 3 million ponds in England that is, in approximately 16% of gardens (Davies et al. 2009). In the Sydney suburbs of New South Wales, garden ponds provide habitats for frogs, some of the 37 species of native amphibians found in the city (Anon 2012b). Additionally, ponds, bird baths and even dishes of water provide not only a source to drink, but also a means to clean their feathers.

The construction and careful siting of artificial nesting boxes can also make a valuable contribution to encouraging wildlife into the garden. For example, it is estimated that there are a minimum of 4.7 million nest boxes in British gardens, that is, one nest box for every six breeding pairs of cavity nesting birds (Davies et al. 2009). Bat houses, constructed similarly to bird houses, except that they have a slit and a crawl-board instead of a hole and a perch provide effective summer roosts and/or hibernation for the pipistrelle bat (*Pipistrellus pipistrellus*), Britain's smallest but most common bat (Harper et al. 1994). Insects too, can be supported through the production of 'bee-quarters', that is a can full of 7 mm in diameter paper straws secured in a crevice in a wall (Harper et al. 1994).

The easiest and therefore the most common means of supporting wildlife is through planting, whether it is planned with that purpose in mind or purely incidental. Some of the hundreds of forms of Buddleia are widely grown throughout the world, because they are so well known in encouraging butterflies into a garden, so much so that *Buddleia davidii* is often nicknamed the 'butterfly bush'. In France, shrubs recommended to encourage butterflies include, varieties of *Berberis, Hedera*, and *Lavandula, Lonicera periclymenum, Rhamnus frangula* and annuals

**Fig. 30.1** A plastic container of water for rainbow lorikeets on the wall of a Sydney suburb

cornflower *Centaurea cyanus*, and cultivars of *Scabious and Scabiosa* (McHoy 2000). In fact the flowers of many shrubs and plants provide nectar and pollen for insects, butterflies, hoverflies and bees (Harper et al. 1994), whether or not that was the intention of the gardener in planting them. Furthermore as Thompson (2006) demonstrated, they do not need to be native species to be effective.

Appropriate planting can similarly encourage birds into a garden and in Barbados the red flowered blossom of the Antigua heath (*Russelia equisetiformis*) is planted to attract the Antillean crested hummingbird (*Orthorhyncus cristatus*). Nectar rich Australian natives, such as *Banksia*, *Grevillea* and *Callistemon* (bottlebrush) are planted in Sydney, Australia to attract the noisy miner (*Manorina melanocephala*), the little wattlebird (*Anthochaera chrysoptera*) and rainbow lorikeets (*Trichoglossus haematodus*), amongst others (Fig. 30.1).

Whilst gardens are often planted with the provision of bird food as a secondary consideration, supplementary feeding of birds is a deliberate action to support wildlife. Feeding wild birds is a common practice among gardeners throughout the western world (O'Leary and Jones 2006) and more recently in many developing countries too. It is estimated for example, that in Australia, 25–57 % of households feed birds, whilst in the USA approximately 43 % of households regularly feed birds (Martinson and Flaspoler 2003) whilst the figure is 48 % of households in England (Davies et al. 2009). Seed is the most provided food with householders in the US and England purchasing 500,000 t of birdseed annually (O'Leary and Jones 2006). Simply scattering the seed loosely on the ground is common. For example, Haikou, in Hainan Province on the southern coast of China, has a subtropical climate and there are opportunities to see a number of endemic bird species such as the White winged magpie (*Urocissa whiteheadi*), which are fed with millet seed by the local residents. In Southern England, a mix of sunflower and smaller seeds attract some of the nation's favourite birds, such as blackbirds (*Turdus merula*), robins (*Erithacus rubecula*), and house sparrows (*Passer domesticus*) as well as the less loved woodpigeon (*Columba palumbus*) and magpie (*Pica pica*). Sunflower seeds are also used to attract ground feeders such as the mourning dove (*Zenaida macroura*) in the USA and southern Canada. Placing the seed on a bird table is useful not only when there is snow on the ground, but also keeps wild birds out of the reach of domesticated animals.

Sunflower seed in bird feeders attracts those birds that feed on the wing such as blue jays (*Cyanocitta cristata*), American robin (*Turdus migratorius*), black-capped chickadees (*Poecile atricapillus*), red-breasted nuthatch (*Sitta canadensis*) and several species of woodpecker including the downy woodpecker (*Picoides pubescens*), northern flicker woodpecker (*Colaptes auratus*) and the pileated woodpecker (*Dryocopus pileatus*) in eastern Canada. Similarly balls of seeds can be hung from the branches of trees to attract birds.

Other popular foods for bird feeding include suet in a feeder (Canada) and cooked long-grain rice scattered on the ground, early in the morning and evening, in Barbados to attract blackbirds (*Quiscalus lugubris*), sparrows (or more accurately the Barbados bullfinch) (*Loxigilla barbadensis*) and wood dove (*Zenaida aurita*). Bread crumbs are popular in England, although as Middleton noted, 'the tamest of all my feathered friends is a cock robin, who sits on the seat beside me, and even on my knee. He is not a vegetarian, and scorns breadcrumbs, but has a great fancy for bits of bacon rind, which I save specially for him' (Middleton 1939, p. 241). Meat is also provided in Australia for Australian magpies (*Gymnorhina tibicen*), the laughing kookaburra (*Dacelo novaeguineae*) (O'Leary and Jones 2006) and the tawny frogmouth (*Podargus strigoides*). It is not just birds that receive supplementary feeding, for example, in England, fat and commercial dog food are put out for mammals including hedgehogs (*Erinaceus europaeus*) and the red fox (*Vulpes vulpes*) respectively.

## Encouraging Wildlife for Horticultural Reasons

Wildlife has been recognised as beneficial in the garden for horticultural reasons, as Sudell (1950) noted. This includes bees for pollination and worms for aerating the soil. Organic gardener, Lawrence Hills, Founder of the Henry Doubleday Research Association (Hills 1989) reported that in an average hectare (two and a half acres) of grassland, 100 t of soil pass through the digestions of 3.75 million earthworms (*Lumbricus terrestris*). In dry climates, ants and termites take on the worm's role (Harper et al. 1994).

Birds also have a horticultural role, for example in England, blue tits (*Cyanistes caeruleus*) consume 'enormous quantities of insects and grubs during the breeding-season' (Sudell 1950, p. 99). Hills suggests hanging 10 cm (4″) square piece of fat above rose bushes, only big enough for two members of the tit family, blue (*Cyanistes caeruleus*), great (*Parus major*) and coal (*Periparus ater*) to feed at a time, encouraging those birds waiting their turn to search the bark at the base of the bushes for greenfly eggs (Hills 1989). Similarly, ramshorn water snails (*Planorbis corneus*) and freshwater winkles (*Paludina vivipara*) consume decayed organic material including any surplus fish food and algae from the sides of the pond (Perry 1955).

## Encouraging Wildlife for Conservation

Generally, people view gardens as an opportunity for encouraging wildlife. Almost half of the respondents in a study in Sheffield, England thought that city gardens contribute to improved environmental quality by creating 'a better environment for wildlife' (Dunnett and Qasim 2000). Gardening for wildlife as described above, provides not only habitats, both permanent and transient, but also a richer variety of habitats and additionally, corridors between habitats (Harper et al. 1994). Furthermore, collaborating "with neighbours to create a 'critical mass' of a particular type of habitat" (Harper et al. 1994, p. 11) or a scarce habitat can be of additional benefit. It is this detailed understanding of wildlife which is of such importance, as large gardens are not better than small for wildlife, nor suburban better than urban, as the Biodiversity in Urban Gardens in Sheffield (BUGS) project showed (Thompson 2006). This programme also confirmed Harper's view that 'Of all the garden developments you can undertake to increase habitat diversity, ponds are probably the most effective and the most gratifying' (Harper et al. 1994, p. 113).

Nonetheless, supplementary bird feeding is widely perceived as a positive activity and is likely to benefit many species, including some of conservation concern, but we still have only a relatively basic understanding of how it affects bird populations. Catterall (2004) observed that planting of eucalypts and nectar-rich native plant species in gardens in Queensland, Australia led to a decrease in the number of species of small-bodied birds, and an increase in numbers of the large, noisy miner (*Manorina melanocephala*). Similarly, Fuller et al. (2008) demonstrated that whilst supplementary feeding increases the total number of birds in an area, it does not increase the number of species. They concluded that 'variation in habitat quality and availability are likely to be much more important drivers of species richness patterns than resource availability, particularly in urban environments' (Fuller et al. 2008, p. 135).

Concerns have also been raised that some species of birds could become reliant on supplementary feeding by people. However, a study of Australian magpies (*Gymnorhina tibicen*) in suburban environments in Queensland, Australia showed that fed birds still obtained 76 % of their food from natural sources. Although their natural behaviour was influenced as they obtained less food items by ground foraging in the morning than unfed magpies and their breeding activities started earlier than the unfed birds. They showed too that in most cases, earlier broods had better survival rates than later ones, enhanced clutch size, hatching success and chick growth rate. The authors determined that the 'magpies were not reliant or dependent on supplementary food provided by wildlife feeders at any time during the breeding season' (O'Leary and Jones 2006, p. 208). However, as it has been shown, feeding influences all aspects of bird behaviour, from daily-survival to large-scale migration. Robb et al. (2008, p. 476) concluded that even 'natural selection is being artificially perturbed, as feeding influences almost every aspect of bird ecology, including reproduction, behaviour, demography, and distribution'.

There are other concerns too, for example, bird feeders have been implicated in the rapid spread of mycoplasmal conjunctivitis through the house finch (*Carpodacus mexicanus*) population in the USA (Fischer et al. 1997). In England, it is suggested that Trichomonosis in greenfinches and chaffinches is similarly spread. 'Disease transmission appears to vary according to the type of feeder used, the number of birds visiting it, and the habitat in which the feeder is located' (Robb et al. 2008, p. 481).

However, there is often little distinction between native and non-native species when information is given about wildlife and it appears that many people neither distinguish between the two, nor in fact care about the distinction. Similarly, which species to encourage has changed over time. In England, Wright (ca. 1902) recommended growing ivy on trees, garden fences and walls to provide a habitat for the common magpie (*Pica melanoleuca*) because they destroy vermin such as mice, voles and young brown rats. Today the European magpie (*Pica pica*) is viewed as a predator as they also collect other bird's eggs and kill nestlings to feed their own young. The British cuckoo (*Cuculus canorus*) is also traditionally disliked, being a brood parasite which lays eggs in the nests of other smaller species of birds, such as meadow pipits (*Anthus pratensis*) and reed warblers (*Acrocephalus scirpaceus*) (Anon 2012c). However the species is on the IUCN Red List, facing a decline in England of 63 % (Anon 2012d).

There can be unintended consequences of conservation efforts by gardeners, too as 'ecology' and 'gardens' have rarely been studied together, probably because of 'their fragmented ownership and essentially private nature' (Thompson 2006, p. 142). As Cannon (1999, p. 287) notes in an opinion piece in Bird Conservation International, 'what is the real global conservation value of a British suburban garden, with its neat little lawns, nut feeders and nestboxes? In my garden, fledgling blue tits (*Parus caeruleus*), a species of no conservation concern, are busy devouring expensive imported peanuts whose production occupied prime agricultural land in a poor country. Pure entertainment, and a sentimental luxury.' However, he then goes on to argue that at the local level, gardens can be of value, citing amongst other examples, the central area of Chile, where natural habitat destruction has heightened the importance of gardens as refuges.

The fact remains that gardens are good for wildlife conservation. Over the past 50 years, the UK has seen the loss of 98 % of wildflower meadows, 50 % of ancient woodlands, 60 % of lowland heathlands, 80 % of downland sheep walks, and 50 % of lowland fens and mires (Baines 2000); all caused by urban sprawl, overgrazing, grubbing out of hedgerows and intensified agriculture. This makes Britain's 22.7 million domestic gardens with a total area of 432,964 ha increasingly important for wildlife conservation (Davies et al. 2009). To this end, wildlife gardening is heavily promoted by the Government and the prime wildlife conservation charities, the Royal Society for the Protection of Birds (RSPB) and the Wildlife Trusts of Great Britain. Over the last 10 years there has been increasing retail space given over to wildlife feeding/ housing in garden centres and nurseries all over the country that profit from the increased demand for wildlife gardening merchandise.

Davies et al. (2009) estimate that in the UK alone, approximately 12.6 million (48 %) households provide supplementary food for birds, 7.4 million of which specifically use bird feeders. Similarly, there are a minimum of 4.7 million nest boxes within gardens. These figures equate to one bird feeder for every nine potentially feeder-using birds in the UK. Gardens also contain 2.5–3.5 million ponds and 28.7 million trees, which is just under a quarter of all trees occurring outside woodlands.

Conservation organisations have also encouraged people to become interested in birds through national events such as the Big Garden Bird Watch which has been running for over 30 years. They are organised by the British Trust for Ornithology (BTO) and the Royal Society for the Protection of Birds (RSPB) (Anon 2012e, f). Designed primarily as an indoor winter activity for children to become interested in birds, it is now undertaken by over half a million people who regularly take part counting the birds that visit their gardens. This has allowed the compilation of 30 years' worth of records detailing garden bird population trends. Indeed most conservation societies in Britain such as The Wildlife Trusts for Great Britain, the RSPB, the BTO, the Butterfly Conservation Society, the Bumblebee Conservation Trust; likewise the National Audubon Society in the USA all promote and provide information on wildlife gardening. Taking part in national surveys and adopting pro-environmental gardening behaviours clearly instill a feel-good factor for gardeners. Understanding how a sense of connection to nature can impact upon people's decisions to seek out nature in their daily lives is important if we wish to encourage the practice of wildlife gardening as a tool to enhance both connection to nature and urban/rural biodiversity.

## Psychological Benefits of Wildlife Gardening and Nature Appreciation

Conservation psychology is a relatively new branch of psychology which looks at the reciprocal relationships between man and nature; notably how people behave toward nature and how people care about or value nature. Part of its focus is to study ways of getting more people involved in, or supporting, conservation with the premise that concrete experiences of nature lead to an emotional affinity towards it and a motivational basis to protect it (Kals et al. 1999).

In his 'biophilia' hypothesis, Wilson (1984) posits that the natural world continues to influence the human condition through our previous close and evolutionary relationship with it. He suggests that technological development has been so rapid that it outpaces our adaptation to modern environments. Therefore inherent in all of us is a need to be with nature through 'an innately emotional affiliation to other living organisms' (Wilson 1993, p. 31). Experimental evidence in support of the theory was provided by a series of conditioning experiments by Öhman (1986). These demonstrated that physiological and emotional responses to natural threats such as snakes and spiders could occur subliminally, despite the participants in the experiments having no conscious recognition of having seen the stimuli before. It was also shown that modern fears such as guns do not invoke similar responses. When it comes to emotional affiliation, environmentalists and nature writers have long since maintained that humans derive psychological and physical benefits from spending time in the natural world (Mayer 2009; Kaplan and Talbot 1983). Research has shown that exposure to nature alleviates aggression, anxiety and depression (Van den Berg 2005), improves mental health and cognitive capacities (Kuo 2001; Wells 2000; Kaplan and Kaplan 1989) aids the healing process (Ulrich 1983) and provides opportunities for reflection (Curtin 2009; Herzog et al. 1997).

There are two important theories that underpin most of the work on the psychological benefits of nature. These are Attention Restoration Theory (Kaplan and Kaplan 1989) and the aforementioned biophilia hypothesis (Wilson 1984). Interest in these formative theories has recently emerged due to a growing unease caused by the recognition of the damage we are doing to the environment and the sociological, physical and psychological challenges of living in modern, affluent, hyper-consumptive societies (Bauman 2001).

The assumption that contact with nature provides people with restoration from stress and fatigue is not a new concept. Experiences in nature have long been seen to have health benefits. The idea you can be mended by the healing currents of the great outdoors goes back to classical times (Mabey 2006). The Romans recommended rambling as a way of resolving emotional tangles (*solvitur ambulando*) and the French philosopher Foucault (2001, p. 62) wrote that the countryside, '*by the variety of its landscapes wins melancholics from their single obsession by taking them away from the cause and the memory of their sufferings*'. The fact that nature reduces stress is predominantly accredited to the Attention Restoration Theory (ART) first espoused by two psychologists Kaplan and Kaplan (1989, 1995) who studied the effects that the natural environment has on the brain. They began this work by looking at levels of concentration.

Their theory proposes that prolonged and/or intensive use of *directed attention* diminishes a person's capacity to ward off distractions which is evidenced by difficulty concentrating, increased irritability and increased rate of errors on tasks which require concentration; thus creating stress because they have less cognitive resources to cope with everyday demands (Kaplan and Kaplan 1995). This is referred to as '*directed attention fatigue*' (Bird 2007). Where a stimulus is weak or uninteresting, it takes greater effort to block out more attractive but less important distractions. This is mentally demanding as the brain uses inhibitory control mechanisms which magnetic resonance imaging (MRI) scans show to be situated in the right cortex of the brain (Kastner et al. 1998); the same part of the brain which is affected in children with deficit hyperactivity disorder (Bird 2007). Examples of directed attention include driving in traffic, studying, working at a computer and making numerous phone calls. Directed attention fatigue is prevalent in people who are stressed, overworked, bereaved or sleep deprived and is a widespread condition of modern life which is overloaded with information, communication and multiple stimuli that either demand our attention or need to be blocked out.

In contrast to directed attention, *involuntary attention* or 'fascination' is effortless and is naturally held when a person finds the activity such as wildlife gardening interesting and absorbing. Recovery from directed attention fatigue requires restorative environments and activities which do not use the tiring inhibitory control mechanism. Attention restoration involves clearing the mind, a recovery from fatigued directed attention, the opportunity to think about personal and unresolved problems and the chance to reflect on life's larger questions such as direction and goals. Clearing the mind and recovery from fatigue is called *attentional recovery* whereas dealing with personal problems and thinking about philosophical viewpoints is *reflection*. Together, reflection and attentional recovery completes the *restorative process*. The outdoor environment is usually restorative but according to Kaplan and Kaplan (1995) it is only so if it:

1. Involves **being away**, i.e. be in a physically distinct location.
2. Has **extent**, i.e. the location must be absorbing and somewhere which is distinct where a person can settle into and where there is enough to see, experience and think about.
3. Is **fascinating** to behold, i.e. effortless attention allows the inhibitory fibres to relax, since they no longer have to block out distractions. Fascination can be divided into **hard fascination** (e.g. watching sport, television and computer games) which holds attention effortlessly but does not allow enough space for reflection, and **soft fascination** (e.g. looking at nature, exploring countryside and gardens) which holds one's attention to allow **attentional recovery** but also allows time and space for personal **reflection** and time to stand and stare.
4. Is **compatible** with our expectations, i.e. the setting must be able to provide what the seeker requires of it without it being a struggle (Hartig et al. 1991).

The activity of gardening and the enjoyment of a garden as a place of sanctuary meet much of the above criteria and produce aesthetic, spiritual and psychological benefits that extend beyond the growing of plants (Dunnett and Qasim 2000). There have been several studies which have explored the benefits of gardening to human well-being; particularly urban gardening. In their study of 376 UK city residents, Dunnett and Qasim (2000) found that creating a pleasant relaxing environment was the most prominent individual value (76 %) and being close to nature was the sixth (44 %). Gardens were viewed as a necessary relief and contrast to the hard elements of the city. Garden wildlife was universally welcomed. Whilst in their survey of garden owners, Bhatti and Church (2004) found that the garden is an important site for privacy, sociability and sensual connections to nature and these activities can be understood as negotiations and practices to address the social and environmental paradoxes of late modern life, i.e. a space for mental and spiritual restoration. Similarly Eigner (2001, pp. 191–192) studied how participants involved in the voluntary maintenance of a local natural site found that working with nature induced 'an amazing feeling of happiness'; 'an inner sort of calm' and a feeling of really being satisfied, more relaxed and more themselves'.

A report for the Health Council for the Netherlands (Anon 2004) proposes that there are five ways that experiences in nature are psychologically beneficial. These are recovery from stress and fatigue (as above); encouragement to exercise; facilitating social contact; encouraging optimal development in children and providing opportunities for personal development and a sense of purpose and belonging. With regards to the latter, Roszak (1995) argued that this sense of belonging extends beyond our social and city limits to include a sense of belonging to the natural world; to feel connected to it. This 'connectedness to nature' depends on how people see themselves in relation to the natural world (Clayton and Opotow 2003).

As well as a way to find solace, wildlife gardening can reflect a self-identity rooted in such feelings of connection. Stets and Biga (2003, p. 406) define environmental self-identity as 'the meanings that one attributes to the self as they relate to the environment'. The relationship between connectedness to nature and self-identity is complex and inter-related. It is also more to do with affective rather than cognitive responses, i.e. the emotions that a particular subject, in this case love of wildness, arouses. In environmental psychology the general consensus is that we tend to identify with what we care about, i.e. the stronger the environmental identity, the more positive the attitudes towards the environment. Gardening is a highly personal activity and therefore it follows that wildlife gardeners represent their love and care of wildlife through their discernible gardening practices and the wildlife places they create. A nature lover's garden becomes a distinct place with which its owner/creator identifies.

Teisl and O'Brien (2003) conclude that people who enjoyed outdoor forms of recreation tended to display greater pro-environmental attitudes and behaviours than those people who do not engage in those activities. It has also been suggested that it is the emotional attachments that people form through experiential encounters that are instrumental in developing commitments to nature (Milton 2002). Thus it follows that the more time spent engrossed in outdoor activities such as gardening, the greater the emotional attachment to it, and the greater this emotional affinity the stronger the environmental self-identity (Hinds and Sparks 2009). For a growing number of people and organisations, this emotional affinity extends to wildlife. In his book, the Philosophy of Gardens (2006), Cooper discusses the manifestation of care and concern that is induced by the cultivation of a garden. Care arises when the garden becomes inhabited by the self alongside the caring of significant others such as plants, insects, birds and mammals.

In her study of wildlife tourism Curtin (2010) discovered a direct relationship between an interest in wildlife watching on holiday and attracting wildlife to their gardens at home; evidencing a distinct cross-over between holiday and home interests. Having designed a space where wildlife is welcome, the participants in her qualitative study revealed that seeing things in their own garden was just as thrilling as, and sometimes even more significant than, seeing wildlife on tour. In part this thrill is due to the nature of the encounter, in that they themselves have been successful in creating an environment which attracts wildlife, and the caring and nurturing emotions it provokes. There is a tenable sense of responsibility and relationship with these regular garden visitors and this is what makes it so important to their everyday world and everyday self. It highlights the protectionist value orientations people have towards wildlife (Kellert and Berry 1987) and was especially apparent for women whose children had left home and whose careers or jobs had become part-time instead of full-time as their financial prosperity had improved. Time becomes available to re-engage their interest in gardens and nature which fulfils an emotional need to tend to other living things. Finally, people often set their calendar by natural wildlife events, for example, in England, the arrival of barn swallows (*Hirundo rustica*) in the spring, the sound of the cuckoo (*Cuculus canorus*) in summer and so forth.

Bentrupperbaumer (2005) suggests that the timely arrival of wildlife represents the 'miner's canary' of the ecosystem; a barometer of life itself (Knopf 1987) and reassurance of a viable and functioning natural environment despite the destruction that man causes. Whilst industrial and urban settings are not in keeping with traditional and romantic views of nature and wildlife, the emotional significance of seeing wildlife here is somewhat heightened by the wonderment and reassurance it arouses.

## Visiting Horticultural Attractions

Key information sources regarding the state of current wildlife and the conservation of wildlife in gardens comes primarily from popular media. However other vital sources of information and inspiration come in the form of horticultural attractions and retail outlets. The latter consist of plant nurseries and garden centres and the former of visitor attractions both permanent, such as gardens, and temporary, such as horticultural events and festivals.

Garden centres have developed from plant nurseries and have an expanded range of products for the home and garden. Many are owned independently, but in the USA, Europe and Australia, home improvement chains have also introduced gardening departments. Additionally, some gardens such as the Royal Horticultural Society gardens at Wisley, England have opened garden centres as an additional revenue stream. In these centres information about wildlife in the garden is always displayed prominently as a promotion for the garden products on sale. Other garden centres provide for light refreshments in an environment surrounded by wildlife.

Whilst there is a growing body of academic literature on wildlife in domestic gardens, there is little when it comes to gardens that are visitor attractions, such as botanic gardens (collection-based institutions) or other gardens open to the public. Nonetheless the wildlife in these gardens is acknowledged and lists of the species that have been seen in the gardens can be found, for example at the Royal Botanic Gardens, Sydney, Australia (Anon 2012g). Occasionally, interpretation or 'living exhibits', such as the butterfly border at Birmingham Botanical Gardens, England (Anon 2012h) are developed; sometimes this is taken further with the inclusion of wildlife viewing infrastructure such as hides, wildlife interpretation boards and real-time television footage.

Figure 30.2 provides an example of how The Lost Gardens of Heligan (Anon 2012i) in Cornwall, England, have developed a wildlife hide alongside webcam technology to provide their visitors with live coverage of nesting birds and visiting mammals. This site has also featured on a wildlife television programme, not only promoting wildlife conservation and wildlife gardening but also the sustainable, eco-centric land management principles of Heligan. Another unique location

'The Lost Gardens of Heligan, near Mevagissey in Cornwall, are one of the most popular botanical gardens in the UK. Heligan's aim is to maximise biodiversity within a patchwork of habitats found throughout the 200 acres of historic Cornish estate and garden. Ancient woodland, hay meadows, grazed pasture, wetlands are all sustainably managed to encourage local wildlife. This is achieved using a variety of traditional methods including coppicing, charcoal burning, hay making, and low intensity grazing with our herd of Dexter cattle'.
'Horsemoor Hide lies at the heart of our estate, and offers the perfect location to enjoy Heligan's wildlife. There is a large wildlife viewing area; along with live and recorded footage, interactive displays, photographs and information gathered by our dedicated Wildlife Team. We use traditional land management techniques to benefit a wide range of wildlife for you to see here. We hope to encourage you to explore the fascinating natural world with us'.
Photographs provided with the courtesy of Lost Gardens of Heligan, 2012

**Fig. 30.2** Horticultural attractions and wildlife

is the Wildlife Botanic Gardens at Bush Prairie, Washington USA. The 'Gardens are devoted to demonstrating and teaching gardening concepts which attract birds, butterflies, hummingbirds and other wildlife to residential gardens' (Anon 2012j). Managed by Naturescaping, a non-profit, all volunteer, educational organisation, the 9th garden was completed in 2008 and is devoted to hummingbirds and the native butterflies of the Northwest of America (Anon 2012k).

Fox and Edwards (2009) describe the development of horticultural shows and festivals from exhibitions of chrysanthemums in Japan about 900A.D. to the first European show in Belgium in 1809 and the Philadelphia Flower Show in the USA, two decades later. This is now held in 33 acres of the Pennsylvania Convention Centre making it the largest indoor Flower Show in the world. They distinguish between three types of horticultural show; first, large national/international shows with show gardens, celebrities and media coverage. The second type, they referred to as professional shows as they are regional events based on professional exhibitors selling plants and gardening accessories. The third are small local shows, labelled by them as amateur shows, at which gardeners compete for prizes for their flowers and vegetables etc. Both the two largest forms of show often contain exhibits on garden wildlife from conservation organisations and commercial sales stands.

Floriade World Horticulture Expo is an international exhibition of flowers and gardening that is held every 10 years in the Netherlands. The 2012 event received 2 million visitors and had as its overall theme 'Be part of the theatre in nature, get closer to the quality of life'. It included a 'Feel Good Garden' as part of a 'Relax & Heal' theme and a butterfly garden entitled, "Footkiss for Butterflies". The garden is in the shape of a huge, sloping leaf from the Japanese Ginkgo Biloba tree. A hedge consisting of large numbers of indigenous flowers marks out the periphery of the entry and is highly attractive to butterflies' (Anon 2011).

Wildlife festivals 'promote a variety of social, educational, economic, recreational, and community development goals' (Hvenegaard 2011). In North America, there were over 240 events in 2002, but they are significantly smaller than many horticultural shows, attracting only hundreds or a few thousand visitors. Rarely, however, do they relate to the wildlifc in people's gardens or yards and as Hvenegaard questions, 'Does educating visitors about wildlife and their habitats at the festivals translate into environmentally friendly behaviour?' (Hvenegaard 2011, p. 382). Ultimately if we are to move to a more sustainable future, changing consumer behaviour towards more eco-friendly practices and consumption is fundamental. Wildlife gardening is perhaps a first important step towards the recognition of biodiversity and its intrinsic value to humankind.

## Conclusions

The ethos of creating a space where wildlife is a vital, holistic part of the overall concept is becoming more fashionable. Wildlife is predominantly attracted to the garden by appropriate planting schemes but also through the provision of water, food and nest boxes. Whilst the inclusion of these is generally positive for conservation, it is not without some drawbacks such as a potential over-reliance on provisioned food, the unsustainable nature of production and transportation of bird food and the spread of diseases from feeders and bird baths. Despite this, the notion of symbiosis between wildlife and gardening appeals to modern concerns and sensibilities of nature, and is somewhat counter to more traditional forms of gardening where pest control and protecting prize produce was key. This gradual shift in values may, in part, be due to the constant reminders from popular media of the damage that man ultimately causes to the natural world as well as more intrinsic motivations such as the need to reconnect with nature through the process of biophilia.

The world's population predominantly lives in large urban areas but whilst this is a more convenient way of life, modern cities can induce high levels of mental fatigue caused by noise, traffic, people and an overload of mental stimuli (Waliczek et al. 2005). There is much research to suggest that all people need at least some interaction with nature (Ulrich 1983). Allowing the 'wild' into our lives has several psychological advantages; notably mental and spiritual restoration, a reconnection to the natural rhythms of life and the happiness and peace derived from slowing down, observing more and reflecting.

Wildlife thus has the potential to add greater meaning and sensual appeal to gardens. It also satisfies the human need to tend to care for other living things. Gardening by its very nature imparts a sense of time and seasonal changes brought not only by the weather but by the natural cycle of fauna and flora. The anticipation of what might arrive and the resultant theatre, beauty and movement it brings engender an emotional attachment to the experience. Psychology suggests that it is the emotional attachments that people derive through experiential encounters with nature that are instrumental in the desire to care for it. Over time this desire to care for wildlife underpins a social self-identity of ‘being a wildlife gardener’ and the stronger that identity becomes, the greater the emotional attachment to its philosophy.

However, it is clear that not all wildlife is equally valued or welcomed with some species being undesirable (e.g. rodents), highly desirable (songbirds) and dubious depending on personal preference (e.g. bats). For a truly holistic approach to wildlife gardening there is arguably some work to be done with regards to the promotion of valuing all species rather than just the aesthetically pleasing or useful. As it is, the attraction of songbirds is the most sought. Horticultural attractions and events have an important part to play in the promotion of valuing biodiversity. Working alongside conservation organisations, some key attractions have taken a strong stance with their philosophy of wildlife gardening and this should only be encouraged.

To date very little research has been undertaken specifically to understand the consumer behaviour and experience of wildlife gardening or the importance and appeal of including wildlife in the design of horticultural attractions. Yet understanding how a sense of connection to nature can impact upon people’s decisions to seek out nature in their daily lives is important if we wish to encourage the practice of wildlife gardening as a tool to enhance both connection to nature and urban/rural biodiversity. It will be interesting to see the content and results of such studies.

## References

Anon (2004) Nature and health: the influence of nature on social, psychological and physical well-being (publication no. 2004/09E: RMNO publication no. A02ae). Health Council of the Netherlands and RMNO, Netherlands

Anon (2011) Floriade 2012. http://www.floriade.com/. Accessed 27 Sep 2012

Anon (2012a) The tomato had to go abroad to make good. Texas AgriLife Extension Service, Texas. http://aggie-horticulture.tamu.edu/archives/parsons/publications/vegetabletravelers/tomato.html. Accessed 26 Sep 2012

Anon (2012b) Wildlife of Sydney. http://australianmuseum.net.au/Wildlife-of-Sydney. Accessed 27 Sep 2012

Anon (2012c) Cuckoo. http://www.rspb.org.uk/wildlife/birdguide/name/c/cuckoo/index.aspx. Accessed 21 Dec 2012

Anon (2012d) About the cuckoo project. http://www.bto.org/science/migration/tracking-studies/cuckoo-tracking/about. Accessed 21 Dec 2012

Anon (2012e) Garden BirdWatch. http://www.bto.org/volunteer-surveys/gbw. Accessed 21 Dec 2012

Anon (2012f) Big Garden Bird Watch. http://www.rspb.org.uk/birdwatch. Accessed 20 Oct 2012

Anon (2012g) Wildlife. http://www.rbgsyd.nsw.gov.au/welcome/royal_botanic_garden/gardens_and_domain/wildlife. Accessed 28 Sep 2012

Anon (2012h) Wildlife in the garden. http://www.birminghambotanicalgardens.org.uk/gardens/wildlife-areas/wildlife-in-the-garden. Accessed 28 Sep 2012

Anon (2012i) The lost gardens of Heligan, Pentewan, St.Austell, Cornwall, United Kingdom, PL26 6EN. http://www.heligan.com. Accessed 22 Nov 2012

Anon (2012j) Wildlife botanical gardens. http://www.prairiewa.com/wildlife.htm. Accessed 28 Sep 2012

Anon (2012k) Wildlife botanical garden. http://www.naturescaping.org/aboutus.php. Accessed 28 Sep 2012

Baines C (2000) How to make a wildlife garden. Francis Lincoln Limited, London

Bauman Z (2001) Consuming life. J Consum Cult 1(1):9–29

Bentrupperbaumer J (2005) Human dimensions of wildlife interactions. In: Newsome D, Dowling RK, Moore SA (eds) Wildlife tourism. Channel View Publications, Clevedon, pp 82–112

Bhatti M, Church A (2004) Home, the culture of nature and the meaning of gardens in late modernity. Hous Stud 19(1):37–51

Bird W (2007) Natural thinking: investigating the links between the natural environment, biodiversity and mental health. A report for the Royal Society for the Protection of Birds. RSPB, Sandy

Cannon A (1999) The significance of private gardens for bird conservation. Bird Conserv Int 9:287–297

Catterall CP (2004) Birds, garden plants and suburban bush lots: where good intentions meet unexpected outcomes. In: Burgin S, Lunney D (eds) Urban wildlife: more than meets the eye. Royal Zoological Society of NSW, Mosman, pp 21–31

Clayton S, Myers G (2009) Conservation psychology: understanding and promoting human care for nature. Wiley, Chichester

Clayton S, Opotow S (2003) Introduction: identity and the natural environment. In: Clayton S, Opotow S (eds) Identity and the natural environment: the psychological significance of nature). MIT Press, Cambridge, pp 1–24

Cooper DE (2006) The philosophy of gardens. Oxford University Press, Oxford

Curtin SC (2009) Wildlife tourism: the intangible, psychological benefits of human-wildlife encounters. Curr Issue Tourism 12(5):451–474

Curtin SC (2010) The self-presentation and self-development of serious wildlife tourists. Int J Tourism Res 12(1):17–33

Daglish EF (1928) The book of garden animals. Chapman and Hall, London

Davies ZG, Fuller RA, Loram A, Irvine KN, Sims V, Gaston KJ (2009) A national scale inventory of resource provision for biodiversity within domestic gardens. Biol Conserv 142(4):761–771

Dunnett N, Qasim M (2000) Perceived benefits to human well-being of urban gardens. HortTechnol 10(1):40–45

Eigner S (2001) The relationship between "protecting the environment" as a dominant life goal and subjective well-being. In: Schmuck P, Sheldon KM (eds) Life goals and well-being: towards a positive psychology of human striving. Hogrefe and Huber, Gottingen, pp 182–201

Ellis ET (ca. 1935) The garden for expert and amateur. Daily Express Publications, London

Fischer JR, Stallknecht DE, Luttrell MP, Dhondt AA, Converse KA (1997) Mycoplasmal conjunctivitis in wild songbirds: the spread of a new contagious disease in a mobile host population. Emerg Infect Dis 3(1):69–72. http://wwwnc.cdc.gov/eid/article/3/1/97-0110.htm. Accessed 24 Sep 2012

Foucault M (2001) Madness civilization: a history of sanity in the age of reason. Routledge, London

Fox D, Edwards JR (2009) A preliminary analysis of the market for small, medium and large horticultural shows in England. Event Manag 12(3/4):199–208

Fuller RA, Warren PH, Armsworth PR, Barbosa O, Gaston KJ (2008) Garden bird feeding predicts the structure of urban avian assemblages. Divers Distrib 14:131–137
Hadfield M (1936) A gardener's bibliography. In: Hadfield M (ed) The gardener's companion. J. M. Dent and Sons Ltd, London, pp 547–605
Harper P, Madsen C, Light J (1994) The natural garden book: a holistic approach to gardening. Simon and Schuster, London
Hartig T, Mang M, Evans GW (1991) Restorative effects of natural environment experience. Environ Behav 23(1), 3–26
Herzog TR, Black AM, Fountaine KA, Knotts DJ (1997) Reflection and attention recovery as distinctive benefits of restorative environments. J Environ Psychol 17(2):165–170
Hills LD (1989) Month-by-month organic gardening. Thorsons Publishers Ltd, Wellingborough
Hinds J, Sparks P (2009) Investigating environmental identity, well-being and meaning. Ecopsychology 1(4):181–186
Hvenegaard GT (2011) Potential conservation benefits of wildlife festivals. Event Manag 15(4):373–386
Inglehart R, Baker WE (2000) Modernization, cultural change, and the persistence of traditional values. Am Sociol Rev 65(1):19–51
Kals E, Schumacher D, Montada L (1999) Emotional affinity toward nature as a motivational basis to protect nature. Environ Behav 31(2):78–202
Kaplan R, Kaplan S (1989) The experience of nature: a psychological perspective. Cambridge Press, New York
Kaplan R, Kaplan S (1995) The restorative benefits of nature: Toward an integrative framework. J Environ Psychol 15(3):169–182
Kaplan S, Talbot JF (1983) Psychological benefits of a wilderness experience. In: Altman I, Wohlwill JF (eds) Human behaviour and the environment: advances in theory and research: behaviour and the natural environment, vol 6. Plenum Press, New York, pp 163–203
Kastner S, De Weerd P, Desimone R, Ungerleider LG (1998) Mechanisms of directed attention in the human extrastriate cortex as revealed by functional MRI. Science 282(5386):108–111
Kellert R, Berry JK (1987) Attitudes, knowledge and behaviours towards wildlife as affected by gender. Wildl Soc Bull 15:363–371
Knopf R (1987) Human behaviour, cognition, and affect in the natural environment. In: Stokols D, Altman I (eds) Handbook of environmental psychology, vol 2. Wiley, New York, pp 783–826
Kuo FE (2001) Coping with poverty. Aggression and violence in the inner city. Environ Behav 1(33):5–34
Mabey R (2006, September) A brush with nature. BBC Wildlife Magazine, p 13
Manfredo MJ, Teel TL, Bright AD (2003) Why are public values toward wildlife changing? Hum Dimens Wildl 8:287–306
Martinson TJ, Flaspohler DJ (2003) Winter bird feeding and localized predation on simulated bark-dwelling arthropods. Wildl Soc Bull 31:510–516
Mayer S (2009) Why is nature beneficial? Environ Behav 41(5):607–643
McHoy P (2000) Jardinier en toute saisons (Gardener in all seasons). Manise, Geneva
Middleton CH (1939) With C. H. Middleton in your garden. George Allen and Unwin Ltd, London
Milton K (2002) Loving nature. Routledge, London
Öhman A (1986) Face the beast and fear the face: animal and social fears as prototypes for evolutionary analysis of emotion. Psychophysiology 23:123–143
O'Leary R, Jones DN (2006) The use of supplementary foods by Australian magpies Gymnorhina tibicen: implications for wildlife feeding in suburban environments. Austral Ecol 31(2):208–216
Perry F (1955) The woman gardener. Hulton Press, London
Robb GN, McDonald RA, Chamberlain DE, Bearhop S (2008) Food for thought: supplementary feeding as a driver of ecological change in avian populations. Front Ecol Environ 6(9):476–484
Rolston H (1987) Beauty and the beast: Aesthetic experience of wildlife. In: Decker DJ, Goff GR (eds) Valuing wildlife: economic and social perspectives. Westview Press, Boulder, pp 187–196

Roszak T (1995) Where psyche meets Gaia. In: RoszaK T, Gomes ME, Kanner AD (eds) Ecopsychology: restoring the earth, healing the mind. Sierra Club Books, San Francisco
Ryrie C (2003) Wildlife gardening. Cassell Illustrated, London
Smith RM, Thompson K, Hodgson JG, Warren PH, Gaston KJ (2006) Urban domestic gardens (IX): composition and richness of the vascular plant flora, and implications for native biodiversity. Biol Conserv 129:312–322
Stets JE, Biga CF (2003) Bringing Identity theory into environmental sociology. Sociol Theory 21:398–423
Sudell R (1950) The new illustrated gardening encyclopaedia. Odhams Press, London
Teisl MF, O'Brien K (2003) Who cares and who acts? Outdoor recreationists exhibit different levels of environmental concern and behaviour. Environ Behav 35(4):506–522
Thomas J (2012) Myna fightback. ABC Science. http://www.abc.net.au/science/articles/2004/04/08/2044900.htm. Accessed 26 Sep 2012
Thompson K (2006) Ecology of the garden. In: Taylor P (ed) The Oxford companion to the garden. Oxford University Press, Oxford, pp 142–143
Turner EL (1935) Every garden a bird sanctuary. Witherby, London
Ulrich RS (1983) Aesthetic and affective response to natural environment. In: Altman I, Wohlwill JF (eds) Human behaviour and the environment: Advances in theory and research: behaviour and the natural environment, vol 6. Plenum Press, New York, pp 85–125
Van den Berg AE (2005) Health impacts of healing environments: a review of the benefits of nature, daylight, fresh air and quiet in healthcare settings. Foundation 200 years University Hospital Groningen, Groningen
van den Duim R, Caalders J (2002) Biodiversity and tourism: ompacts and Interventions. Ann Tourism Res 29(3):743–761
Waliczek TM, Zajicek JM, Lineberger RD (2005) The Influence of Gardening Activities on Consumer Perceptions of Life Satisfaction. HortScience 40(5):1360–1365
Wells NM (2000) At home with nature. Effects of greenness on children's cognitive functioning. Environ Behav 32(6):775–795
Wilson EO (1984) Biophilia. Harvard University Press, Cambridge
Wilson, EO (1993) Biophilia and the conservation ethic. In: Kellert SR, Wilson EO (eds) The Biophilia Hypothesis. Island Press, Washington DC, pp 31–41
Wright WP (ed.) (Ca. 1902) Cassell's dictionary of practical gardening, vol 1. Cassell and Co, London
Wyman D (1971) Wyman's gardening encyclopaedia. The Macmillan Company, New York

# Chapter 31
# Horticultural Science's Role in Meeting the Need of Urban Populations

**Virginia I. Lohr and P. Diane Relf**

**Abstract** Horticultural products and services impact the health and well-being of urban populations. This is an extremely important group for horticultural scientists and researchers to serve: more than half of all people worldwide already live in urban areas, and more than two-thirds will do so by 2050. In this chapter we address the past, current, and future roles that horticultural science plays in the major issues of concern to public welfare: public health, environmental health, food security, and economic stability. Urban horticulture has important impacts on the health of the individual and the community, two concerns of public health. Documented individual health benefits include less depression and improved pregnancy outcomes from walking in or living near urban green spaces. Community gardens, parks, and other urban vegetation enhance community health by improving social interactions, such as family dynamics, and public safety, such as protection from crimes. Uses of plants to improve the urban environment include temperature modification, air pollution reduction, and water quality improvement. Impacts on biological diversity are mixed. Other negative impacts include the introduction of invasive species. Urban food security requires food in sufficient, nutritious, and affordable quantities. Providing this for all people is one of the greatest challenges for horticultural science. Potential solutions include increasing small-scale food production in urban areas by providing more community gardens or converting vacant lots. Horticulture contributes directly to urban economics through the production and sales of horticultural products by urban businesses. Indirect contributions from plants include higher property values and more productive employees. The increasing urbanization and aging of the human population is happening in conjunction with rising environmental destruction from global warming and climate change. Combining the traditional horticultural concern of feeding the world with an expanded understanding of the additional functions provided by horticultural products, the needs of urban people, and the opportunities to partner with professionals in other disciplines will be essential in the unpredictable future.

V. I. Lohr (✉)
Department of Horticulture, Washington State University, Pullman,
WA 99164-6414, United States of America
e-mail: Lohr@wsu.edu

P. D. Relf
Department of Horticulture, Virginia Tech, VPI & SU, Blacksburg,
VA 24061, United States of America
e-mail: pdrelf@vt.edu

G. R. Dixon, D. E. Aldous (eds.), *Horticulture: Plants for People and Places, Volume 3*,
DOI 10.1007/978-94-017-8560-0_31, 

**Keywords** Biodiversity · Climate change · Community gardens · Economics · Ecosystem services · Environment · Food security · Global warming · Human health · Horticultural therapy · Human population · Public safety · Urbanization

## Introduction

In this chapter we address the impacts of horticulture on urban populations and their physical and psychosocial needs. We use the word, *urban*, in a broad sense, including related, human-dominated landscapes, such as suburban and peri-urban areas. We also address the applications of horticultural science in urban agriculture. For the purpose of the discussion, we draw on the broad definition of *urban agriculture* given by the Council for Agricultural Science and Technology: "Urban agriculture is a complex system encompassing a spectrum of interests, from a traditional core of activities associated with the production, processing, marketing, distribution, and consumption, to a multiplicity of other benefits and services that are less widely acknowledged and documented. These include recreation and leisure; economic vitality and business entrepreneurship, individual health and well-being; community health and well-being; landscape beautification; and environmental restoration and remediation" (Butler and Maronek 2002). Horticulture is clearly included in this definition of urban agriculture, and for the purposes of this chapter we focus on that aspect by using the term urban horticulture.

In this chapter, we will consider the impact on urban populations on the full range of horticultural plants, including but certainly not limited to trees, shrubs, flowers, turf, indoor plants, fruits, vegetables, native and introduced species, cut and potted flowers, and medicinal plants. These products of horticulture are essential for a healthy urban population. At the same time, topics related to horticultural services are considered. These include meeting the needs of human beings and addressing their quality of life through contact with plants, utilization or consumption of plant parts, or involvement in the cultivation of plants. The information explored by horticultural science that is included in this concept ranges from ecosystems to green-care farms, from landscape design to healthcare gardens, from economics and marketing to school gardens, and multitudes of other human concerns. Opportunities to grow plants, to nurture the life of the plant, and to feel personal responsibility for caring for life in urban environments are also essential for the health and well-being of urban populations. While our focus is primarily on the benefits from horticultural crops and services, we also present some detrimental aspects, including environmental destruction from the introduction of invasive species and property damage from poorly sited or maintained vegetation.

Many of the items in this chapter are addressed in other chapters, but we include them here to emphasize that their role has a strong urban aspect and that professionals responsible for the management of urban areas need to be aware of the importance of horticulture and horticultural science as a part of the urban complex. Among the urban professionals that we, as horticulturists, need to address are: urban planners, engineers, public health officials, business leaders, politicians, educators, non-profit staff, and volunteers. In addition, our focus is on the role that horticultural science and horticultural scientists play in the urban natural environment in partnership with urban foresters, agronomists, landscape architects, urban ecologists, and others concerned with linking people and nature.

As the world continues its rapid urbanization, with projections that 67 % of the human population will be concentrated in cities and their surroundings by 2050 (United Nations 2012) and as the negative impacts of climate change become increasingly apparent (Lelieveld et al. 2012), the importance of horticultural science in conducting research and disseminating information to address world issues effectively becomes more evident. In this chapter we address the role of horticultural science in the major issues of concern to public welfare: public health, environmental health, food security, and economic stability. Needless to say, these are tightly interrelated, but for the purpose of this chapter we will consider specific elements of each separately.

## Horticultural Science's Role in Public Health

Public health is a major social and economic issue that will continue to grow as the human population expands and ages. It is also an environmental issue: practitioners in public health have long focused on the reduction of health problems by remediation of causal factors including degraded environments (see section on: Urban Environmental Health). Certainly urban horticulture has important impacts on the health of the individual, the community, and the environment, the three areas that constitute public health concerns. Health of the individual and health of the community are discussed in this section with an emphasis on the role of horticulture in both.

To meet the needs of healthcare and public health practitioners and develop strong utilization of horticulture in public health, to meet the needs of the rapidly growing and aging population and the professionals that serve this group, and to provide knowledge and skills to help build healthy communities, we need to develop a long-term plan for research and outreach. Among the tasks to be completed in cooperation with researchers from other universities, colleges, and departments is to identify the most critical issues to be researched. Long-term cooperative research will demand external funds from sources not usually approached by researchers in Colleges of Agriculture. This cross disciplinary work with researchers in medicine, education, social sciences, urban planning, and others will open doors to different private foundations, corporate foundations, and government organizations that must be identified and communicated with in terms that clearly address their goals.

## Health of the Individual

Horticulture has important impacts on the health of individuals through direct interaction with plants and the natural environment. Horticulture promotes individual health through exercise, stress reduction, social interaction, and mental stimulation. Gardening is recommended by such groups as the American Heart Association (2013) as a technique to improve general physical health and thus prevent many human diseases. In addition urban horticulture can plan a role in improved health through access to high quality fresh produce either locally produced or self-produced (see discussion below on *Food Security*).

How we use plants in our cities, whether indoors or out, can have strong influences on health and well-being (Lohr 2010, 2011). Increasing research is identifying strong links between plants in our surroundings and positive health outcomes. In fact, documentation of such links have become so strong that the medical and public health communities are promoting the expansion of green spaces, parks, green roofs, and community gardens and the planting trees to reduce the incidence of human diseases, including heat-related deaths, respiratory illnesses, and cardiovascular diseases (Younger et al. 2008; O'Neill et al. 2009; Cheng and Berry 2012). Some examples of connections between human health and the use of plants in our surroundings are presented below, focused on areas that have significant potential for further involvement by horticultural scientists. Additional examples are presented in other chapters. Such information can be useful in promoting the need for municipalities to spend money on establishing and maintaining plants in urban areas.

*Trees, walkable communities, and human health.* According to the World Health Organization, obesity has more than doubled worldwide since 1980 and being overweight is the fifth leading risk factor for deaths worldwide (World Health Organization 2012). Common health problems associated with being overweight or obese include heart disease and cancer. The Centers for Disease Control and Prevention (2012) attribute the lack of physical activity in the United States of America (U.S.), in part, to current patterns of land use and transportation. Studies have shown that walkable outdoor spaces with trees (Fig. 31.1) are correlated with lower rates of obesity (Lachowycz and Jones 2011). Lovasi et al. (2012) found that residents in New York City who lived in areas with more walkable streets with trees had lower body mass indices than people living in areas without street trees. In Tokyo, five-year survival rates for citizens in their 70s and 80s were greater if they had space for walking near their residences and if there were parks and street trees near them (Takano et al. 2002). Even children's health can be improved through appropriate incorporation of plants in urban areas. One study showed that children had better body mass index scores if they lived in communities with more vegetation nearby compared to children in areas with low amounts of green (Bell et al. 2008). As a result of such studies, public health professionals and even the American Planning Association are emphasizing the need to plant more trees and build more parks (Younger et al. 2008; Ricklin et al. 2012). Communities are also responding. For example, Fort Worth, Texas has a plan to increase neighborhood and community park space by more than 1 acre per 1,000 people by 2025 (Ricklin et al. 2012).

Other documented health benefits from walking in parks and green spaces include reduced stress and less depression. In the 1980's, researchers began to document evidence of reduced human stress when passively viewing plants or nature (Moore 1981; Ulrich and Simon 1986). Benefits from actively walking among trees or in gardens were documented in subsequent studies. Cimprich (1993) studied

**Fig. 31.1** Walkable street with trees in Madrid, Spain. Streets with shade trees increase the likelihood that people will walk along them and gain health benefits, including reduced obesity. (Photo by V. I. Lohr)

women undergoing surgery for breast cancer. They typically suffer from mental fatigue and show signs of depression. Half of the subjects were given only routine treatment, while the others were also asked to perform a mentally restorative activity. Most women decided to walk in a garden. Depression began to lift within 3 months of surgery for those who walked in gardens, while depression worsened for those in the control group. A more recent study, using medical records from people in The Netherlands, found that rates of anxiety disorders and depression were lower when the amount of green space within 1 km of people's residences were greater (Maas et al. 2009). In another study, Berman et al. (2010) worked with people with major depressive disorder. The moods and mental capacities of the subjects were better on days when they walked in an arboretum than on days when they walked in an urban area. Gardening has similar benefits to walking in nature: gardeners who performed a stressful, non-gardening task and then gardened for 30 min had lower levels of salivary cortisol, a hormone associated with human stress, than those who read absorbing material for 30 min following the task (Van den Berg and Custers 2011).

*Green areas and human health.* A range of positive and perhaps surprising health outcomes have been documented in urban areas with increasing amounts of green space. One study, looked at mortality rates for people in England based on income deprivation and found that overall mortality rates dropped as the amount of vegetation near their residence increased (Mitchell and Popham 2008). Two recent studies have shown a positive link between pregnancy outcomes and green space near the mother's home. One conducted in Barcelona, Spain, documented an increase in birth weight from mothers in the lowest educational group as the amount of vegetation within 500 m their residences increased or as the distance to a major green open space shortened (Dadvand et al. 2012). A study in Portland, Oregon, U.S. documented a reduction in the number of babies who were small for their gestational age as tree canopy cover within 50 m of the mother's residence increased (Donovan et al. 2011).

Other studies have looked at the effects of nature on exercise and sports. Barton and Pretty (2010) showed that activities performed in areas with plants, such as walking, cycling, or gardening in urban green areas, could improve mood and self-esteem within just 5 min. A recent study found that track and field athletes received their best performance marks at sites with more green landscaping and their worst scores at venues with the least amount of vegetation (DeWolfe et al. 2011).

A wide range of health benefits that accrue from being in urban areas with high amounts of vegetation are known, and some researchers are now examining the level of access to vegetation for different populations. Environmental inequities are wide-spread. In Montreal, Canada, disparities based on income have been found, with lower income people having less green in their neighborhoods than more affluent people (Pham et al. 2012). Similar relationships have been documented in many places, including six cities in Australia (Kirkpatrick et al. 2011), the urban and peri-urban area of Phoenix, Arizona, U.S. (Hope et al. 2003), and the municipality of Campos Dos Goytacazes in the state of Rio de Janeiro, Brazil (Pedlowski et al. 2002). This relationship was not found in Paris, France (Cohen et al. 2012a). Mitchell and Popham (2008) showed how critical green areas for low income people could be: they documented that the increase in mortality associated with income deprivation in urban areas could be overcome with more exposure to vegetation. This knowledge is critical to incorporate in urban and peri-urban planning (Frumkin and McMichael 2008; Rydin et al. 2012).

The role of horticultural science in contributing to walkable communities and green health includes selection and testing of plants for the specific areas and intent; development of maintenance techniques to withstand the environmental stresses placed on urban trees; cooperative research with other professionals to determine the sustainability of plants in the research that is conducted (for example, are the trees healthy and appropriate for their location and does that matter as far as the effect on humans); and appropriate plant materials to withstand stresses related to efforts to rectify environmental inequality. Ideally these roles involve horticultural scientists working with public health professionals and other social scientists to conduct research and provide educational materials to the urban officials who make the decisions regarding funding.

*Horticulture and healthcare.* Horticulture has a long history of use as a treatment for individuals within the healthcare system (Warner and Baron 1993). As part of the professional area of healing landscapes, horticulture and plants play roles in providing an atmosphere that is conducive to recovery (Relf 2005). In the professional areas of social horticulture and therapeutic horticulture and as part of greencare farming (Hassink and van Dijk 2006), horticulture is valued as an activity to stimulate, motivate, and rehabilitate the gardener with health issues whether or not they are in a structured, goal directed program (Relf 2006). In the profession of horticultural therapy, horticulture has use and significant potential as a structured treatment tool for individuals with diagnosed health issues in defined treatment programs with specific achievable goals utilizing living plants under the direction of a trained professional therapist (Relf 2005).

Extensive information is available on the design and use of healing landscapes for the positive ambiences they provide. Research focused on post-occupancy surveys of hospital and other healthcare landscapes have been conducted and reported (Marcus and Barnes 1999). There is access to links and descriptions of healing landscapes sites on-line (Anon 2012b). Therapeutic gardens, as sites for the implementation of therapeutic horticulture and horticultural therapy programs, are less well researched, as they require the integration of therapist, patients, and programs to test their efficacy.

The majority of the research on the utilization of horticulture as an activity in the healthcare area has looked at social and therapeutic horticulture, with particular focus on aging (Gigliotti et al. 2004; Wichrowski et al. 2005; Collins and O'Callaghan 2008). Activities with children have also been studied (Kim et al. 2012). Theoretical models have been put forth to serve as guides for research and to stimulate query (Relf 2006; Shoemaker and Lin 2008) and methods for research have been discussed (Shoemaker et al. 2000), but the limited number of potential researchers within academic horticulture has resulted in a lack of adequate studies that look at the efficacy of horticultural therapy as a treatment regime eligible for third party reimbursement. Anecdotal evidence and growth of horticultural practices among activity therapists and other healthcare professionals justifies an expanded role for horticultural scientists working cooperatively with healthcare professionals to conduct the research and teaching to ensure this professional area reaches its potential.

It is widely recognized that a rapidly aging population worldwide has significant implications for health issues. Horticulture has an important role to play in enhancing and/or ameliorating health factors (Rappe and Kivelä 2005). In recent years connections have been found between gardening and a delay in Alzheimer's disease (Simons et al. 2006). Two major causes of health problems among the elderly are lack of exercise and loneliness. Horticulture addresses both of the issues for many people. However, many people unnecessarily stop gardening as they age due to impacts of arthritis, back problems, heart disease, and other ailments often associated with aging. This translates into landscapes in both private and public areas lying idle, which, if properly designed, could serve to meet recreational and health needs of this large population. Despite the widely accepted value of gardening, only limited research has been done to understand how gardening affects elderly individuals socially, psychologically, physically or intellectually (Park et al. 2011). The research that has been conducted provides important indicators of the value of continued work in this area. In two separate studies of intergenerational programming in horticulture with adults who still lived at home either independently or with care givers, researchers (Kerrigan and Stevenson 1997; Predny and Relf 2000) found that horticultural activities that focused on growing plants resulted in greater interaction than those activities that involved craft-type work. For seniors in intermediate care, Mooney (1994), using three different psychological measuring tools, found a pattern of improvement after the treatment was implemented and decline when the therapy was withdrawn to be a "classic" pattern for the experimental group. For elderly adults with cognitive impairment such as Alzheimer's disease, studies indicated that a properly designed outdoor environment reduced incidents of aggressive behavior (Mooney and Nicell 1992) and agitation (Detweiler et al. 2008).

This type of research data is important in justifying the expenditure of dollars for gardening facilities in public parks and gardens, retirement communities, public housing, and hospitals as well as residential and healthcare sites for elderly. Research in this area would also have implications for making gardening easier and more rewarding for the well-elderly and for the expansion of horticulture in the recreation and tourist industry.

## Health of Communities

A third significant area in which horticulture impacts public health is through the interaction and dynamics of a healthy community. Research has indicated that community gardens, street trees, parks, and other urban vegetation play a role in reduction of crime, including child abuse, and that gardens can be a central focus for community development and neighborhood partnerships (Kuo et al. 1998; Kuo and Sullivan 2001a). Charles Lewis (1996) wrote that if an area is dilapidated or vandalized, has trash-filled vacant lots, or is sterile steel and concrete, it sends messages that those in charge (the city government, the owner, the employers) do not place value on the area or the people there. It implies that the people have no intrinsic worth and no control over their environment. It tells outsiders that this is not a good place to be. The opposite is also true; for example, as a consequence of businesses and neighborhoods beautifying their surroundings as part of the Philadelphia Garden Blocks program, other areas followed suit, a phenomena reported as early as the 1960's.

The daily contact with nature that takes place in the landscapes around homes is important to people's welfare. For example, in a National Gardening Survey in the U.S., 37 % of respondents said gardening gave a sense of peace and tranquility (Relf et al. 1992). Forty percent reported that being around plants helped them feel calmer and more relaxed, and 46 % said that nature was essential to their well-being. In Uruguay, people in urban areas with trees were more likely to report being happy and having an improved social life than people in areas without trees (Gandelman et al. 2012). In this section we look at the impact of horticulture by expanding on two factors critical to healthy communities: social functioning, including neighbor cohesion and family dynamics, and public safety, including reduced crime.

*Social Functioning.* Positive connections between plants and social functioning have been documented for decades. An early study by Brogan and Douglas

(1980) in Atlanta, Georgia, U.S. examined the association between the psychosocial health of the community and the physical environment (e.g., landscaping and nearby land use) and sociocultural environment (e.g. population density and income). They found that the characteristics of physical and sociocultural environments were about equally important in explaining the variations in the psychosocial health of the community.

A wide range of positive connections between plants and social functioning have been revealed by researchers comparing residents in randomly assigned Chicago, Illinois, U.S. public housing units with differing amounts of vegetation. They found that greener landscapes led to stronger social integration and a stronger sense of community for older adults (Kweon et al. 1998), better parental functioning (Taylor et al. 1998), and less verbal aggression (Kuo and Sullivan 2001a). Their work also indicated that green outdoor spaces were associated with a lower incidence of "incivilities" including litter and graffiti (Kuo and Sullivan 2001b). One study, particularly relevant to community cohesion, examined the role of plants in the formation of neighborhood social ties in neighborhood common spaces and found that levels of vegetation predicted both the use of common space and the strengths of the ties (Kuo et al. 1998).

Studies of the impact of community gardens and gardening on community cohesion have also shown positive trends. Kidd and Brascamp (2004) surveyed gardeners in New Zealand who reported that gardening was peaceful and almost never frustrating. Female gardeners were especially likely to value stress reduction from gardening, while men were more likely to value gardening as a shared activity with others. In another study, community garden project leaders reported positive social impacts from community gardens, including the promotion of neighborhood cohesion and trust and an increase in civic participation and diversity in neighborhood associations (Feenstra et al. 1999). The potential for community gardens to address social, cultural, and educational needs was revealed by a gardening program started in an immigrant center in Germany in 1995 (Müller 2007). The intercultural garden involved immigrants from many countries with the initial goals of providing meaningful work and healthful food for the families. It was recognized for its impact on intercultural communication and integration into the German community on the basis of a resource-oriented approach that built on the knowledge base of participants.

*Public safety.* Positive impacts of urban nature on public safety can be readily seen in two areas: reduced traffic related injuries and reduced property and violent crimes. Examples of the magnitude of the improvement to public safety in these two areas are presented below.

Traffic calming is widely accepted as a technique by which landscaped circles and chicanes and other environmental designs slow traffic and increase pedestrian and neighborhood safety (Fig. 31.2). In Seattle, Washington, U.S., the city's traffic calming program has reduced accidents by more than 90 % (Mundell and Grigsby 1997). Lockwood and Stillings (2001) reported that traffic calming and streetscaping techniques installed in West Palm Beach, Florida, U.S., managed traffic effectively by altering driver behavior, thereby reducing car speeds and reducing collision frequency and severity.

Mok et al. (2006) analyzed crashes along stretches of urban roads ranging from highways to cities streets in 8 cities in Texas, U.S., before and after landscape im-

**Fig. 31.2** A traffic island in Aix-en-Provence, France. Trees and flowers planted within the traffic island help calm drivers and focus their attention. (Photo by V. I. Lohr)

provements were installed. They documented significant reductions in crash rates, speculating that trees affected driver behavior by improving driver alertness. It is important to note that, while the rate of accidents on roads with trees may be lower than on roads without trees and the actual number of crashes with trees is low, when crashes with trees occur, they are more likely to be fatal than crashes with other vehicles (Wolf and Bratton 2006). Fatal accidents with trees are much less likely in urban areas than in rural areas (Wolf and Bratton 2006). Given the trade-off between the benefits and hazards of trees along roadways, the general consensus is that trees should be used, but with reasonable setbacks along high-speed roadways. In cities, traffic calming landscapes give a psychological indications that drivers should proceed at lower speeds (Ewing and Dumbaugh 2009).

Lockwood and Stillings (2001) reported that one of the results of the traffic calming efforts in West Palm Beach was a reduction in some crimes. Arrests for prostitution dropped 80 % as the streets became safer and more useable. In the same area, incidences involving drugs went down 60 % over the same period. The authors speculated that increases in pedestrian and bicycle traffic lead to better surveillance of the neighborhood contributing to reductions in other crimes. They suggested that, as these elements change, more residents and businesses would improve their property inducing others to move to the community and further reduce crime. Results such as these have been generalized. In a paper co-authored from the U.S., Australia, and the United Kingdom, large-scale evaluations of crime prevention through environmental design were reviewed to compile current knowledge on the evidence of crime prevention through environmental design. It identified a large and growing body of research supporting the claim that crime prevention through environmental design is effective in reducing both crime and fear of crime in communities (Cozens et al. 2005).

Many people believe that vegetation is positively linked to crime, however, findings in urban residential areas indicate that the opposite may be true. Kuo and Sullivan (2001b) examined crime reports in Chicago, Illinois, U.S. and found that fewer property crimes, such as theft, and fewer violent crimes, such as homicides, were reported in public housing with trees than without. Donovan and Prestemon (2012) examined the relationship between trees and crime at single-family homes in Portland, Oregon, U.S. and found similar connections. Property crime rates, such as burglary and car theft, were lower at houses with larger tree canopies on the lot or on the street. Having more, smaller, view-obstructing trees was associated with increased property crime, but was not associated with violent crimes, such as simple assault. The only significant relationship between trees and violent crime was a decrease in violent crimes at homes with larger canopies trees on the lot. In Baltimore, MD, U.S., increases in canopy cover were associated with reductions in outdoor crime, especially on public land (Troy et al. 2012).

In addition to the impact of traffic calming landscapes and trees in urban areas, community gardening has been reported to have a positive impact on community safety. In one Philadelphia, PA, U.S., neighborhood, resident involvement in community greening was the catalyst for a 90% reduction in neighborhood crime (Macpherson 1993). In the Mission District of San Francisco, CA, U.S., residents noted a 28% drop in crime after the first year of their garden project (Malakoff 1995). Feenstra et al. (1999) reported a decrease in vandalism near garden sites. A San Francisco County Sheriff reported that the recidivism rate was cut in half (from 55 to 24%) among prisoners who participated in the prison gardening project in the San Francisco County Jail (Gilbert 2012).

*School and youth gardening education.* There has been a national movement over the last 10 years, encouraged by the National Gardening Association, the American Horticultural Society, and numerous botanical gardens, to integrate gardening into the school curriculum, as evidence of the benefits of such programs grows. Researchers at Texas A&M, for example, have been conducting research in this area for many years (Waliczek and Zajicek 1996; Waliczek et al. 2001; Aguilar et al. 2008).

Studies are determining the efficacy of specific resource materials and to understand what makes a school gardening program effective (Dobbs et al. 1998). In an early survey of teachers who received National Gardening Association gardening grants, DeMarco et al. (1999) found that the most important factor in the successful integration of gardening into the school curriculum was ownership of the concepts and goals by the teachers and students. They also found that the teachers did not use the garden simply to teach gardening, plant science, or environmental attitudes. They also used it to teach language arts, art, and ethics. They reported that their goals when using school gardens were academic, social, recreational, and therapeutic.

Juvenile offenders are a unique subset of youth, and the application of gardening with these individuals often has goals that are different than in the classroom. The

Green Brigade is a community-based program started by the Bexar County Agriculture Extension Service (Finch 1995). Cammack et al. (2002a) reported about a 10 % increase in both horticultural knowledge and environmental attitude scores among offenders in this program. They also found that the Green Brigade program was as effective as traditional probationary programming at reducing the rate and severity of crimes by juvenile offenders (Cammack et al. 2002b). Findings reported by Flagler (1995) on the Rutgers Careers in the Green Industry program, where over 70 % of the youth indicated an increase in experience, skills, contacts with people that could help them, and ideas about future education, indicated that an organized vocational based education program is an effective curriculum for this population. In a study conducted at an alternative education program for youth on probation, McGuinn and Relf (2001) noted that among the small group that they studied, there was strengthening in the delinquent individual's bond with society and the youth were motivated to think more practically about their future and career possibilities. Five of the six students in the study were hired for summer positions by horticulture establishments. These findings further reinforce Flagler's conclusions that horticulture is an effective curriculum focus for vocational training of juveniles on probation and other youth at risk.

## Horticultural Science's Role in Urban Environmental Health

Urban environmental health directly impacts quality of life, public health, economic stability, and food security (see related sections of this chapter). The use of plants to improve the urban environment has been well documented. Plants provide valuable ecosystem services, including air quality improvement, storm water runoff management, carbon sequestration, and temperature modification. McPherson et al. (2005) calculated the annual net benefits from the economic return of trees in cities to be US$ 21 to US$ 38 per tree. Public domain software for individual communities to assess these benefits (i-Tree 2013) is now in use in more than 100 countries (i-Tree 2012). The connections of urban environmental issues to global warming and to plant diversity are so strong that we are focusing on these elements for this chapter.

## Urban Environmental Issues and Climate Change

As the world is getting hotter and the climate more extreme, it is especially important for horticulturists to understand a range of the environmental impacts they may cause and the actions needed to ameliorate these impacts. Land management practices rooted in horticultural science can play a role in landscape design and plant selection for temperature and water management, pollution abatement and remediation, and management of health hazards associated with urban plants stressed by global warming and climate change. Some of the services, values, and negative impacts are presented below; others, such as carbon sequestration, are covered in other chapters.

Green roofs and green walls are examples of ways that horticulturists can contribute to ameliorating the issues discussed below. These horticultural uses of plants are increasingly valued as ways to create more places for vegetation in cities. They contribute to improved air quality, reduced city temperatures, and improved water quality (Alexandri and Jones 2008; Yang et al. 2008; Berndtsson 2010). They can also provide healthy space for wildlife and people. Rain gardens and bioswales are also becoming popular ways to improve water quality (Dietz and Clausen 2005; Xiao and McPherson 2011). The role of horticulturists in planning, designing, and selecting plant material for these is essential (Dylewski et al. 2011). Too many rain gardens, for example, are created by well-meaning ecologists who do not consider design aesthetics or maintenance requirements when selecting plants for such urban environments. They may have no idea how the plants in these will grow in urban landscapes over time and select plants that are native, but look like weeds to the neighbors. This can encourage negative attitudes toward the use of non-traditional landscape designs (Morzaria-Luna et al. 2004). For full acceptance of horticultural solutions to urban problems, the solutions must serve their ecological functions, be viable for long-term survival and maintenance, and appeal to people. In a similar manner xeriscaping, as means of limiting use of water in dry periods, must be approached from a horticultural science perspective to insure sustainability from a socio-cultural as well as ecological perspective.

*Cooling*. Horticultural science contributes to an understanding of the role of vegetation in affecting temperatures and energy use in cities and the selection and maintenance of plant systems that reduce the demands for energy in cities. Trees and other vegetation can be used to reduce temperatures indirectly though evaporative cooling and directly through shading and directing wind. The need for cooling in urban areas is increasing (Fig. 31.3). Average temperatures and high temperature extremes are rising around the world as a result of global warming (Lüdecke et al. 2011; Lelieveld et al. 2012). The temperatures are exacerbated in cities, where the urban heat island effects from the excess of hardscape compared to surrounding vegetated areas can magnify temperatures by 10 °C (Kim 1992; Akbari et al. 2001). As cities grow in size and population, the number of people affected by these often-fatal heat waves increases (O'Neill et al. 2009; Egondi et al. 2012). The potential to provide cooling through plantings in cities has been documented around the globe. In Manchester, UK, planting new trees has the potential to reduce maximum surface temperatures between 0.5 °C and 2.3 °C (Hall et al. 2012). In Tel Aviv, Israel, urban parks can be up to 3.8 °C cooler than urban street canyons (Cohen et al. 2012b). In Washington, D.C., models showed that decreasing the width of urban streets to increase planting space and add 50 % tree canopy coverage would drop temperatures of the air, building walls, and road surfaces by 4.1 °C, 8.9 °C, and 15.4 °C, respectively (Loughner et al. 2012). In Hong Kong, models predicted that planting trees or grass on the roofs of tall buildings could reduce temperatures at pedestrian level by 0.2 °C to 0.6 °C (Ng et al. 2012); they also showed that trees were more effective than grass.

**Fig. 31.3** Urban area of Hiroshima Japan. Hardscape, such as buildings and roads, contribute to the urban heat island effect, while the addition of trees and grass help counteract that by reducing temperatures on hot days. (Photo by V. I. Lohr)

A direct benefit of using vegetation to reduce temperatures is the effect on energy usage. If energy usage drops and if the energy source is carbon-based, such as coal is, then greenhouse gas emissions are also reduced (Akbari 2002). Huang et al. (1987) predicted that increasing tree cover around a home by 25 % could save 25 % in energy for cooling in Phoenix, Arizona, and 40 % in Sacramento, California, U.S. They showed that there were additional benefits in lowering the peak energy loads. McPherson and Rowntree (1993) used computer simulations to show that planting a single, small deciduous tree could reduce annual heating and cooling costs for a typical residence by 8 to 12 %.

*Air Quality*. Planting trees, establishing green roofs and wall plantings, and creating community gardens are ways that horticulturists can contribute to improved air quality and human health. Air pollution in urban areas is a problem worldwide

(Fenger 1999; Duh et al. 2008). Trees improve air quality through the physical removal of chemical and particulate air pollutants, such as ozone, carbon monoxide, sulfur dioxide, and dust (Nowak et al. 2006; Popek et al. 2013). McDonald et al. (2007) used modeling to show that trees could contribute to the reduction of particulate matter of< 10 μm in diameter, which are associated with adverse human health effects. By understanding the results of scientific studies in this area, horticulturists can increase the quantity of pollutants removed from the air by selecting specific trees and planting systems. For example, trees with smaller leaves, such as conifers, and hairier leaves are more effective than other trees at removing particulate matter (Beckett et al. 2000; Xie et al. 2011). Horticultural scientists can also be involved in such research, using their extensive knowledge of plants to suggest and study other features for their effectiveness (Xie et al. 2011; Popek et al. 2013), including the use of vines, shrubs, and green walls of herbaceous plants for faster, short-term impacts.

*Water quality*. Clean water in cities for human and ecosystem health are critical, yet water quality in cities is often compromised by the traditional handling of storm water (Whitehead et al. 2009). Many cities have relied on concrete culverts and ditches to drain storm water away from developed areas (Roy et al. 2008). As cities become more urbanized, the amount of impervious surface area increases. Climate change is bringing more and heavier rain events to some cities (Whitehead et al. 2009). Together, these are exacerbating flooding and reducing water quality. Plantings contribute to improved water quality by slowing water run-off and reducing soil erosion (Xiao et al. 1998). They also filter sediments and chemical pollutants from the run-off (Davis et al. 2006). There is an important role for horticultural science to play in understanding what can be done and what plants and planting systems are most effective for improving water quality. This may prove to be particularly important in developing countries as urban populations continue to grow faster than the infrastructure to provide water or remove excesses. Integration of water quality projects into multifunction uses including green roofs, green walls, community gardens, market gardens, parks, and recreational and tourism sites will maximize their sustainability and the potential for implementation where the urban planner, politicians, and taxpayers can better understand and appreciate the planted areas.

*Hazards and safety issues presented by plants*. All interactions between humans and plants in cities are not positive. Weak limbs can break and damage vehicles or injure people. Utility wires can come down with fallen trees. Trees with invasive root systems can seriously damage sidewalks. Improperly selected or managed urban vegetation can become a serious liability. Trained horticultural professionals are essential to address these hazards and safety issues. Climate change means greater wind speeds, more intense rainstorms, and more flooding, so we need to have trees that are pruned properly and transplanted correctly. Improper pruning leads to poorly attached branches and decay (Shigo 1985; Dahle et al. 2006). Improper transplanting leads to girdling roots and tree failure (Gouin 1983; Maleike and Hummel 1992).

In the U.S., courts of law have recognized that land managers and owners have a legal obligation to maintain vegetation in a healthy and safe condition. A 1978 court case, Husovsky vs. United States (described in detail in Anderson and Eaton 1986), involved a driver who suffered permanent paralysis when a limb dropped onto his vehicle in a park in Washington, D.C. The defendants (in this case the District of Columbia and the National Park Service) were found negligent for not inspecting for defects and, therefore, liable for damages of nearly US$ 1 million. This case demonstrates the serious legal ramifications of poor urban landscape management.

## Urban Environmental Issues and Biodiversity and Genetic Diversity

As the world has become more and more urbanized, both biodiversity (here used to mean diversity of different species, species richness, or the number of different species) and genetic diversity (diversity within a single species) are being affected. Urbanization can create highly degraded landscapes that are unsuitable for most species, thus negatively impacting diversity. There are also positive aspects related to urbanization and diversity. Understanding what is happening to diversity in urban areas is of significant concern to horticultural science in order to develop methods to address the actual problems.

*Enhanced biodiversity in urban areas.* When properly managed, urban parks and forests can afford landscape diversity and rich habitats for many different organisms (Alvey 2006). Residential gardens can also provide a rich diversity of species (Kendall et al. 2012). While the dense urban cores of most cities have little space for plants of any sort, other urban areas can have greater biodiversity than surroundings non-urban areas; this has been documented in both developed and developing countries (Kühn et al. 2004; Alvey 2006). This may be due to the introduction of different plants by people from around the world and the intense cultivation of land where plants will be grown. Many urban areas include parks, zoological gardens, botanic gardens, and similar sites dedicated to collection of different species. In addition, a number of local native species find appropriate habitat in cities. For example, in northern Belgium, 30 % of the local wild plant species and even greater percentages of wild birds, butterflies, and amphibians can be found within city parks (Cornelis and Hermy 2004). Larger parks, not surprisingly, had greater biodiversity due to greater diversity of habitats, such as forests, ornamental gardens, and hedges. Small parks still have value for biodiversity and can be critical for wildlife in increasingly urban areas. Ikin et al. (2013) showed that small parks can provide needed habitat for bird species, especially when there is green space in the surrounding urban blocks. Tree size also has an impact: larger trees have increased bird species richness (number of different species), abundance, and breeding (Stagoll et al. 2012).

*Reduced biodiversity and genetic diversity in urban areas.* Many urban areas currently have greater biodiversity than surrounding areas (Kühn et al. 2004; Alvey 2006), however they may have less biodiversity than they did historically. Gregor et al. (2012) looked at changes in the number of plant species in Frankfurt, Germany, over 200 years: from 1800 to 1900, the number dropped by 2.6 %, while from 1900 to 2000, it dropped an additional 7.75 %. On degraded urban landscapes, highly adapted, early succession species are often required for successful establishment (McKinney 2006). To survive these conditions, plants must grow under high light and temperatures and in droughty soils with high pH (Wittig and Becker 2010). There are a limited number of such species, which contributes to reduced biodiversity in such areas. Selecting and cloning cultivars of landscape plants, such as street trees that are well-adapted for urban conditions and human preferences, lead to reduced genetic diversity (Morton and Gruszka 2008). In a similar fashion, seed companies throughout the world distribute the same cultivars of flowers and vegetables.

At a time when global warming and climate change are subjecting urban plants to more environmental extremes and increased pest problems, the risks from greatly reduced diversity are growing (Lohr 2013). Elm trees (*Ulmus* spp.) provide an example of the problems that can arise from reduced diversity. They were widely planted as street trees in cities in Europe and America more than a century ago. Then Dutch elm disease (*Ophiostoma ulmi*), first reported in 1921, began to decimate the elms and cities where they had been planted (Wilson 1975). There was a problem from over-reliance on a few species of elms (little biodiversity), all of which had some susceptibility to the disease. The close proximity of the trees made the spread of the disease, once it was introduced, common. Within cities on both continents, many avenues with large, continuous canopies of cool shade became barren as the elms died. The spread of the disease continues today, but a handful of resistant American elms and hybrids created by crossing with resistant Asian species are now available for planting (Santini et al. 2002; Townsend et al. 2005). As plant breeders were searching for resistant trees, the disease was evolving. In 1972, a new more virulent strain of the disease (*Ophiostoma novo-ulmi*) was found (Potter et al. 2011). It is only a matter of time until other strains of the disease or other problems impact the new, currently resistant cultivars. In fact, Dutch elm disease has recently been found in trees in Japan on species that had been considered resistant (Masuya et al. 2010).

If we continue to overplant a few species and a limited numbers of cultivars in cities, other catastrophes like those from Dutch elm disease will continue to occur. In fact, similar catastrophes are readily evident today. Ash trees (*Fraxinus* spp.) were frequently planted (over-planted) to replace the decimated elms. Now emerald ash borer (*Agrilus planipennis*) in the U.S. and ash dieback (*Hymenoscyphus pseudoalbidus*) in Europe are wreaking havoc on urban areas at a similar scale to Dutch elm disease, but at an alarmingly faster pace (Poland and McCullough 2006; Bakys et al. 2009). Horticultural scientists need to understand the problems and risks from the worldwide homogenization of species in urban areas and work with the nursery industry to diversify their crops while maintaining market and profit if we are to maintain healthy green spaces in cities in the decades to come (Lohr 2013).

*Invasive species.* A weed is a plant that is growing in a place where it is not wanted. Historically, weeds have often been native plants, such as ragweed (*Ambrosia artemisiifolia*) in North America and dandelion (*Taraxacum officinale*) in Europe and Asia, that have invaded cultivated gardens. Today, non-native plants, such as African boxthorn (*Lycium ferocissimum*) in North America and Australia, common rhododendron of southern Europe and southwest Asia (*Rhododendron ponticum*) in Great Britain and New Zealand, neem from India (*Azadirachta indica*) in West Africa, and yarrow from the northern hemisphere (*Achillea millefolium) in Australia*, invade native ecosystems and have become weeds of increasing concern. These are considered invasive species. According to U.S. Federal law, an invasive species is: "an alien species whose introduction does or is likely to cause economic or environmental harm or harm to human health" (NISIC 2012). These plants may be introduced accidentally or intentionally. In the U.S., about 5,000 alien plant species have become established outside of cultivation (Pimentel et al. 2005). Not all non-natives become problems. For example, in Florida, only about 1 in 25 alien introductions have become invasive (Pimentel et al. 2005). While the percent may seem small, the amount damage to ecosystems and the economy may not be. The estimated annual cost in the U.S. for European purple loosestrife (*Lythrum salicaria*) (Fig. 31.4) alone is estimated at US$ 45 million (Pimentel et al. 2005). Ecosystem damage from the same plant includes loss of native plant species, increased wetland evapotranspiration rates, and reduced bird habitat (Blossey et al. 2001).

Increasingly, the horticultural industry and gardeners are being seen as a major source of the problem (Niemiera and Von Holle 2009; Barbier et al. 2011). There are a number of different responses to this within and outside of the horticultural industry. Some groups organize parties to remove invasive plants (WNPS 2012), while others recommend avoiding invasives and using alternatives (Coats et al. 2011). Bans on the sales of particular plants occur (Dehnen-Schmutz and Touza 2008). Some breed low fertility versions of otherwise desirable invasive species (Ranney et al. 2006). Some people work to predict invasive potential and recommend avoiding particular plants (Kueffer and Loope 2009). Some have recommended concentrating on natives (Missouri Prairie Foundation 2013). Others propose the implementation of regulation and taxes (Barbier et al. 2011). Disturbed, unnatural conditions in human dominated landscapes make urban areas inhospitable to many native plant species. Global warming and climate change contribute to the inhospitability. Some researchers are now suggesting that alien plants must be brought into cities to maintain green space (Hitchmough 2008). But we must maintain a balance between those that can tolerate the urban conditions and those that would dominate and exclude other species.

## Horticultural Science's Role in Urban Food Security

Food security is a complex concept with complex causes and no simple solutions. According to the Food and Agriculture Organization (FAO) of the United Nations, food security is defined as existing: "…when all people, at all times, have physical and economic access to sufficient, safe and nutritious food that meets their dietary needs and food preferences for an active and healthy life" (FAO 2013). This is exceedingly difficult to obtain at a time when the human population of our planet exceeds 7 billion people (U.S. Census Bureau 2013). The concern for world food security is not simply related to the size of the population, but also to the concentration in urban locations, where the problems causing food insecurity are

**Fig. 31.4** European purple loosestrife (*Lythrum salicaria*). This beautiful plant, which had been available in nurseries, has invaded vast areas of North America, causing the loss of native plant species and increased evapotranspiration in wetlands. (Photo by V. I. Lohr)

compounded. More than half of the world's people live in urban areas (United Nations 2012). In 1970, there were two megacities (more than 10 million inhabitants) in the world, Tokyo and New York City; in 2011, there were 23 megacities (United Nations 2012). Rates of urbanization are not uniform worldwide. Africa's urban population is growing faster than that of any other region, while Asia is close behind (United Nations 2012). This has the potential to further exacerbate food insecurity in those areas.

## Causes of Food Insecurity

Most causes for food insecurity ultimately translate into three key factors: lack of available nutritious food, lack of money to purchase food, and lack of the ability to grow food. The quantity of food is impacted by issues including climate change, sustainability, and affluent consumption of food products (e.g. corn for meat and ethanol). Availability of food is impacted by the political will to provide for all citizens; the use of agricultural commodities in developed and developing countries; the impact of war and climate change related disasters; infrastructure for storage, transport, and delivery; and household resources for transportation, preparation, and storage. Even if food is available, if the quality is insufficient, then food security will not been achieved. Particularly among the urban poor, there is a significant need, for example, for sources of fresh food, which is beyond their limited purchasing power. Funds for the purchase of food is embroiled in all of the causes of poverty including health, education, gender freedoms, and other societal factors. Inability to grow food for the family is related to the urbanization of the world's population and the lack of policy and procedure to enable people who are food insecure to access urban land for cultivation.

## Role of Industrial Food Production

The insurance of safe and nutritionally adequate food for all people is perhaps the greatest challenge for horticultural science worldwide. This issue is being addressed extensively from many directions—political, economic, social, environmental, and medical. But at the core of these discussions is still the optimum production techniques for horticultural crops to contribute to meeting the nutritional needs of populations lacking in food security, both in developed and developing countries that can be implemented within the constraints of the social, environmental, economic, and political realties of the population. This brings into question the other possible applications of horticulture and higher education to find sustainable solutions to intractable problems, particularly in the light of climate change and increased risk of natural disasters.

It is argued that commercial production of horticultural crops as currently practiced in the U.S. and other developed countries is not suitable for supplying the bulk of the human population with the fresh fruits and vegetables needed to provide a nutritionally adequate diet. Looking at food security from a public health perspective, Dixon et al. (2009) argue that "functional foods", one area of high interest and involvement for horticultural science, are produced by an industrialized agricultural model that will not be able to address the needs of the urban poor and are counter to long term sustainability. While functional foods are an efficient means to supply essential micronutrients, they have several problems on a worldwide basis. Involving large agrifood and pharmaceutical corporations, it is a highly profitable sector that is outside the financial reach of most of the world's population, it fails to recognize and address the realities of production with climate change, it fails to provide the nutritional balance in a culturally acceptable way that the World Health Organization (2012) recognizes as essential, and it reduces biodiversity at the same time as altering genomes. The authors point out that it is often overlooked that the green revolution contributed to marked rural inequalities and despoiled traditional agricultural environments (Dixon et al. 2009).

An additional problem with industrial-scale food production as the sole source of nutritious diets for the urban poor is that the method for getting the food to the populous is costly and wasteful, with large amounts of packaging, transportation, and fossil fuels involved. Despite this, with urban growth and increasing demand for prepared or easily prepared and storable foods, there has been a significant emergence of supermarkets and fast food chains worldwide (Kennedy et al. 2004). The demand for large quantities of uniform product has contributed to erosion of food culture and reduction in biodiversity, as well as the loss of a major sector of employment in agriculture and horticulture. Furthermore, packaged products in the supermarket contain large quantities of salt and sugar, and fast food is often very high in fat, salt, and sugar, all contributing to the worldwide health epidemic. All of these factors increase the cost of the food, often beyond the finances of the poor.

## Potential Solutions

It was not until 1996 at the International Conference on Human Settlements, that the United Nations formally recognized gardening and urban agriculture for their potential contribution to the health and welfare of urban populations worldwide (United Nations 1996). In 1999, the FAO pointed out that when done in a safe and secure manner, the local production of food in urban and peri-urban regions can be successful for several reasons: increased quantity of food available, increased quality of food, and increased income from jobs or from successful marketing (FAO 1999b). According to FAO this source of food proves to be important during times of crisis and severe scarcity (civil war, widespread drought, currency devaluations, inability to import) as well as personal upheaval (illness, health, sudden unemployment).

The FAO reports case study data indicating that both food availability and incomes in poor farming households are significantly higher compared to households of non-farmers (FAO 1999b). However, it is important to note that urban gardeners are not typically the poorest residents, but rather those families that have lived long enough in the city to secure land and water and become familiar with the market channels for selling. The quality of the perishable food produced near the consumer, whether at home or in peri-urban farms, is often higher as it is fresher than when it travels long distances along the supply chain often lacking cooling in storage or transport thus quality and nutritional value deteriorates rapidly (Fig. 31.5).

As a source of income, food production offers few barriers to employment. The FAO estimates that worldwide, 800 million urban residents are involved in food producing/income earning activities with one-quarter to two-thirds of urban and peri-urban households involved in agriculture and horticulture, depending on location (FAO 1999b). In many cases, women are the primary food producer as they combine gardening with childcare and other home based responsibilities.

To overcome some of the limitations of previous anecdotal and qualitative data, a recently created dataset bringing together comparable, nationally representative

**Fig. 31.5** Produce stand at the Green Market in Almaty, Kazakhstan.Locally produced, fresh crops, which are available for sale to urban populations, is not transported long distances along a supply chain. (Photo by V. I. Lohr)

household survey data for 15 developing or transition countries, was used by Zezza and Tasciotti (2010) to analyze the importance of urban agriculture for the urban poor and food insecure and provide a comparative international perspective. This study pointed out that "the potential for urban agriculture to play a substantial role in urban poverty and food insecurity reduction should not be over-emphasized, as its share in income and overall agricultural production is often quite limited." However its value cannot be dismissed as a source of income for the urban poor. And there was evidence of a positive statistical association between participation in urban agriculture and positive indicators of dietary adequacy.

The greatest growth of urban farming has been in the developing countries, but in recent years it has also become a factor of importance in the developed world as more affluent individuals choose to produce their own food to address quality rather than quantity, desiring greater freshness and control over pesticides and other inputs. The bulk of the information on-line regarding urban farming in developed countries is presented by advocates and may be biased regarding the projected efficacy in meeting significant nutritional requirements of urban populations. However the total positive impact of urban horticulture must be taken into the discussion of food security. Food grown in home and community gardens supplements other food sources and provides excess for trade or informal markets. Even with little or no land, high valued vegetables can be produced in containers on floors or walls. Other gardening sites include parks, utility rights-of-way, bodies of water, rooftops, walls and fences, balconies, basements, and courtyards (Brown et al. 2003).

Through their web-pages and numerous books, conferences and papers the FAO (Anon 2013b) and the Resource Centres on Urban Agriculture and Food Security (Anon 2012a) provide a tremendous amount of information regarding urban population growth, food security needs, and urban agriculture practice and policy development, accompanied by a significant amount of anecdotal and qualitative research to document its efficacy (FAO 1995, 1999a; Drescher et al. 2000; FAO 2001, 2008, 2012; FAO et al. 2012). In addition, an interesting source of information is Mission 2014: Feeding the World (Anon 2013d), which provides current data and recommendations addressing world food needs developed by Terrascope, a student-run class offered to MIT freshmen that focuses on solving complex world problems through the collaboration of students, faculty, and alumni. This outstanding teaching model is indicative of the views of future generations of problem solvers and how they learn.

## Additional Benefits of Urban Food Production

Health professionals are recognizing the value of farm- and garden-scale urban agriculture (Baumgartner and Belevi 2001; Bellows et al. 2004). Numerous reasons are discussed for growing food and non-food crops in and near cities. Urban food production contributes to healthy communities by engaging residents in work and recreation that improves individual and public well-being, providing exercise, enhancing mental health, as well as improving social and physical urban environments.

Based on surveys and case studies from 31 countries, the FAO (2012) recommended that African policymakers act now to refocus urbanization toward a greener, healthier environment. The current path is unsustainable as the urban populations continue rapid growth. The FAO highlights a key component of sustainable urban development: that peri-urban and urban horticulture needs to ensure food and nutrition security, decent work and income, and a clean environment for all their citizens. Studies of urban agriculture in South Africa found it is important to women of low-income households in ways less directly related to monetary gain (Slater 2001); for women, urban agriculture promotes empowerment, establishes social networks, symbolizes a sense of security, and encourages community development.

Urban food production has multiple functions, playing a role in urban poverty alleviation, social inclusion, urban food security, urban waste management, and urban greening (Hoekstra 2006). Smit and Nasr (1992) discussed the potential for transforming urban centers from consumers of agricultural products to centers that conserve resources, improve health, and produce foods in a sustainable fashion with particular emphasis on conversion of urban wastes, use of vacant areas, and other improvements of the environment. According to Bon et al. (2010), challenges for urban agriculture include obtaining inputs, such as fertilizers, and growing fresh and nutritious food in polluted environments. They advocate the reuse of city wastes to help alleviate these challenges, but recognize that those alone will not be sufficient to achieve needed yields.

Urban food production has strong proponents worldwide who feel that it contributes to mitigating the two most intractable problems facing Third World cities – poverty and waste management (Baumgartner and Belevi 2001). They recognize that urban food production is simply one of several food security options for individuals and family groups. Urban agriculture and horticulture complement, rather than supplant, rural supplies and imports of food and will continue to do so. Cities will continue to depend largely on rural food production for bulkier, less perishable foodstuffs (Mougeot 2000). Similarly, it is only one of many tools for making productive use of urban open spaces, treating urban waste, saving or generating income and employment, and managing freshwater resources more effectively. Many professionals in the field also highlight its importance public health and sustainable resource management.

## Challenges for Urban Food Security

There are a number of significant issues that challenge the success of peri-urban and urban food production to provide food security. They include:

- safety of the soils in which crops might be grown, particularly in brown fields and other reclaimed industrial sites;
- utilization of scarce resources such as water and heat for protected spaces;
- potential for health hazards including attracting rats and other pests; pesticides; polluted flood water;
- difficulty of obtaining resources such as seeds and fertilizer;
- problems in finding and keeping land for gardening;
- impact that food production for personal consumption and marketing has on family and community dynamics; and
- lack of training and skills to implement urban horticulture issues.

Research issues for horticultural science include:

- the identification of the best crops to grow given the environmental and sociocultural demands of an area to optimize the nutrition value of the resources utilized;
- affordable techniques for producing crops under the environmental and economic constrains of urban poverty, both for market and personal consumption;
- use of horticultural crops to ameliorate some of the problems such as soil reclamation, revegetation, and other bio-system issues;
- food gardening for health to include psychosocial, and physical benefits as well as nutrition as presented in other chapters in this book;

- food gardening for community development, economic opportunity, and personal growth; and
- development of educational tools and techniques particularly using computers and smartphones to improve communication and technology especially in poorer, developing countries.

## Horticultural Science's Role in Urban Economics

Ultimately much of the application of horticulture to the urban environment comes down to economics and the basic concept of cost/benefit analysis. One question is: "Does the use of urban land and the installation and maintenance of plants cost more than the resulting 'profit'?" This profit may take many forms such as the savings from a reduced cost of energy or less flood damage; actual profit from increased worker productivity or property values; financial savings to individuals from improved nutrition and health; or, less tangible, thus more difficult to measure, profits from the perception of improved quality of life. In this section we will look briefly at the following areas related to economic issues: urban businesses based on horticultural crops and services; non-horticultural businesses that utilize plants as part of their business plan; and potential savings from cost of environmental issues being ameliorated by plants.

## Urban Businesses Based on Horticultural Crops and Services

A majority of urban and peri-urban horticultural commodities in both developed and developing countries include vegetables, fruits, herbs, potted and cut flowers, commercial turfgrass, and horticultural crops cultivated for indoor and outdoor landscapes. In addition to production, horticultural businesses are part of the retail and service industries. Urban farm and garden enterprises employ a variety of marketing models in urban settings including: direct sales to grocery outlets, restaurants, schools, hospitals, and other institutions; community supported agriculture; cooperatives; value-added processing and sales; and sale at farmers markets and farm-stands in a neighborhood (Feenstra et al. 1999). Environmental horticulture crops are sold through nurseries, garden centers, landscape contractors, and mass marketers (Hayes et al. 2007). Flowers and related floral products are commonly found in grocery stores, quick-stop stores, roadside stalls, and corner flower stands as well as full service florists. Horticultural service businesses, including interior and exterior landscape design, construction, and maintenance, continue to grow as the urban population expands and the demand for public and private landscape increases, particularly at recreational and tourism sites. The expansion of rooftop and wall gardens offers significant opportunity for growth of the industry in urban settings. Potential for specialized small scale and essentially self-maintaining landscapes are presented by the increase in apartment dwellers with neither space for landscapes nor skills and time for maintenance. At the same time, in peri-urban settings, demands for well-designed small-scale gardens for intense caring and cultivating are increasing in demand for retired hobbyists.

Horticultural businesses provide entry level and unskilled jobs in both the production area of the industry and sales and service, which rely heavily on individuals with relatively little education or experience, thus providing jobs for population such as disenfranchised youth and immigrants. This in turn opens opportunities for training programs (Justen et al. 2009). Further demand for education among consumers opens opportunities for writers and media professionals, education consultants, and other entrepreneurs to profit from urban horticulture. In addition, urban farmers markets provide opportunities for entrepreneurial businesses for people who manage the site as well as small-scale farmers, home gardeners, producers of specialty crops such as mushrooms, and organic farmers who profit from the direct sales.

To give a glimpse of the money involved for just a small percent of the worldwide market, one study estimated that the economic impacts of the U.S. environmental horticulture industry were about US$ 148 billion in output, 2 million in jobs, US$ 95 billion in value added, US$ 64 billion in labor income, and US$ 7 billion in indirect business taxes (Hall et al. 2006). In addition, the study evaluated the value and role of urban forest trees (woody ornamental trees); the total output of tree production and care services was valued at US$ 15 billion, which translated into US$ 21 billion in total output impacts, 260,000 jobs, and US$ 4 billion in value added. Another example comes from the small country of The Netherlands, which in 2010 alone accounted for 24% of global trade in horticultural products, 50% in world trade of floricultural products, 80% of the world market in flower bulbs (Anon 2011). In 2011, their total horticultural production amounted to € 8.6 billion, and in 2010, they exported € 4.2 billion worth of vegetables. Exports and re-exports of Dutch horticultural crops amounted to € 16.2 billion in 2011 (Anon 2011). FloraHolland is the world's largest auctioneer for cut flowers and plants. In 2011, it employed 4,000 people and sold 12.5 billion cut flowers and plants for € 4.2 billion (Anon 2013a).

## Non-Horticultural Organizations Utilizing Plants as Part of Their Business Plan

From tourist sites to shopping malls to corporate offices, horticultural crops and services play an important role in the profitability of many urban businesses. According to Wolf (2003) the character of a place is important to business communities as it influences consumer choices and ultimately the profitability of retail business. Among benefits were an increase in return visits, a message of care, and a perception of higher quality merchandise; negative aspects included reduced usable parking space, increased waste from tree debris. There were also higher positive

**Fig. 31.6** A bridge in Frederiksberg Have in Copenhagen, Denmark. This popular urban park provides opportunities for nature-based recreational experiences, such as walking along paths among mature trees. (Photo by V. I. Lohr)

perceptions of business districts and a willingness to pay more for merchandise at businesses that had street trees and other landscape improvements (Wolf 2009).

Plants are an essential part of the urban tourism experience. Half of the respondents to a U.S. survey conducted by the Gallup Organization indicated that plants and flowers at theme parks, historic sites, golf courses, and restaurants were important enjoyment of visiting there (Relf et al. 1992). At Opryland in Nashville, TN the "greatscapes" contributed to higher occupancy and room rates for those rooms overlooking the gardens, which yielded US$ 7 million in additional room revenue annually (Evans and Malone 1992). Tyrväinen (2001) found that people want green nature-based recreation areas in their cities (Fig. 31.6), and 82 % of users were willing to pay for the recreational experiences these sites provide. A cost-benefit analysis revealed that revenues could be as much as 25 times more than the costs. A Canadian study suggested that tourist locations that feature plants attract an older, wealthier, and better-educated clientele (Lang Research 2001). This group of tourists has the interest and resources to visit public and private gardens and related tourist sites such as historical sites, natural wonders, museums, art galleries, zoos, aquariums, and planetariums and to take scenic bus tours.

Demand for recreational activities dependent upon the products and services of environmental horticulture (e.g. athletic fields, parks, golf courses) continue to in-

**Fig. 31.7** Floriade in Canberra, Australia. This annual spring event is an example of a festival celebrating horticultural products that brings tourists to an area and generates income for businesses nearby. (Photo by V. I. Lohr)

crease as population increases. In Europe, new forests are also being established with public recreation and tourism very much in mind, often close to large centers of urban population (Bell et al. 2007). Tourism based on natural environments is an increasing international industry with major economic, social, and environmental consequences at both local and global scales (Buckley 2003). Festivals celebrating horticultural products are another popular form of horticultural tourism (Gen-Song et al. 2012). There are large events, such as the Flower Festival in Chiang Mai, Thailand, the Infiorata Flower Festival in Genzano, Italy, La Tomatina in Buñol, Spain, Leboku (New Yam Festival) in Ugep, Nigeria, and the Cherry Blossom Festivals in Washington D.C. and across Japan, and small ones, such as the Dogwood Festival in Lewiston, Idaho, U.S., and the Fête du Melon in Cavaillon, France. An example of the impact of such an event comes from Floriade a month-long festival held every year to celebrate spring in Canberra, Australia (Fig. 31.7). In 2012, nearly ½ million people visited Floriade (Barr 2013). More than 1/4 of the visitors were national or international travelers who came to Canberra because of Floriade and spent nearly AU$ 30 million. The City estimated that news stories about the event reached almost 34 million people, and they estimated that to be worth more than AU$ 5 million.

Employees are the single greatest expense for any business. Even small increases in job satisfaction, productivity, and health can have significant impact on the net profit of a business. Plants contribute to all of these. Workers with views of nature, such as trees and flowers, have been shown to experience less job pressure, be more satisfied with their jobs, and have fewer ailments than those who could only see built elements from their windows (Kaplan et al. 1988). Productivity on a computer task was shown to be higher in a room with interior foliage plants compared to one without plants (Lohr et al. 1996). Subsequent studies have further explored the potential impact on employees and anecdotal evidence from corporations utilizing plants further substantiates their value (Thomsen et al. 2011). Workers in an office with foliage plants reported fewer physical symptoms including coughing, hoarse throat, and fatigue than when no plants were present, translating into more productive workers (Fjeld 2000). A major cost to employers is employee's sick time. Additional studies showed improvements in indoor air quality through reductions in air pollution (Burchett et al. 2005) and dust (Lohr and Pearson-Mims 1996) and an increase in relative humidity (Lohr 1992), all of which could have a positive impact on health of employees. Employees appear to understand the need for plants, as those who work in windowless offices have been found to be 5 times more likely to have brought plants in their offices than those with windows (Bringslimark et al. 2011).

## Value of Plants to Real Estate

The real estate industry is impacted both by the quality of the landscape of the property on the market and the proximity to parks, botanic gardens, and other urban green. Crompton (2001) reports positive impact of parks, open spaces, and water features on residential property values. Behe et al. (2005) reported that landscape plant material, size, and design sophistication increase the perceived home value from 5 to 11 % for homes with good landscaping. In another study comparing homes with the same square footage and other characteristics, Stigarll and Elam (2009) reported that homes that improved landscaping from average quality to good or excellent quality increased selling price by 5.7 % and 10.8 %, respectively. HomeGain (Anon 2013c) surveyed nearly 600 real estate agents nationwide to determine the top 10 low cost, do-it-yourself home improvements for people getting their home ready to sell. They reported an average investment in landscaping of US$ 540 gave an average US$ 1,932 price increase for a 258 % return on investment. Culp (2008) reported a study of on-site inspections of 3,088 home sites that showed that time on market is reduced and price is increased by a variety of green features, such as trees, landscaping, open spaces, and parks.

Another way in which plants have significant impact on real estate value and the economics of real estate to urban governments is as an alternative method of dealing with vacant properties. In Baltimore MD U.S., where it is estimated that it can cost city services between US$ 2,000 and US$ 4,000 per year to clean a problem lot, the city established a new program "Vacants to Value" (Sernovitz 2011) aimed at reducing the estimated 40,000 vacant lots or abandoned row homes through sale to homeowners for planting gardens and lawns. Converting vacant lots to community gardens is another approach to significant sanitation savings. In Sacramento, California (CA) researchers compared community gardens to city-managed parks and found that the community garden was 20 times cheaper to create and 27 times cheaper to maintain each year (Francis 1987). Urban green spaces from rooftop plantings to community gardens to parks and greenways reduce storm water runoff and improve water quality and thus the related service costs. Reduced noise, glare, and wind and temperature moderation all have impact on energy costs, human stress and health, and other factors that translate into dollar savings for the urban governments and businesses.

## Personal Perspectives on the Urbanization of Horticultural Science and Conclusions

The role of horticultural science in the urban environment is an issue of definition and perception by researchers, academics, and others who shape the knowledge and attitudes about a scientific arena and the profession that it supports. The transition in attitude over the last 50 years has been significant and promises to accelerate as the human population continues to urbanize. In 1963, co-author Relf, as a Horticulture Science freshman in a land-grant university in the U.S., was told that horticulture was the intensive production of crops and encompassed the skills and knowledge for growing fruits, vegetables, cut and potted flowers, and woody plants. The commonly held definition of horticulture as an art as well as a science was acknowledged via studies in landscape design and construction. While insects, diseases, and weeds were essential parts of the science of horticulture; related topics such as extension, marketing, and education were peripheral subjects to be studied by non-horticulturists. Environmental impact was not discussed, and organic gardening was heresy and a return to the mythology and ignorance of the past. Women and urban students were an anomaly. As the top graduating senior in horticulture in 1967, Relf was told by a noted horticultural production firm that they would hire a man who was classified 1A by the draft board and on his way to Vietnam before they would hire a woman. In 1975, co-author Lohr was told that it was great to have women in horticulture departments, because they were good at flowering arranging due to their manual dexterity.

In the 1970's the environmental movement began to influence people's thinking. In the U.S., the demand for information from the public and the influx of urban students wishing to change the world by growing a garden forced horticulture departments to become aware of the urban population and to expand their course offerings to a wider non-production audience, thus indoor plants and home horticulture were offered to non-majors. As more urban students joined horticulture departments, they become more diversified in race and gender. The changes in size and diversity of the students forced open doors for additional course work to be included in horticultural science degrees. Kansas State University established a degree in Horticultural Therapy and shortly thereafter Michigan State University began a program. John Carew, their department head, fully demonstrated the challenges in attitude to be faced in redefining horticultural science while at the same time indicating a willingness to try to adapt to new paradigms when he commented to Relf, "I am glad you invented this horticultural therapy. It will give my little girls something to do."

By the late 70's and early 80's the academic acceptance of human-oriented, non-traditional horticulture increased (Cotter et al. 1978; Relf 1982). *Home Gardening* had become *Consumer Horticulture* and the formation of HortTechnology as a refereed journal by the American Society for Horticultural Science (ASHS) opened more doors for an expanded definition of horticultural sciences. The specific needs of urban communities came to the forefront in the U.S. with the establishment of the Urban Horticulture Institute at Cornell in 1980 and the Center for Urban Horticulture at University of Washington in 1983.

As horticultural science moved closer to the 21st Century, researchers including Relf, Lohr, Shoemaker, and others called for greater recognition of the relationship between horticultural science and the urban environment with a greater focus in horticultural science on the psychosocial aspects being researched by urban forestry, environmental psychology, landscape architecture, social ecology, anthropology, sociology, geography, communications, and other fields (Relf 1990; Relf 1992b; Lohr and Relf 1993). Relf also proposed a more complete definition of horticulture (Relf 1992a).

> Horticulture—the art and science of growing flowers, fruits, vegetables, trees and shrubs resulting in the development of the minds and emotions of individuals, the enrichment and health of communities and the integration of the 'garden' in the breadth of modern civilization. By this definition, horticulture encompasses PLANTS, including the multitude of products (food, medicine, O2) essential for human survival; and PEOPLE, whose active and passive involvement with 'the garden' brings about benefits to them as individuals and to the communities and cultures they comprise.

Today, this broader understanding of horticulture is widely embraced. The International Society for Horticultural Science has had a Commission on Landscape and Urban Horticulture for many years. The broader understanding of horticulture became the theme of the XXVIth International Horticultural Congress held in Toronto, Canada in 2002: Horticulture—Art and Science for Life. The opening plenary colloquium of that Congress focused on human issues in horticulture, recognizing "...that the horticultural arts and sciences exist to nourish and enrich the human body and the human soul" (Lohr et al. 2004). It is widely evident in the chapters in this book.

With the increasing urbanization of the world and the daunting global environmental changes instigated by human activity and related global warming and climate change, it is essential that horticulturists focus on the needs of urban populations. We have illustrated some of the issues and opportunities for horticulture. For example, environmental justice demands that all individuals have access to a healthy environment. This is a particular problem for disadvantaged and disenfranchised individuals with strong racial and income implications that can be addressed through properly researched and designed horticultural initiatives. Howard Frumkin, from the Centers for Disease Control and Prevention in the U.S., and Anthony McMichael, from the National Centre for Epidemiology and Population Health in Australia, called on readers to recognize the value of designing and maintaining a healthy natural environment as a essential preventative measure integral to public health (Frumkin and McMichael 2008). By combining awareness of traditional horticultural issues, i.e. feeding the world, with expanded awareness of functions and services provided by additional horticultural crops and awareness of the needs of people, horticulturists should become indispensable members of teams addressing urban populations.

## References

Aguilar OM, Waliczek TM, Zajicek JM (2008) Growing environmental stewards: the overall effect of a school gardening program on environmental attitudes and environmental locus of control of different demographic groups of elementary school children. HortTechnology 18:243–249

Akbari H (2002) Shade trees reduce building energy use and CO2 emissions from power plants. Environ Pollut 116:199–126

Akbari H, Pomerantz M, Taha H (2001) Cool surfaces and shade trees to reduce energy use and improve air quality in urban areas. Sol Energy 70:295–310

Alexandri E, Jones P (2008) Temperature decreases in an urban canyon due to green walls and green roofs in diverse climates. Build Environ 43:480–493

Alvey AA (2006) Promoting and preserving biodiversity in the urban forest. Urban For Urban Green 5:195–201

American Heart Association (2013) Why we garden. http://www.heart.org/HEARTORG/Getting-Healthy/HealthierKids/TeachingGardens/Why-We-Garden_UCM_436620_SubHomePage.jsp. Accessed Feb 13 2013

Anderson LM, Eaton TA (1986) Liability for damage caused by hazardous trees. J Arboric 12:189–195

Anon (2011) Horticulture. http://www.hollandtrade.com/sector-information/horticulture/?bstnum=4928. Accessed Feb 13 2013

Anon (2012a) RUAF Foundation: Resource Centres on Urban Agriculture & Food Safety. http://www.ruaf.org. Accessed Dec 17 2012

Anon (2012b) Therapuetic Landscapes Network. http://www.healinglandscapes.org. Accessed Feb 26 2013

Anon (2013a) FloraHolland. http://www.floraholland.com/en/about-floraholland/who-we-are-what-we-do/facts-and-figures. Accessed Feb 12 2013

Anon (2013b) Food and agricultural organization of the United Nations. http://www.fao.org. Accessed Feb 18 2013

Anon (2013c) HomeGain 2011 home improvement national survey. http://www.eximus.com/blog/homegain-2011-home-improvement-national-survey-results.aspx. Accessed Feb 26 2013

Anon (2013d) Mission 2014: Feeding the World. http://12.000.scripts.mit.edu/mission2014. Accessed Feb 18 2013

Bakys R, Vasaitis R, Barklund O, Ihrmark K, Stenlid J (2009) Investigations concerning the role of *Chalara fraxinea* in declining *Fraxinus excelsior*. Plant Pathol 58:284–292

Barbier EB, Gwatipedza J, Knowler D, Reichard SH (2011) The North American horticultural industry and the risk of plant invasion. Agr Econ 42 supplement:113–129

Barr, A (2013) Floriade delivers second biggest result on record. Available from http://www.cmd.act.gov.au/open_government/inform/act_government_media_releases/barr/2013/floriade_delivers_second_biggest_result_on_record. Accessed Mar 5 2013

Barton J, Pretty J (2010) What is the best dose of nature and green exercise for improving mental health? A multi-study analysis. Environ Sci Technol 44:3947–3955
Baumgartner B, Belevi H (2001) A systematic overview of urban agriculture in developing countries. EAWAG/SANDEC, Dübendorf. Int J Environ Tech Manag 3:193–211
Beckett KP, Freer-Smith PH, Taylor G (2000) Particulate pollution capture by urban trees: Effect of species and windspeed. Glob Change Biol 6:995–1003
Behe B, Hardy J, Barton S, Brooker J, Fernandez T, Hall C,... Schutzki R (2005) Landscape plant material, size, and design sophistication increase perceived home value. J Environ Hortic 23(3):127–133
Bell JF, Wilson JS, Liu GC (2008) Neighborhood greenness and 2-year changes in body mass index of children and youth. American J Prev Med 35:547–553
Bell S, Tyrväinen L, Sievänen T, Pröbstl U, Simpson M (2007) Outdoor recreation and nature tourism: A European perspective. Living Rev Landsc Res 1(2):1–46
Bellows AC, Brown K, Smit J (2004) Health benefits of urban agriculture. Community Food Security Coalition's North American Initiative on Urban Agriculture. Portland
Berman MG, Kross E, Krpan KM, Askren MK, Burson A, Deldin PJ, Kaplan S, Sherdell L, Gotlib IH, Jonides J (2010) Interacting with nature improves cognition and affect for individuals with depression. J Affect Disord 140:300–305
Berndtsson JC (2010) Green roof performance towards management of runoff water quantity and quality: A review. Ecol Eng 36:351–360
Blossey B, Skinner LC, Taylor J (2001) Impact and management of purple loosestrife (*Lythrum salicaria*) in North America. Biod and Cons 10:1787–1807
Bon HD, Parrot L, Moustier P (2010) Sustainable urban agriculture in developing countries. A review. Agron Sustain Dev 30:21–32
Bringslimark T, Hartig T, Patil GG (2011) Adaptation to windowlessness: Do office workers compensate for a lack of visual access to the outdoors? Environ Behav 43:469–487
Brogan DR, Douglas JL (1980) Physical environment correlates of psychosocial health among urban residents. Am J Commun Psychol 8:507–522
Brown KH, Carter A et al. (2003) Urban Agriculture and community food security in the United States: Farming from the city center to the urban fringe. Urban Agriculture Committee of the Community Food Security Coalition (CFSC). CFSC Report. Feb 2003. p. 30
Buckley R (2003) The practice and politics of tourism and land management. In Buckley R, Pickering C, Weaver DB (eds) Nature-based tourism, environment and land management. Paper presented at the 2001 Fenner Conference on Nature Tourism and the Environment, in Canberra, Australia, CABI Publishing, Wallingford, Cambridge
Burchett M, Wood R, Orwell R, Tarran J, Torpy F, Alquezar R (2005) How and why potted-plants really do clean indoor air summary. http://www.interiorplantscape.asn.au/Downloads/M_B_Papers/mburchett_transcript_040305.pdf. Accessed Feb 15 2013
Butler LM, Maronek DM (eds) (2002) Urban and agricultural communities: Opportunities for common ground. CAST Task Force Report 138. Council for Agricultural Science and Technology, Ames, 124 p
Cammack C, Waliczek TM, Zajicek JM (2002a) The Green Brigade: The educational effects of a community-based horticultural program on the horticultural knowledge and environmental attitude of juvenile offenders. HortTechnology 12:77–81
Cammack C, Waliczek TM, Zajicek JM (2002b) The Green Brigade: The psychological effects of a community-based horticultural program on the self-development characteristics of juvenile offenders. HortTechnology 12:82–86
Centers for Disease Control and Prevention (2012) Healthy places: Physical activity. http://www.cdc.gov/healthyplaces/healthtopics/physactivity.htm. Accessed Dec 30 2012
Cheng JJ, Berry P (2012) Health co-benefits and risks of public health adaptation strategies to climate change: A review of current literature. Online Int J Publ Health DOI:101007/s00038–012-0422-5 Available from http://link.springer.com/article/10.1007%2Fs00038-012-0422-5#page-1. Accessed Dec 30 2012
Cimprich B (1993) Development of an intervention to restore attention in cancer patients. Cancer Nurs 16:83–92

Coats VC, Stack LB, Rumpho ME (2011) Maine nursery and landscape industry perspectives on invasive plant issues. Invasive Plant Sci Manag 4:378–389

Cohen M, Baudoin R, Palibrk M, Persyn N, Rhein C (2012a) Urban biodiversity and social inequalities in built-up cities: New evidences, next questions. The example of Paris, France. Landsc Urban Plan 106:277–287

Cohen P, Potchter O, Matzarakis A (2012b) Daily and seasonal climate conditions of green open spaces in the Mediterranean climate and their impact on human comfort. Build and Comf 51:285–295

Collins CC, O'Callaghan AM (2008) The impact of horticultural responsibility on health indicators and quality of life in assisted living. HortTechnology 18:611–618

Cornelis J, Hermy M (2004) Biodiversity relationships in urban and suburban parks in Flanders. Landsc Urban Plan 69:385–401

Cotter DJ, Gomez RE, Lohr VI (1978) Enhancing ASHS efforts at the plant people interface. HortScience 13:216

Cozens PM, Saville G, Hillier D (2005) Crime prevention through environmental design (CPTED): a review and modern bibliography. Prop Manag 23:328–356

Crompton J L (2001) The impact of parks on property values: A review of the empirical evidence. J Leisure Res 33:1–31

Culp RP (2008) Predicting days on market: The influence of environmental and home attributes. New York Econ Rev 39:70–84

Dadvand P, de Nazelle A, Figueras F, Basagaña X, Su J, Amoly E, Jerrett M, Vrijheid M, Sunyer J, Nieuwenhuijsen MJ (2012) Green space, health inequality and pregnancy. Environ Int 40:110–115

Dahle GA, Holt HH, Chaney WR, Whalen TM, Cassens DL, Gazo R, McKenzie RL (2006) Branch strength loss implications for silver maple (*Acer saccharinum*) converted from round-over to v-trim. Arboric Urban For 32:148–154

Davis AP, Shokouhian M, Sharma H, Minami C (2006) Water quality improvement through bioretention media: Nitrogen and phosphorus removal. Water Environ Res 78:284–293

Dehnen-Schmutz K, Touza J (2008) Plant invasions and ornamental horticulture: pathway, propagule pressure and the legal framework. In da Silva JAT (ed) Floriculture, Ornamental and Plant Biotechnology, vol V. Global Science Books, Isleworth, pp. 15–21

DeMarco LW, Relf D, McDaniel A (1999) Integrating gardening into the elementary school curriculum. HortTechnology 9:276–281

Detweiler MB, Murphy PF, Myers LC, Kim KY (2008) Does a wander garden influence inappropriate behaviors in dementia residents? Am J Alzheimer's Dis Other Demen 23:31–45

DeWolfe J, Waliczek TM, Zajicek JM (2011) The relationship between levels of greenery and landscaping at track and field sites, anxiety, and sports performance of collegiate track and field athletes. HortTechnology 21:329–335

Dietz ME, Clausen JC (2005) A field evaluation of rain garden flow and pollutant treatment. Water Air Soil Poll 167:123–138

Dixon JM, Donati KJ, Pike LL, Hattersley L (2009) Functional foods and urban agriculture: two responses to climate change-related food insecurity. N S W Public Health Bull 20(2):14–18

Dobbs K, Relf D, McDaniel A (1998) Survey on the needs of elementary education teachers to enhance the use of horticulture or gardening in the classroom. HortTechnology 8:370–373

Donovan GH, Michael YL, Butry DT, Sullivan AD, Chase JM (2011) Urban trees and the risk of poor birth outcomes. Health Place 17:390–393

Donovan GH, Prestemon JP (2012) The effect of trees on crime in Portland, Oregon. Environ Behav 44:3–30

Drescher AW, Nugent R, de Zeeuw H (2000) Final report: Urban and periurban agriculture on the policy agenda. http://www.fao.org/docrep/MEETING/003/X6091E.HTM. Accessed Feb 15 2013

Duh J-D, Shandas V, Chang H, George LA (2008) Rates of urbanisation and the resiliency of air and water quality. Sci Total Env 400:238–256

Dylewski KL, Wright AN, Tilt KM, LeBleu C (2011) Effects of short interval cyclic flooding on growth and survival of three native shrubs. HortTechnology 21:461–465
Egondi T, Kyobutungi C, Kovats S, Muindi K, Ettarh R, Rocklo J (2012) Time-series analysis of weather and mortality patterns in Nairobi's informal settlements. Global Health Action 5. DOI:19065 http://dx.doi.org/10.3402/gha.v5i0.19065. Accessed Jan 3 2013
Evans MR, Malone H (1992) People and plants: a case study in the hotel industry. p.220 In: Relf D(ed) The role of horticulture in human well-being and social development: A National Symposium. Timber Press, Portland.
Ewing R, Dumbaugh E (2009) The built environment and traffic safety: A review of empirical evidence. J Plan Lit 23:347–367
FAO (1995) Improving nutrition through home gardening: A training package for preparing field workers in Southeast Asia, 171 pp. English—Job Number V5290E
FAO (1999a) Field programme management- Food, nutrition and development. 244 pp. English—ISBN 92-5-104387-6
FAO (1999b) Urban and peri-urban agriculture. http://www.fao.org/unfao/bodies/COAG/COAG15/X0076e.htm. Accessed 15 Feb 2013
FAO (2001) Urban and peri-urban agriculture A briefing guide for the successful implementation of urban and peri-urban agriculture in developing countries and countries of transition. SPFS/DOC/27.8 Revision 2 Handbook Series Volume III. http://www.fao.org/fileadmin/templates/FCIT/PDF/briefing_guide.pdf. Accessed Feb 15 2013
FAO (2008) Briefing paper: Hunger on the rise. http://www.fao.org/newsroom/common/ecg/1000923/en/hungerfigs.pdf. Accessed Nov 22 2010
FAO (2012) Growing greener cities in Africa. http://www.fao.org/docrep/016/i3002e/i3002e.pdf. Accessed Feb 28 2013
FAO (2013) Definitions of key bioenergy and food security terms. http://www.fao.org/energy/befs/definitions/en/. Accessed Feb 15 2013
FAO, WFP, IFAD (2012) The state of food insecurity in the world 2012. Economic growth is necessary but not sufficient to accelerate reduction of hunger and malnutrition. http://www.fao.org/docrep/016/i3027e/i3027e00.htm. Accessed Dec 17 2012
Feenstra G, McGrew S, Campbell D (1999) Entrepreneurial community gardens: Growing food, skills, jobs and communities. Agricultural and Natural Resources Publication 21587, Univ Calif Davis
Fenger J (1999) Urban air quality. Atmos Environ 33:4877–4900
Finch CR (1995) Green Brigade: Horticultural learn-and-earn program for juvenile offenders. HortTechnology 5:118–120
Fjeld T (2000) The effect of interior planting on health and discomfort among workers and school children. HortTechnology 10:46–52
Flagler J (1995) The role of horticulture in training correctional youth. HortTechnology 5:185–187
Francis M (1987) Some different meanings attached to a city park and community gardens. Landsc J 6:101–112
Frumkin H, McMichael AJ (2008) Climate change and public health. Am J Prev Med 35:403–410
Gandelman N, Piani G, Ferre Z (2012) Neighborhood determinants of quality of life. J Happiness Stud 13:547–563
Gen-Song W, Wang J, Ming S, Yang W-R, Qi-Xiang Z (2012) Developing the concept of mei flower culture themed tourism—A case study of resources development of She County's mei flower in Huangshan City. Acta Hort 937:1201–1208
Gigliotti CM, Jarrott SE, Yorgason J (2004) Harvesting health effects of three types of horticultural therapy activities for persons with dementia. Dementia 3:161–180
Gilbert E (2012) Five urban garden programs that are reaching inmates and at-risk populations. http://blogs.worldwatch.org/nourishingtheplanet/five-urban-garden-programs-that-are-reaching-inmates-and-at-risk-populations. Accessed Feb 28 2013
Gouin FR (1983) Girdling by roots and ropes. J Environ Hort 1:50–52
Gregor T, Bönsel D, Starke-Ottich I, Zizka, G (2012) Drivers of floristic change in large cities—A case study of Frankfurt/Main (Germany). Landsc Urban Plan 104:230–237

Hall CR, Hodges AW, Haydu JJ (2006) The economic impact of the green industry in the United States. HortTechnology 16:345–353
Hall JM, Handley JF, Ennos, AR (2012) The potential of tree planting to climate-proof high density residential areas in Manchester, UK. Landsc Urban Plan 104:410–417
Hassink J, van Dijk M (eds) (2006) Farming for health: Green-care farming across Europe and the United States of America. Springer Dordrecht, Wageningen
Haynes C, Van Der Zanden AM, Iles JK (2007) A survey of the ornamental horticulture industry in Iowa. HortTechnology 17:513–517
Hitchmough J (2008) New approaches to ecologically based, designed urban plant communities in Britain: do these have any relevance in the United States? Cities and the Environment 1(2), Article 10, 15 pp
Hoekstra F (2006) Cities farming for the future—Urban agriculture for green and productive cities. RUAF Foundation, IDRC and IIRR
Hope D, Gries C, Zhu W, Fagan WF, Redman CL, Grimm NB, Nelson AL, Martin C, Kinzig A (2003) Socioeconomics drive urban plant diversity. Proc Natl Acad Sci 100:8788–8792
Huang, YJ, Akbari, H, Taha, H, Rosenfeld, AH (1987) The potential of vegetation in reducing summer cooling loads in residential buildings. J Clim Appl Meteorol 26:1103–1116
i-Tree (2012) International milestone—100 countries of i-Tree. i-TreeNewsletter, February 1–2
i-Tree (2013) What is i-Tree? http://www.itreetools.org. Accessed 11 Jan 2013
Ikin K, Beaty MR, Lindenmayer DB, Knight E, Fischer J, Manning AD (2013) Pocket parks in a compact city: how do birds respond to increasing residential density? Landscape Ecology 28:45–56
Justen EAK, Haynes C, VanDerZanden AM, Grudens-Schuckfile N (2009) Managers of Latino workers in the Iowa horticulture industry want educational programs to bridge language and cultural barriers. HortTechnology 19:224–229
Kaplan S, Talbot JF, Kaplan R (1988) Coping with daily hassles: The impact of nearby nature on the work environment. Project Report. USDA Forest Service, North Central Forest Experiment Station, Urban Forestry Unit Cooperative Agreement 23–85–08
Kendal D, Williams NSG, Williams KJH (2012) Drivers of diversity and tree cover in gardens, parks and streetscapes in an Australian city. Urban For Urban Green 11:257–265
Kennedy G, Nantel G, Shetty P (2004) Globalization of food systems in developing countries: a synthesis of country case studies. In: Globalization of food systems in developing countries: Impact on food security and nutrition, p 1–25. FAO Food and Nutrition Paper 83. http://www.fao.org/docrep/007/y5736e/y5736e00.htm. Accessed Feb 28 2013
Kerrigan J, Stevenson NC (1997) Behavioral study of youth and elders in an intergenerational horticultural program. Act Adapt Aging 22:141–153
Kidd JL, Brascamp W (2004) Benefits of gardening to the well-being of New Zealand Gardeners. Acta Horticulturae 639:103–112
Kim BY, Park SA, Song JE, Son KC (2012) Horticultural therapy program for the improvement of attention and sociality in children with intellectual disabilities. HortTechnology 22:320–324
Kim HH (1992) Urban heat-island. Int J Remote Sens 13:2319–2336
Kirkpatrick JB, Daniels GD, Davison A (2011) Temporal and spatial variation in garden and street trees in six eastern Australian cities. Landsc Urban Plan 101:244–252
Kueffer C, Loope L (eds) (2009) Prevention, early detection and containment of invasive, non-native plants in the Hawaiian Islands: Current efforts and needs. Pacific Cooperative Studies Unit Technical Report 166, University of Hawai`i at Manoa, Department of Botany, Honolulu, HI
Kühn I, Brandl R, Klotz S (2004) The flora of German cities is naturally species rich. Evol Ecol Res 6:749–764
Kuo FE, Sullivan WC (2001a) Aggression and violence in the inner city: Effects of environment via mental fatigue. Environ Behavior 33:543–571
Kuo FE, Sullivan WC (2001b) Environment and crime in the inner city: Does vegetation reduce crime? Environ Behav 33:343–367
Kuo FE, Sullivan WC, Coley RL, Brunson L (1998) Fertile ground for community: Inner-city neighborhood common spaces. Amer J Community Psychol 26:823–851

Kweon BS, Sullivan WC, Wiley A (1998) Green common spaces and the social integration of inner-city older adults. Environ Behav 30:832–858
Lachowycz K, Jones AP (2011) Greenspace and obesity: a systematic review of the evidence. Obes Rev 12:e183–e189
Lang Research (2001) TAMS Horticultural tourism report. http://www.ontla.on.ca/library/repository/mon/3000/10298621.pdf. Accessed Jan 12 2013
Lelieveld J, Hadjinicolaou P, Kostopoulou E, Chenoweth J, El Maayar M, Giannakopoulos C, Hannides C, Lange MA, Tanarhte M, Tyrlis E, Xoplaki E (2012) Climate change and impacts in the Eastern Mediterranean and the Middle East. Clim Change 114:667–687
Lewis CA (1996) Green nature/human nature: the meaning of plants in our lives. University of Illinois Press, Urbana, Ill
Lockwood IM, Stillings T (2001) Traffic calming for crime reduction and neighborhood revitalization. http://www.ite.org/traffic/documents/AHA98A19.pdf. Accessed Jan 25 2013
Lohr VI (1992) The contribution of interior plants to relative humidity in an office. In: Relf D (ed) The role of horticulture in human well-being and social development. Timber Press, Portland, pp 117–119
Lohr VI (2010) What are the benefits of plants indoors and why do we respond positively to them? Acta Horticulturae 881(2):675–682
Lohr VI (2011) Greening the human environment: The untold benefits. Acta Horticulturae 916:159–170
Lohr VI (2013) Diversity in landscape plantings: Broader understanding and more teaching needed. HortTechnology 23:126–129
Lohr VI, Pearson Mims CH (1996) Particulate matter accumulation on horizontal surfaces in interiors: Influence of foliage plants. Atmospheric Environ 30:2565–2568.
Lohr VI, Pearson-Mims CH, Goodwin GK (1996) Interior plants may improve worker productivity and reduce stress in a windowless environment. J Environ Hort 14(2):97–100
Lohr VI, Relf D (1993) Human issues in horticulture: Research priorities. HortTechnology 3:106–7
Lohr VI, Relf PD, Looney NE (2004) A focus on human issues in horticulture: An introduction to the opening plenary colloquium—Applying the art and science of horticulture to improving human life quality. Acta Horticulturae 642:69–70
Loughner CP, Allen DJ, Zhang D-L, Pickering KE, Dickerson RP, Landry L (2012) Roles of urban tree canopy and buildings in urban heat island effects: Parameterization and preliminary results. J Appl Meteorol Climatol 51:1775–1793
Lovasi GS, Bader MDM, Quinn J, Neckerman K, Weiss C, Rundle A (2012) Body mass index, safety hazards, and neighborhood attractiveness. Am J Prev Med 43:378–384
Lüdecke H-J, Link R, Ewert F-K (2011) How natural is the recent centennial warming? An analysis of 2249 surface temperature records. Int J Mod Physics C 22:1139–1159
Maas J, Verheij RA, de Vries S, Spreeuwenberg P, Schellevis FG, Groenewegen PP (2009) Morbidity is related to a green living environment. J Epidemiology Com 63:967–973
Macpherson M (1993) Benefits of urban greening. Merck Family Fund, Milton, Mass
Malakoff D (1995) What good is community greening? American Community Gardening Association Monograph. Pennsylvania Horticultural Society, Philadelphia PA
Maleike R, Hummel RL (1992) Planting landscape plants. Arboric J 16:217–226
Marcus CC, Barnes M (1999) Healing gardens: Therapeutic benefits and design recommendations. Wiley, New York
Masuya H, Brasier C, Ichihara Y, Kubono T, Kanzaki N (2010) First report of the Dutch elm disease pathogens *Ophiostoma ulmi* and *O. novo-ulmi* in Japan. Plant Pathol 59:805
McDonald AG, Bealey WJ, Fowler D, Dragosits U, Skiba U, Smith RI, Donovan RG, Brett HE, Hewitt CN, Nemitz E (2007) Quantifying the effect of urban tree planting on concentrations and depositions of $PM_{10}$ in two UK conurbations. Atmos Environ 41:8455–8467
McGuinn C, Relf PD (2001) A profile of juvenile offenders in a vocational horticulture curriculum. HortTechnology 11:427–433
McKinney ML (2006) Urbanization as a major cause of biotic homogenization. Biol Cons 127:247–260

McPherson EG, Rowntree RA (1993) Energy conservation potential of urban tree planting. J Arboric 19:321–331
McPherson G, Simpson JR, Peper PJ, Maco SE, Xiao Q (2005) Municipal forest benefits and costs in five US cities. J Forestry 103:411–416
Missouri Prairie Foundation (2013) Grow native. http://grownative.org/. Accessed Jan 15 2013
Mitchell R, Popham F (2008) Effect of exposure to natural environment on health inequalities: an observational population study. Lancet 372:1655–1660
Mok J-H, Landphair HC, Naderi JR (2006) Landscape improvement impacts on roadside safety in Texas. Landsc Urban Plan 78:263–274
Mooney PF (1994) Assessing the benefits of a therapeutic horticulture program for seniors in immediate care. In: Francis M, Lindsey P, Rice JS (eds) The healing dimensions of people-plant relations. Ctr for Design Res, Davis, California, pp 173–194
Mooney PF, Nicell PL (1992) The importance of exterior environment for Alzheimer's residents: Effective care and risk management. Healthcare Mgt Forum 5(2):23–29
Moore EO (1981–1982) A prison environment's effect on health care service demands. J Environ Sys 11:17–34
Morton CM, Gruszka P (2008) AFLP assessment of genetic variability in old vs. new London plane trees (*Platanus* × *acerfolia*). J Hort Sci Biotechnol 83:532–537
Morzaria-Luna HN, Schaepe KS, Cutforth LB, Veltman RL (2004) Implementation of bioretention systems: A Wisconsin case study. J Am Water Resour Assoc 40:1053–1061
Mougeot JAL (2000) The hidden significance of urban agriculture. Trialog 65:8–13
Müller C (2007) Intercultural gardens: Urban places for subsistence production and diversity. German J Urban Stud 46(1):55–65
Mundell JE, Grigsby D (1997) Neighborhood traffic calming: Seattle's traffic circle program. http://ite.org/traffic/documents/Seattle/SeattlesTrafficCircleProgram.pdf. Accessed 15 Jan 2013
Ng E, Chen L, Wang Y, Yuan C (2012) A study on the cooling effects of greening in a high-density city. Build Environ 47:256–271
Niemiera AX, Von Holle B (2009) Invasive plant species and the ornamental horticulture industry. In Inderjit A (ed) Management of invasive weeds. Springer Science, Dordrecht, The Netherlands, pp 167–187
NISIC (2012) Federal laws and regulations: Executive order 13112. http://www.invasivespeciesinfo.gov/laws/execorder.shtml. Accessed 15 Jan 2013
Nowak DJ, Crane DE, Stevens JC (2006) Air pollution removal by urban trees and shrubs in the United States. Urban For Urban Green 4:115–123
O'Neill MS, Carter R, Kish JK, Gronlund CJ, White-Newsome JL, Manarolla X, Zanobetti A, Schwartz JD (2009) Preventing heat-related morbidity and mortality: New approaches in a changing climate. Maturitas 64:98–103
Park SA, Lee KS, Son KC (2011) Determining exercise intensities of gardening tasks as a physical activity using metabolic equivalents in older adults. HortScience 46:1706–1710
Pedlowski MA, Silva VACD, Adell JJC, Heynen NC (2002) Urban forest and environmental inequality in Campos dos Goytacazes, Rio de Janeiro, Brazil. Urban Ecosyst 6:9–20
Pham T-T-H, Apparicio P, Séguin A-M, Gagnon M (2012) Spatial distribution of vegetation in Montreal: An uneven distribution or environmental inequity? Lands Urban Plan 107:214–224
Pimentel D, Zuniga R, Morrison D (2005) Update on the environmental and economic costs associated with alien-invasive species in the United States. Ecol Econ 52:273–288
Poland TM, McCullough DG (2006) Emerald ash borer: Invasion of the urban forest and the threat to North America's ash resource. J Forestry 104:118–124
Popek R, Gawrońska H, Wrochna M, Gawroński SW, Sæbø A (2013) Particulate matter on foliage of 13 woody species: Deposition on surfaces and phytostabilisation in waxes—a 3-year study. Int J. Phytoremediation 15:245–256
Potter C, Harwood T, Knight J, Tomlinson I (2011) Learning from history, predicting the future: The UK Dutch elm disease outbreak in relation to contemporary tree disease threats. Phil Trans R Soc B 366:1966–1974
Predny M, Relf PD (2000) Interactions between elderly adults and preschool children in a horticultural therapy research program. HortTechnology 10:64–70

Ranney T, Touchell D, Olsen R, Eaker T, Lynch N, Mowrey J (2006) Progress in breeding non-invasive nursery crops. SNA Res Conf 51:597–598
Rappe E, Kivelä SL (2005) Effects of garden visits on long-term care residents as related to depression. HortTechnology 15:298–303
Relf D (1992a) Human issues in horticulture. HortTechnology 2:159–171
Relf D (ed) (1992b) The role of horticulture in human well-being and social development. Timber Press, Portland, 254 pp
Relf D, McDaniel AR, Butterfield B (1992) Attitudes toward plants and gardening. HortTechnology 2:201–204
Relf PD (1982) Consumer horticulture: A psychological perspective. HortScience 17:317–319
Relf PD (1990) Psychological and sociological response to plants: Implications for horticulture. HortScience 25:11–13
Relf PD (2005) The therapeutic values of plants. Pediatr Rehabil 8:235–237
Relf PD (2006) Theoretical models for research and programme development in agriculture and health care. In: Hassink J, Dijk M (eds) Farming for health. Springer, Dordrecht, pp 1–20
Ricklin A et al. (2012) Healthy Planning: an evaluation of comprehensive and sustainability plans addressing public health. American Planning Association, Chicago
Roy AH, Wenger SJ, Fletcher TD, Walsh CJ, Ladson AR, Shuster WD, Thurston HW, Brown RR (2008) Impediments and solutions to sustainable, watershed-scale urban stormwater management: Lessons from Australia and the United States. Environ Manage 42:344–359
Rydin Y, Bleahu A, Davies M, Dávila JD, Friel S, De Grandis G, Groce N, Hallal PC, Hamilton I, Howden-Chapman P, Lai K-M, Lim CJ, Martins J, Osrin D, Ridley I, Scott I, Taylor M, Wilkinson P, Wilson J (2012) Shaping cities for health: Complexity and the planning of urban environments in the 21st century. Lancet 379:2079–2108
Santini A, Fagnani A, Ferrini F, Mittempergher L (2002) 'San Zanobi' and 'Plinio' elm trees. HortScience 37:1139–1141
Sernovitz DJ (2011) Baltimore approves plan to shed vacant lots. http://www.bizjournals.com/baltimore/news/2011/08/17/baltimore-approves-plan-to-shed-vacant.html. Accessed Feb 26 2013
Shigo AL (1985) Compartmentalization of decay in trees. Sci Am 252:96–103
Shoemaker CA, Lin MC (2008) A model for healthy aging with horticulture. Acta Horticulturae 775:93–98
Shoemaker CA, Relf PD, Lohr VI (2000) Social science methodologies for studying individuals' responses in human issues in horticulture research. HortTechnology 10:87–93
Simons LA, Simons J, McCallum J, Friedlander Y (2006) Lifestyle factors and risk of dementia: Dubbo Study of the elderly. Medical J Australia 184(2):68–70
Slater RJ (2001) Urban agriculture, gender and empowerment: an alternative view. Dev South Afr 18(5):635–650
Smit J, Nasr J (1992) Urban agriculture for sustainable cities: using wastes and idle land and water bodies as resources. Environ Urban 4(2):141–152
Stagoll K, Lindenmayer DB, Knight E, Fischer J, Manning AD (2012) Large trees are keystone structures in urban parks. Conserv Lett 5:115–122
Stigarll A, Elam E (2009) Impact of improved landscape quality and tree cover on the price of single-family homes. J Environ Hortic 27:24–30
Takano T, Nakamura K, Watanabe M (2002) Urban residential environments and senior citizens' longevity in megacity areas: The importance of walkable green spaces. J Epidemiology and Community Health 56:913–918
Taylor AF, Wiley A, Kuo FE, Sullivan WC (1998) Growing up in the inner city: Green spaces as places to grow. Environ Behav 30:3–27
Thomsen JD, Sønderstrup-Anderse HKH, Müller R (2011) People–plant relationships in an office workplace: Perceived benefits for the workplace and employees. HortScience 46:744–752
Townsend AM, Bentz SE, Douglass LW (2005) Evaluation of 19 American elm clones for tolerance to Dutch elm disease. J Environ Hort 23:21–24
Troy A, Grove JM, O'Neil-Dunne J (2012) The relationship between tree canopy and crime rates across an urban–rural gradient in the greater Baltimore region. Landsc Urban Plan 106:262–270

Tyrväinen L (2001) Economic valuation of urban forest benefits in Finland. J Environ Manag 62:75–92
U. S. Census Bureau (2013) U.S. & World Population Clocks. http://www.census.gov/main/www/popclock.html. Accessed Feb 15 2013
Ulrich RS, Simons RF (1986) Recovery from stress during exposure to everyday outdoor environments. In: Wineman J, Barnes R, Zimring C (eds) The costs of not knowing: Proceedings of the 17th Annual Conference of the Environmental Research and Design Association, Washington, DC, pp 115–122
United Nations (1996) Report of the United Nations Conference on Human Settlements (Habitat II). http://daccess-ods.un.org/access.nsf/Get?Open&DS=A/CONF.165/14&Lang=E. Accessed Feb 15 2013
United Nations (2012) World urbanization prospects: The 2011 revision, highlights. United Nations Department of Economic and Social Affairs, Population Division, New York
Van den Berg AE, Custers MHG (2011) Gardening promotes neuroendocrine and affective restoration from stress. J Health Psychol 16:4–11
Waliczek TM, Bradley JC, Zajicek JM (2001) The effect of school gardens on children's interpersonal relationships and attitudes toward school. HortTechnology 11:466–468
Waliczek TM, Zajicek JM (1996) The effect of school gardens on self-esteem, interpersonal relationships, attitude toward school, and environmental attitude in populations of children. HortScience 31:608
Warner SB Jr, Baron JH (1993) Restorative gardens: Green thoughts in a green shade. British Medical J 306:1080–1081
Whitehead PG, Wilby RL, Battarbee RW, Kernan M, Wade AJ (2009) A review of the potential impacts of climate change on surface water quality. Hydrological Sciences J 54:101–123
Wichrowski M, Whiteson J, Haas F, Mola A, Rey MJ (2005) Effects of horticultural therapy on mood and heart rate in patients participating in an inpatient cardiopulmonary rehabilitation program. J Cardiopulm Rehabil 25:270–274
Wilson CL (1975) The long battle against Dutch elm disease. J Arboriculture 1:107–112
Wittig R, Becker U (2010) The spontaneous flora around street trees in cities—A striking example for the worldwide homogenization of the flora of urban habitats. Flora 205:704–709
WNPS (2012) Ivy OUT. http://www.ivyout.org/index.html. Accessed Jan 15 2013
Wolf KL (2003) Public response to the urban forest in inner-city business districts. J Arboriculture 29:117–126
Wolf KL (2009) Strip malls, city trees, and community values. Arboric Urban For 35:33–40
Wolf KL, Bratton N (2006) Urban trees and traffic safety: Considering U.S. roadside policy and crash data. Arboric Urban For 32:170–179
World Health Organization (2012) Obesity and overweight, fact sheet No 311. http://wwwwhoint/mediacentre/factsheets/fs311/en/. Accessed Dec 30 2012
Xiao QF, McPherson EG (2011) Performance of engineered soil and trees in a parking lot bioswale. Urban Water J 8:241–253
Xiao QF, McPherson EG, Simpson JR, Ustin SL (1998) Rainfall interception by Sacramento's urban forest. J Arboriculture 24:235–244.
Xie Q, Zhou Z, Chen F (2011) Quantifying the beneficial effect of different plant species on air quality improvement. Environ Engineer Manag J 10:858–963
Yang J, Yu Q, Gong P (2008) Quantifying air pollution removal by green roofs in Chicago. Atmos Environ 42:7266–7273
Younger M, Morrow-Almeida HR, Vindigni SM, Dannenberg AL (2008) The built environment, climate change, and health: Opportunities for co-benefits. Am J Prev Med 35:517–526
Zezza A, Tasciotti L (2010) Urban agriculture, poverty, and food security: Empirical evidence from a sample of developing countries. Food Policy 35(4):265–273

# Chapter 32
# Education and Training Futures in Horticulture and Horticultural Science

**David E. Aldous, Geoffrey R. Dixon, Rebecca L. Darnell and James E. Pratley**

**Abstract** Horticultural knowledge and skills training have been with humankind for some 10,000 to 20,000 years. With permanent settlement and rising wealth and trade, horticulture products and services became a source of fresh food for daily consumption, and a source of plant material in developing a quality environment and lifestyle. The knowledge of horticulture and the skills of its practitioners have been demonstrated through the advancing civilizations in both eastern and western countries. With the rise of the Agricultural Revolutions in Great Britain, and more widely across Continental Europe in the seventeenth and eighteenth centuries, as well as the move towards colonisation and early migration to the New Worlds, many westernised countries established the early institutions that would provide education and training in agriculture and horticulture. Today many of these colleges and universities provide undergraduate, postgraduate and vocational and technical training that specifically targets horticulture and/or horticultural science with some

---

Professor David E. Aldous – deceased 1st November 2013

---

J. E. Pratley (✉)
Charles Sturt University, School of Agricultural and Wine Sciences, Locked Bag 588,
Wagga Wagga, NSW 2678, Australia
e-mail: jpratley@csu.edu.au

D. E. Aldous
School of Land, Crop and Food Science, The University of Queensland, Gatton Campus, Lawes,
Queensland 4343, Australia

G. R. Dixon
School of Agriculture, University of Reading, Earley Gate, Reading,
Berkshire RG66AR, United Kingdom
e-mail: geoffrdixon@btinternet.com

GreenGene International, Hill Rising, Horsecastles Lane, Sherborne,
Dorset DT9 6BH, United Kingdom

R. L. Darnell
Horticultural Sciences Department, 1131 Fifield Hall,
PO Box 110690, Gainsville, FL 32611-0690, USA
e-mail: rld@ufl.edu

G. R. Dixon, D. E. Aldous (eds.), *Horticulture: Plants for People and Places, Volume 3*,
DOI 10.1007/978-94-017-8560-0_32, 

research and teaching institutions also providing extension and advisory services to industry. The objective of this chapter is to describe the wider pedagogic and educational context in which those concerned with horticulture operate, the institutional structures that target horticulture and horticultural science education and training internationally; examine changing educational formats, especially distance education; and consider strategies for attracting and retaining young people in the delivery of world-class horticultural education. In this chapter we set the context by investigating the horticultural education and training options available, the constraints that prevent young people entering horticulture, and suggest strategies that would attract and retain these students. We suggest that effective strategies and partnerships be put in place by the institution, the government and most importantly the industry to provide for undergraduate and postgraduate education in horticulture and horticultural science; that educational and vocational training institutions, government, and industry need to work more effectively together to improve communication about horticulture and horticultural science in order to attract enrolments of more and talented students; and that the horticulture curriculum be continuously evaluated and revised so that it remains relevant to future challenges facing the industries of horticulture in the production, environmental and social spheres. These strategies can be used as a means to develop successful programs and case studies that would provide better information to high school career counsellors, improve the image of horticulture and encourage greater involvement from alumni and the industries in recruitment, provide opportunities to improve career aspirations, ensure improved levels of remuneration, and promote the social features of the profession and greater awareness and recognition of the profession in the wider community. A successful career in horticulture demands intellectual capacities which are capable of drawing knowledge from a wide field of basic sciences, economics and the humanities and integrating this into academic scholarship and practical technologies.

**Keywords** Pedagogic philosophies · Horticultural education · Vocational education and training · Community college · Higher education

## Introduction—The Challenge

The origins of agriculture and horticulture date back at least to the lands of Mesopotamia, irrigated and enriched by the Tigris and Euphrates Rivers some 10,000 to 20,000 years ago (Mudge et al. 2009). Humanity's change from nomadic hunter-gatherers to pastoral village cultures necessitated the evolution of husbandry knowledge and skills. From these earliest times, knowledge, ingenuity, invention, and adaptation required for growing crops were transmitted across generations. As individual and civic wealth were created, horticulture developed a broader relevance providing civilised surroundings populated with flowering trees and shrubs and enabling the cultivation of plants of medicinal, perfumery and stimulatory natures.

Possibly this was first evident with the Hanging Gardens of Babylon. Certainly by the time of the earliest Egyptian dynasties, plant husbandry supplying both food and pleasurable leisure is recorded. The essential knowledge and skills probably passed largely by word of mouth through the Persian, Greek, Roman and Arab civilizations. At about the same time or possibly earlier, China was developing its own horticultural knowledge and skills (Jin 1994). China has deservedly been called the "Mother of Gardens" for her rich resources of wild and cultivated flowers and her long history of landscape gardening which was passed on into Japan.

In each of these great civilisations, the development of towns, cities and states was only possible if surrounding rural areas could regularly supply fresh produce into the markets for daily consumption. Concurrently, the wealthier classes would be demanding pleasing landscapes where they could walk, talk and entertain. All of this required horticultural knowledge and skills which were transmitted across generations of "gardeners". The initial attempts at recording horticultural knowledge and skills may have come from the apothecaries who codified the medicinal properties and cultural requirements of annual and perennial herbs. As Christian Europe developed, monasteries emerged as centres of medical and horticultural knowledge, leading to the production of the great hand written herbals. Gardens were cultivated within the castles as places of rest, relaxation and intrigue. Areas capable of high quality horticulture were associated with wealth and the creation of art of the highest quality. It can be no accident that the Loire Valley, which has some of Europe's most horticulturally productive soils, also produced one of its greatest cultural gems. In 1373, King John (Le Bon) commissioned the Apocalypse Tapestry from the painter Hennequin de Bruge and weaver Nicholas Bataille. This was the largest tapestry ever woven in Europe with a length of 140 m covering a total area of 850 $m^2$ (Delwasse 2008). The background of each scene in the surviving 104 m shows flowers that would have been commonly cultivated around Angers Castle, illustrating a high level of horticultural knowledge and gardening skill. As horticulture flourished so did trade, which led to the founding of guilds such as the Worshipful Company of Fruiterers of the City of London, which regulated the quality of fruit and vegetable brought into London. Centres of specialist expertise developed, such as parts of the Spanish Netherlands (today parts of The Netherlands and Belgium) where expertise in fruit tree propagation was developed. In parallel, botanical and medical gardens began appearing first in Padua, Italy followed by the Chelsea Physic Garden in London. The knowledge of horticulture and skills of its practitioners is well illustrated by medieval and Renaissance art, most notably from the Flemish and Dutch painters in the period 1400–1600.

Increasing global exploration from the fifteenth century onwards brought great wealth into Europe accompanied by an increasing flow of new plants. These provided both novel forms of food and exciting flowers and foliage for decoration and landscaping. Europe saw the rise of hugely ambitious landscape projects, as typified by Louis XIV's Versailles and the re-creation of Catherine de Medici's Tuileries by Le Notre, followed by English landscape art created by Kent, Brown and Repton. These were horticulturists of considerable social power and influence who recorded and illustrated their knowledge, ensuring that it was available for future generations. Education and the passage of horticultural knowledge through the generations became increasingly important throughout the seventeenth and eighteenth centuries. Each of the great centres of learning, wealth, trade and industry founded new botanical gardens. Their function was identification of the tidal waves of new plants arriving from around the world and the development of their husbandry such that they might be exploited. Useful plants could be employed either at home or in creating further wealth, as plantations in the colonies of Great Britain and other European powers producing commodities for worldwide trading. John Ray and Linnaeus, the visionary taxonomists, required cultivated botanical gardens in order to understand and classify plants of worldwide origins. Dynasties grew of nurserymen who exploited the new plants for trade, filling the gardens of the *nouveau riche* who desired the latest novelty and would pay handsomely for it. Meanwhile, gardens such as the Royal Botanic Garden at Kew began receiving public, tax-raised, funding as governments started to appreciate the economic value of research and education. In Germany for instance, Humboldt developed his idea of the unity of research and teaching which has become a model for universities world-wide (Bokelmann 2007).

The Enclosure Acts of the seventeenth and eighteenth centuries, which swept away medieval systems of land tenure, laid the foundations for Agricultural Revolutions in Great Britain and more widely in Continental Europe. Once individual ownership was secured, land lords began experimenting with crop and animal production, increasing productivity in order to supply the food needed for the burgeoning numbers of factory workers in the rapidly expanding nineteenth century industrial towns. With colonisation and early migration to the New Worlds, which included the Orient, the Americas, and Australia, many westernised countries established colleges and universities that provided training from vocational to higher education levels in agriculture and horticulture. The British greatly influenced the countries they colonised with knowledge of growing methods in agriculture and horticulture (Hendrick 1950; Aldous et al. 2011). This is seen most clearly in the United States where visionary politicians appreciated the need for education and training for the settlers as they moved westwards across the continent. The Morrill Act of 1862 founded the land grant colleges, which focused on agricultural and mechanical arts. Recognition of the need to underpin education with scientific research resulted in the Hatch Act of 1887, which provided federal funding for state agricultural experiment stations. Moving research and education into the population was further promoted by the Smith-Lever Act of 1914, which founded the co-operative extension service. These acts produced an effective and efficient system linking education, research and extension, and allowed the United States to develop its vast land resources, feed its own population and ultimately become the largest provider of food worldwide.

Much of Australia's early horticultural education commenced with the establishment of the Burnley Horticultural College in Victoria, Australia. On January 1 1863, the Horticultural Society of Victoria, opened the gardens and became home to a new horticultural college that taught both production and ornamental horticulture

**Fig. 32.1** Boater hats and late Victorian clothing styles with seed collection from the English firm, Suttons. (Source: Burnley Archives)

**Fig. 32.2** Ladies turning the soil circa 1890s. (Source: Burnley Archives)

(Aldous 1990; Winzenried 1991). In 1891 the site was transferred to the Victorian Department of Agriculture. Today the site is the Burnley campus of The University of Melbourne (Figs. 32.1–32.4).

The founding of research stations in Europe was spearheaded by the establishment of the Rothamsted Experimental Station (now Rothamsted Research) in England. This was followed by establishment of the John Innes Institute, East Malling Research Station, and the Long Ashton Research Station, all of which became world leading centres of excellence in horticultural science. Each was founded around the turn of the nineteenth century as self-governing trusts established by innovative farmers and growers. Similar research institutions were developed as either public or privately funded organisations worldwide. Universities and colleges took responsibility for providing undergraduate, postgraduate and vocational and technical training that specifically targeted horticulture and/or horticultural science. Some of these

**Fig. 32.3** Students learning pruning techniques circa 1890s. (Source: Burnley Archives)

**Fig. 32.4** Mowing Burnley Gardens. Copy of illustration in "Prize Essays," by A. E. Bennett, published 1894. (Source: Burnley Archives)

research and teaching institutions also sought to provide extension and advisory services to industry. Two world wars fought within a generation during the first half of the twentieth century left governments determined to defeat famine on a global scale. The developed world successfully applied agricultural and horticultural scientific research, education and extension/advisory services on an unprecedented scale. The Green Revolution spearheaded by Norman Borlaugh applied the science of genetics to cereal breeding and produced basic food self-sufficiency for China and India. Education, training and extension/advisory services in agricultural chemistry, plant breeding and crop nutrition through the late 1980s resulted in rising living standards and an increasing level of world trade and wealth. Regrettably, as demonstrated elsewhere in these volumes, insufficient attention was given to the effects of massive over exploitation of the Earth's basic resources of land, water and its atmosphere. Climate change, a devastated environment, and much diminished natural biodiversity are the results. These problems present the current generation of horticultural scientists and future students with the biggest challenges their science has ever faced.

Today the global human population is over 7 billion and is expected to reach 9 billion by mid twenty-first century. More than half of this population is now urban-based and lives and works in cities. The World Bank indicates that the global middle class comprises about 8 % of the world's population, or 600 million people, and this percentage is projected to double in the coming decades. With increases in population, urbanisation, and affluence, there are increased expectations for the availability of a safe, affordable, healthy, and consistent food supply. Improving the quality, quantity and security of food produced becomes an imperative, achieved with better environmental safeguards. In the context of environmental, ecological, and socioeconomic sustainability there is an increasing desire for products that are "natural", "organic", and/or "local". For this to be achieved, there is an increasing requirement for an intellectually able, highly educated and trained agricultural and horticultural workforce which is capable of utilising the opportunities offered by scientific research and translating them carefully into sustainable husbandry.

The objective of this chapter is to: (1) establish the context within which horticultural education is part; (2) describe the institutional structures that target horticulture and horticultural science education and training internationally; (3) examine changing educational formats, especially distance education; and (4) consider strategies for attracting and retaining young people in the delivery of world-class horticultural education.

## The Wider Context of Horticultural Education

National and international educational policies set the context within which all aspects of horticultural teaching, learning, scholarship, research, training and outreach take place. In common with many other disciplines, there is often a mistaken tendency to view the provisions for horticulture in isolation from the surrounding educational world. While imbuing students with professional pride in their chosen discipline is creditable and productive, encouraging them to consider that they are in some way different from others within their peer group who have chosen alternative disciplines is a grave and often very damaging mistake. Horticulture is a relatively small discipline for which isolation diminishes students' comprehension of their wider context and limits the vision of those responsible for course provision.

Possibly the largest single and most radical educational change witnessed over the past half century has been an acceptance by governments that education should be spread as widely as possible within their populations. This recognised the need to encourage in students aspirations that will take individuals as far as possible along the educational road. This may now seem self-evidently sensible and economically advantageous but that was not the case before the provision of universal secondary education or extending opportunities for entry into universities. Cognitive ability theory shows that increasing peoples' educational competencies is decisive in raising a community's economic wealth. Education and wealth creation go hand in hand. Encouraging achievement by higher ability groups is important since it stimulates growth through artistic and scientific-technological progress (Rindermann 2012). Devising pedagogic processes that permit achievement is the role of educators.

Educators have sought to fulfil this responsibility by changing from syllabus-based teaching to competency-based learning. Syllabuses set out broad areas of teaching that provide students with an understanding of their subject in its totality and their understanding was tested largely by end-of-session examinations. Competency-based learning moved responsibility for learning towards the student by dividing the subject into sub-sets, termed modules or units. Each of these provided knowledge of an aspect of the subject either delivered by formal face-to-face teaching or from studying prescribed documents. Learning capacity or competence for each aspect was tested by a series of instruments of assessment. These varied from simple 'yes/no' questions and answers to extended essay responses. Once students could demonstrate acceptable competency they moved on to other similar or more advanced aspects of the subject. Modularisation was seen to offer ease of access by students of mixed ranges of ability to individual components of courses; flexible accumulation of qualifications over several years; ease of updating in response to technological and social change; the identification and provision of common elements in different courses and the motivation and clarity of purpose that can be engendered by short-term targets.

Opinion is divided now over the value of modularisation. Some educators believe it can lead to fragmentation and trivialisation of learning. It can also contribute to over-assessment and assessment-driven learning, and necessitates an emphasis on internal assessment which, though welcome in certain circumstances, ought to be only one element in a balance of assessment modes (McVittie 2008). Coherence between and within cohorts of students becomes difficult to sustain where modules are employed. As a result, previously successful programmes that allowed student options for intercalated periods of work-experience in industry lost their educational attraction and became administratively difficult to operate, with the inevitable consequence that they were discontinued. That process was accelerated since they were financially expensive in staff time and apparently did little to embellish the reputation of the educational provider. Some authorities consider that describing educational goals in terms of competencies instead of knowledge levels has played a crucial role in modernizing the higher education systems for example in The Netherlands (Trip et al. 2004). Dutch teachers were generally positive about the use of modules but they did not believe that they would improve students' results. Employers were neutral with respect to the change to competency models, doubted the relevance of half of the formalized competencies, but were positive about the other related changes such as developing closer connections between education, the job market and the international focus. Students see the relevance of the majority of formalized competencies and consider they will be better equipped for working in teams than their predecessors.

Those who favour changes to competency-based studies suggest that they have entirely laudable foundations. They reflect the ideas of early German educators such as Wagenshein, a mathematician who believed in "open learning", that is, the need to understand and argue with knowledge not simply accumulate facts (Jung 2012). It could be suggested that a syllabus-based system also achieves these objectives. No doubt the quality of the results achieved by either system will depend on the level of resources, including available staff. This was highlighted by one of few attempts at analysing pedagogic thinking as it relates to horticulture by an authority from outside the discipline (Parker 2005). She identified key themes developed in Australia, Great Britain and the U.S.A. Major efforts have added focus to the critical evaluation of student work. Better design of assessments which identify improvements in student knowledge, skills, attitudes and values has been a major advance as has greater accountability for both institutions of learning and teachers. Parker (2005) makes the plea, however, that such accountability should not stifle independent and individual initiatives by teachers, termed "lone-ranger" approaches as these may ultimately be adopted much more widely. She also identifies the need for diversity amongst teaching institutions and staff. This is a feature of the manner by which horticultural education has evolved since about 1990.

The opportunities provided by modularised approaches have been analysed in the U.S.A. Recent applications of competency-based curricula involving the definition of knowledge, skills, and values for horticultural education are described by Basinger et al. (2009). This national Delphi study is the first concise list of competencies described for a horticulture curriculum in the U.S.A. Twenty-two expert horticultural educators within the United States were selected as experts in education and curriculum improvement. The final compilation of competencies describes 108 specific learning outcomes comprising 41 horticulture technical competencies, 34 life science technical competencies, and 33 professional competencies. Table 32.1 provides an abridged version of horticultural competencies.

Internationalisation of education has developed in parallel with greater opportunities for travel and communication. European partnerships arranged through European Union (EU) have instituted programmes aim at increasing staff and student mobility. One of the earliest to be established was the Erasmus (*E*u*R*opean Community *A*ction *S*cheme for the *M*obility of *U*niversity *S*tudents), and still encourages study in other relevant international centres. The philosophical model for this programme is the medieval student who journeyed between centres of excellence gaining knowledge from expert professors. Transnational collaboration offers substantial benefits for the staff who are required to develop a network of peers throughout Europe and perhaps more widely. The students participating in these programmes gain valuable understandings of their discipline viewed from alternative perspectives, interact with others from differing cultural heritages and develop better language skills. This has proven to be especially valuable for horticultural students coming as they do from small cohorts in their home institutions (G R Dixon, personal communication). Opportunities for Erasmus exchanges have improved as coherence in the delivery of disciplines has been encouraged by the European Union (EU) Bologna and Copenhagen Agreements.

In a response to increasing worldwide competition for the benefits of a "knowledge society" the EU has attempted to develop comprehensive European education models (Powell et al. 2012). This is influenced by systems historically developed

**Table 32.1** Abridged table from competencies for a U.S. horticulture undergraduate major: a National Delphi Study (Basinger et al. 2009) with study topic area, Delphi round of acceptance, competency, description, and acceptance rate of topic and competency

| Topic area that includes the competency | Delphi round number when accepted | Competency description example | Acceptance rate (%)[a] |
|---|---|---|---|
| Plant identification | | | 100 |
| | 1 | Identify plants commonly used in the discipline by scientific and common name | 100 |
| Field and greenhouse management | | | 95 |
| | 1 | Define and select plant methods for different horticulture crops | 100 |
| Technology | | | 95 |
| | 1 | Understand new technologies within the horticulture discipline | 100 |
| Biological science | | | 90 |
| | 1 | Demonstrate fundamental knowledge of biological science | 100 |
| Plant genetic/biotechnology | | | 60 |
| | 2 | Recognize how a plant responds to environmental stress | 100 |
| Basic plant breeding | | | 65 |
| | 1 | Recognize the potential impact plant breeding has on agriculture | 90 |
| Abiotic/biotic factors | | | - |
| | 1 | Recognize the common biotic and abiotic stresses | 100 |
| Principles of soils | | | 100 |
| | 1 | Understand the influences of soil pH on plants and fertilization methods | 100 |
| Environmental awareness | | | 75 |
| | 1 | Understand the impact of human activities on the environment | 96 |
| Writing/verbal skills | | | 100 |
| | 2 | Apply critical thinking with respect to analysing and interpreting available information | 100 |
| Research skills | | | 50 |
| | 1 | Understand the scientific method | 96 |
| Problem solving skills | | | 95 |
| | 1 | Work effectively with others on complex issues | 100 |
| Decision making skills | | | 100 |
| | 1 | Make decisions on the basis of thorough analysis of a situation | 91 |

**Table 32.1** (continued)

| Topic area that includes the competency | Delphi round number when accepted | Competency description example | Acceptance rate (%)[a] |
|---|---|---|---|
| Leadership/collaborative skills | | | 90 |
| | 1 | Work effectively in a team situation as a leader or a participant | 91 |
| Professionalism | | | 100 |
| | 1 | Realize the need for continuing education through short courses, seminars, and conferences | 100 |
| Ethical/professional judgement | | | 90 |
| | 1 | Understand ethical issues involved in the formation of professional judgments of horticulture | 91 |
| Lifelong learning skills | | | |
| | 1 | Stay updated on latest developments in horticulture | |
| Visioning | | | - |
| | 1 | Understand the scope of horticulture and its relation to other disciplines | 86 |

[a] Percentage of acceptance of the identified competency after three rounds of surveys. The minimum threshold for inclusion of the competency in the curriculum was 75 %

in Germany, France, Great Britain, and the USA. The result is a bricolage (do-it-yourself) approach that integrates diverse characteristics of the influential models. The key outcome is a change to the bachelors-masters-doctoral degree process with each stage shortened to conform with the basic British-American pattern (3–4 years + 1 year + 3 years) compared with the open-ended approach favoured previously in much of Europe. The pace and effects of the Bologna protocol is reviewed with 10 years perspective in Germany by Mechan-Schmidt (2012). Students have adapted very quickly and with benefit since they can see the advantages of shorter periods of study; some staff have taken longer and with less enthusiasm.

Vocational and skills training should be available throughout a professional's career in what is now termed 'life-long-learning'. This is intended to ensure that professionals continuously improve and upgrade their knowledge and skills in line with scientific discoveries and technological developments. Most professional bodies now offer their members systems of "continuing professional development" (CPD). Not infrequently it is a condition of professional membership that knowledge and skills are demonstrably updated annually. These principles are enshrined by EU protocol in the twin Bologna and Copenhagen processes for higher education and vocational training. A review of provisions and advances resulting from these initiatives as they affect horticulture is provided by Sansavini (2010). Typical of the bricolage approach are the master degrees in horticultural science ranging from those offered by the universities in Great Britain to those obtained from institutions such as the Royal Horticultural Society (RHS) that offer the Master of Horticulture (M.Hort). The latter award testifies to the holder's competence in practice as well as his/her scientific and technical knowledge.

The need for practice based learning is regaining prominence in education and training. Frameworks for developing applied learning are emerging (Pridham et al. 2012) and are of particular relevance to subjects such as horticulture. Whereas previous generations of students were drawn largely from rural backgrounds and carried inherent cultural understanding of the practicalities of crop and animal production, many students are now drawn from urban backgrounds. Horticulture itself has also evolved requirements for much broader and deeper scientific understanding related to expanding environmental and social dimensions of the discipline. Thus the provision of horticultural education and training needs to be fit–for-purpose for such students whose professional careers will extend into the 2050s (Dixon 2005a, b, 2001, 1991; Scott and Dixon 2004).

In the twenty-first century, there has been renewed interest in vocational education and training (VET) amongst the international community (McGrath 2012). Governments in developed countries (e.g. Organisation for Economic Co-operation and Development, OECD countries) consider VET as a means of increasing the education and skills of their labour forces and hence raising wealth whilst diminishing the cost of social welfare. For similar reasons, international aid agencies view the encouragement of VET in developing countries as a means of moving families out of poverty and deprivation. VET is likely to be most successful where there is a partnership between educators and industry (Strawbridge et al. 2011). The horticultural industry in developed countries is increasing its demand for suitably qualified staff but the shortage of appropriately qualified staff limits business development in the sector. Industry requires the new knowledge and skills in technology and environmental management in order to capture opportunities and meet compliance.

## Bridging the Binary Divide

The view that graduates gain their education and qualifications solely from recognised universities is now challenged. Other organisations considered to be less academically rigorous and some vocational course providers are gaining 'degree awarding powers' and, in some instances, university status. These institutions may be drawn from what were previously local or regional government administered sectors or charities and not-for-profit companies. Even more radical is the rise of education provided by 'for-profit' organisations which have gained a substantial foothold in North America. This trend is spreading into Europe, especially Great Britain. The potential pitfalls of radical approaches have been described by Harkin (2012).

Bringing agrarian higher education and vocational education into closer partnerships in Great Britain was advocated by Gill (2007). Progression between these different levels of education offers substantial opportunities for students who may have been unsuccessful earlier in their lives. This process is described by St Hilaire and Thompson (2005) for North American students. Strengthening linkages from 2-year community colleges to 4-year universities has fostered the transition of more students into higher education and enhanced student diversity. Course evaluations have shown that 63 % of students enrolled in the combined class rated the combining of a university and community college class as an above average or excellent model of education. It was concluded that this collaborative approach for teaching landscape horticulture enhanced horticultural education and fostered student development.

Such an approach is especially suited to the discipline of horticulture, because it covers a unique array of extensive and intensive knowledge and skills matched in few other areas of academic and practical activity (Dixon 2005a, b). The discipline may be divided into three interrelating sectors: production, environmental sustainability and social sustainability, with significant intellectual overlap among these components. Thus, scholars require an understanding of a knowledge base emanating from all three. Making concentrations of teaching expertise available within single monotechnic institutions is becoming increasingly expensive and administratively difficult to sustain. Consequently, horticultural education is evolving away from the traditional solo institutional approach at the further vocational education level. Institutions are uniting in manners that exploit their particular strengths and approaches to teaching while retaining their own entities. Educational co-operation and collaboration may exploit partnerships across national boundaries. Within the joint Asian-European project DOCUMAP there is cooperation among three Asian universities where the horticultural and agricultural sector is being introduced in academic study programmes, in an effort to respond to demands on graduates seeking jobs in rapidly changing horticultural and food sectors in China, Indonesia and Vietnam (Li et al. 2011).

Some of the finest examples of the provision of practice based education and training come from botanic gardens and similar institutions. For example the Niagara Parks Commission (Ontario, Canada), established in 1885 (Klose and Whitehouse 2004), created a School of Horticulture in 1936. Hence the Botanical Garden has a role as both an educational institution for study and research, as well as a tourist attraction. This is seen elsewhere such as in the Royal Botanic Garden Kew's diploma course founded 1963. The advantage of these practice based qualifications is that they attest to a high degree of plantsmanship knowledge. Their merit can be increased immensely by linking them with formal science-based degree courses that provide students with a continuum of diploma and degree qualifications, as was devised by the University of Strathclyde, Glasgow and Royal Botanic Garden Edinburgh (Dixon 1993). On completion of these courses, students possessed an array of valuable qualifications attesting to their robust scientific and technical knowledge plus substantial practical abilities in plantsmanship. This approach reached its zenith with Dutch horticultural provision. In the Netherlands, a national agricultural research and education system was established in 1876 (Spiertz and Kropff 2011). Initially, the emphasis was strongly on education and applied research. The higher professional school for teaching agriculture, horticulture and forestry at Wageningen was admitted to the status of technical university ('Hoogeschool') in 1918. Complementary to the university, a wide array of discipline-oriented research institutes and commodity-oriented research stations were founded; especially after World War II. Associating these with university education and providing students with opportunities for study in specialist locations where particular sectors of the Dutch industry were concentrated produced graduates with extensive and intensive practical and academic knowledge that allowed the Dutch horticultural industry opportunities for gaining commercial pre-eminence worldwide. A radical restructuring into one organization for research and education—Wageningen University and Research Centre—occurred in 1998 and provides for the needs of modern scientific approaches while still offering to produce wealth creators for industry.

## The Education Options

Worldwide, horticultural education and training have generally involved one of three sectors: the universities and/or polytechnics, which provide undergraduate and postgraduate education, and where much of the horticultural research is undertaken; the community or technical colleges, where vocational and skills level training are provided; and adult education and training, which can be offered by all sectors for industry and personal use. Across North America, undergraduate education in horticulture has been dominated by academic programs at research-focused land-grant universities, non-land grant universities, community and technical colleges, as well as cooperative extension and other agencies that offer adult, continuing, or lifelong education. In the United Kingdom, education is provided at three levels; the universities, the colleges of further and higher education, and institutions and agencies that offer adult, continuing or lifelong education. In Australasia, baccalaureate and postgraduate training in horticulture and horticultural science has been offered through the university system, with VET institutes and other registered providers offering apprenticeships, national certificates, and diploma programs (McSweeney et al. 2009; Rayner et al. 2009; McEvilly and Aldous 2009). All levels offer adult, continuing, or lifelong education, with universities and the VET institutes offering industry and other professional short courses.

In recent years, the national financial pressures in many countries have resulted in the restructuring and amalgamation of higher education departments and academic programs. This has led to a loss of horticultural identity that compromises undergraduate courses serving the discipline. Of the more than 200 land-grant institutions in the United States, only 20 currently offer an undergraduate degree with the word "Horticulture" or "Horticultural" in the name. Most of the land-grant institutions have consolidated horticulture, agronomy, and in some cases, soil science, landscape architecture and other disciplines into more general plant science departments and/or degrees. In Australia there has been a reduction from 23 to just 9 university campuses that provide agricultural and horticultural education courses (McSweeney et al. 2009; Rayner et al. 2009), with only one university currently retaining an undergraduate degree in horticulture (Pratley 2012b). In Great Britain 16 institutions offer courses which are officially labelled "horticulture" (Anon 2012a). Whilst there is diversity in offerings, few, if any, provide an education directed towards careers in the production horticulture industry. Many are oriented towards landscape design and management. As institutions came under financial pressures there was a regrettable tendency to chase those topics which provided "bums-on-seats" and hence in Great Britain at least the Further Education (FE) colleges which edged into Higher Education (HE) have promoted for example courses in "garden design". The result is a large number of qualification holders chasing a diminishing pool of employment in a very specialised aspect of horticulture and with little intellectual capabilities which would allow them to move elsewhere when the career market alters.

Traditionally, agricultural and horticultural science degrees have been 3 to 4 years in duration, with or without an Honours year, are generally taken on a full and/or part-time basis, and are largely funded through the Federal Government of the country. While some programs have previously led to a degree in horticulture and horticultural science, the Australian university system now offers only a broader agriculture, science or biotechnology program, where horticulture is offered as a major, minor or single unit. In Australia, only 7.8 and 4.6 % of the population have undergraduate level qualifications in agriculture and horticulture respectively, whereas in the United States approximately 25 % of staff in the agricultural industry have BS degrees (Pratley 2012a, b). Postgraduate qualifications in agriculture, horticulture and other allied areas are also available at the Graduate Diploma, Masters, and PhD levels in Australia, and at the Masters and PhD levels in the U. S.

In recent years there has been a range of collaborative efforts by universities, VET providers, state and federal government departments, and industry associations to assist in retaining young people in horticulture and in delivering courses to meet the needs of students and employers. For example, the University of Western Sydney collaborates with a VET provider in the delivery of their landscape horticulture sub-major, the School of Agriculture and Food Sciences at the University of Queensland collaborates internationally with Pennsylvania State University, USA to provide web-based higher education training in turf management, and the Central Queensland University has collaborated with a Queensland State Department in developing a jointly funded Chair in Vegetable Science. The United States Department of Agriculture (USDA) supports a federally-funded Higher Education Challenge grant program that requires inter-institutional collaboration in proposals that focus on increasing the number and diversity of students who pursue and complete a postsecondary degree in the food and agricultural sciences, and to enhance the quality of secondary and postsecondary instruction in order to help meet national food and agricultural science needs. Although not focused solely on horticulture, this program does emphasize systems approaches to sustainable food production and security.

Worldwide, the VET programs are commonly available at technical institutions and community colleges, and strengthen the skills-training associated with the horticultural trades. In Australia in recent years, this sector has undergone several

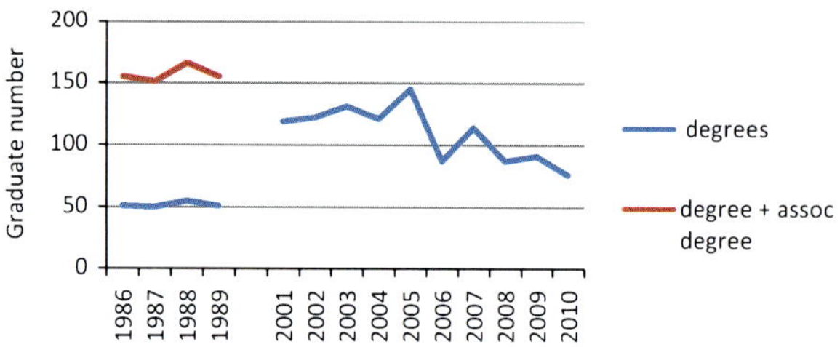


**Fig. 32.5** Graduate completions in horticulture, 1986 to 2010, from higher education institutions in Australia. (Pratley 2012b)

reviews, but generally has experienced growth in program diversity, provider and student numbers across the broad field of agriculture, horticulture and the environment (NCVER 2007; Puflett 2009). The United States has experienced similar increases with their two and four year horticultural programs offered through the non-land grant community and technical college system. Currently there are almost 300 two year horticulture programs offered through these institutions (Ingram 2012). Traditionally the VET system offers apprenticeships in horticulture and traineeships in the more service-oriented vocations in Australia. Apprenticeships typically last for up to four years and traineeships only one or two years in specific vocational areas nursery production, arboriculture, turf management, landscape construction, and parks and gardens management.

Universities, community colleges, VET institutes, and other non-profit agencies can all offer adult, continuing, or lifelong education in horticulture where there is a need to educate better the industry and general public about the scope of horticulture in the community and wider industry. The objective of many of these education providers is to achieve learning outcomes that have been designed for personal enrichment and skills development for both industry and personal use.

There have been considerable shifts in education and training of horticulture and horticultural science since the mid twentieth century, particularly in many western countries. This period contrasts with the previous 100 years, where there was growth in student numbers, the range of career opportunities, and stability in the number of Universities providing undergraduate horticulture training. Since the 1980s, institutions have experienced declining undergraduate and graduate student numbers (Darnell and Cheek 2005; Guisard and Kent 2009; Pratley 2008; Pratley and Copeland 2008; Pratley 2012b), a resultant industry shortfall in graduates (Chapman 2006; Collins and Dunne 2009; Guisard and Kent 2009; Pratley 2012b), a decline in the use of horticulture in course titles and a consolidation of academic programs (Looney 2004; McSweeney et al. 2009; Rayner et al. 2009), departments and faculties (Falvey 1998; Coons 2001; Pratley 2012b).

Figure 32.5 shows the decline in horticulture graduate numbers for Australian universities over the period 1986 to 2010.

Although the number of undergraduate and graduate degrees awarded has been decreasing over the past years, the demand for trained graduates is expected to grow. The USDA estimates that between 2010 and 2015, almost 20% of its staff will retire (Gerwin 2010). Industry is also expecting to increase staff; for exam-

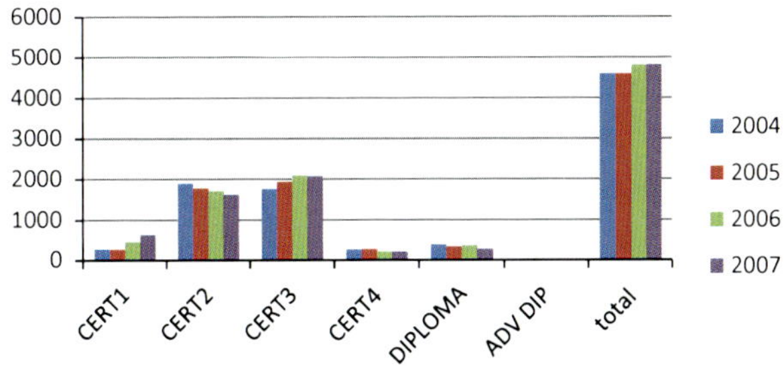


**Fig. 32.6** Completions in vocational horticulture courses in Australia 2004–2007. (Pratley 2012b)

ple, Pioneer Hi-Bred International seed company has recently hired more than 700 scientists and support staff globally (Gerwin 2010). Corporations such as Monsanto are instituting in-house internship programs to aid students in developing their scientific skills, with the ultimate goal of obtaining a permanent position at Monsanto (Gerwin 2010). In particular, job growth in the horticulture industry is expected in fruit and vegetable production, speciality crops that have medical or energy applications, landscape and turf production, and organic farms (Anon 2009). Thus, the need for trained horticulturists is growing and it presents a challenge to educational institutions to respond to that need. In addition the industry needs to promote career opportunities within its industry and encourage students through its respective education and training system.

The VET sector has responded more positively than higher education to industry need in Australia, particularly in the traditional horticultural trades, by introducing competency based workplace training, focussing on one to two qualifications in the general horticulture training package, stimulating demand for training at higher levels within the VET sector, making provision for distance education, external, off-campus delivery of units, and reinvigorating the advanced standing mechanisms into higher education horticulture (McSweeney et al. 2009). The challenges ahead are to find out if these national Australian qualifications meet the needs and demands of the workplace or horticultural businesses and whether the continued financial pressure from the Australian State and Federal Government sources will impact on the future restructuring of the VET sector. Horticultural industry-based training works because of the training system and regulations that are provided to deliver relevant national qualifications that meet the skills needed for a growing economy (Puflett 2009). Data by Pratley (2012b) in Fig. 32.6 indicate that the industry has about 35% of its workforce with vocational training qualifications and that the number of completions is sufficient to maintain that level. The qualifications are not uniform however across the various components of the horticulture industry. An example of collaboration between horticultural academics at the University of Reading, The Horticultural Trades Association (HTA), and the levy raising board Horticultural Development Company (HDC) in Great Britain is providing a suitable mechanism for the provision of VET into hardy nursery plant raising companies. Here academia and industry are working collectively to ensure that staff gain the knowledge and skills needed by particular sectors of the production industry (G R Dixon, personal communication).

## Constraints to Attracting Young People into Horticulture

Coutts et al. (2004), Bogers (2006), and Guisard and Kent (2009), *inter alia*, suggest that horticulture has a poor image, with the negative perceptions associated with poor remuneration relative to levels of profitability in the industry, the extent of manual work, and a lack of awareness of the range of career opportunities (Cecchettini et al. 1992; Matthews and Falvey 1999). This image was cited by academic programme administrators as the most important factor leading to lack of interest in agricultural sciences as a career by USA high school students (Anon 2009). Public understanding of horticulture is poor and many people are unaware of where their food comes from and the multi-faceted nature of the horticulture industry. Horticulture was listed as the second most "useless degree" by Newsweek magazine (LaWell 2011). Similar experience has occurred in Great Britain where there has been a tendency for career advisory staff to treat horticulture (in their eyes "gardening") as suitable for the mentally challenged and less able students. This damage has been compounded by multitudes of television programs which simply 'dumb-down' the whole discipline propelled by naive "celebrities". As a result, parents and students see horticulture as a "low-pay-low-skill" job, even though it requires considerable knowledge interwoven with practical abilities for those who wish to make progress beyond the most basic grades.

Falvey (1998) suggested that the decline in number of students entering horticulture programs was associated also with a decline in demand for science more generally. Although horticulture will always retain a "hands on" component, modern horticulture has become more intimately associated with the sciences and now requires a more highly skilled, innovative and versatile workforce to manage the technological, financial and marketing systems along the production and environmental horticulture supply chain. Looney (2004) suggests finding a way to position horticulture more effectively as an essential life science in modern society that requires further investment in education and training to answer these challenges. If these economic and social trends continue and enrolments in horticulture do not improve, there will be continued compromise in programs and likely cuts in staffing levels, all of which will seriously damage the economic development and future of innovative horticultural industries.

The definition of horticulture also contributes to the educational decline (Darnell 2006), as the traditional description of the "art or science of cultivating gardens" is now outdated. Horticulture has been better described as the science, technology and business of intensive plant cultivation for human use (Anon 2012d). Alternatively horticulture could be described as the manipulation of plant growth and reproduction for the sustainable generation of profit in industry, the environment or social context. This lack of definition and the inability to communicate the meaning of "horticulture" may be contributing to the lack of knowledge about the fields of study and career opportunities. Many students today not only want careers that are financially rewarding, but perhaps increasingly, they want careers that are rewarding in terms of their personal values. These values may include work-life balance, opportunities for creativity, and environmental stewardship (Anon 2009d). The conflict in understanding what "horticulture" is and the relationship between "horticulture" and "gardening" has contributed to the poor image of the former. It is becoming evident however that mature, returning, second-career students can make a very effective contribution to the discipline. They bring knowledge and experience in other spheres concerned with science, business and technology which they wish to deploy into a new plant-based occupation.

## Strategies for the Attraction and Retention of Students

Strategies to attract and retain students should include (1) development of smarter partnerships with industry, institutions, and government to promote horticulture and horticultural science, (2) continuous evaluation, revision and marketing of curricula by academic institutions, (3) increased use of different delivery formats, i.e. distance education, and (4) development of a strong advocacy agency that engages with industry, institutions and government on horticultural education and training agenda issues.

Thompson and Russell (1993), Russell et al. (2004) and Russell et al. (2006) found that specific agricultural science programs in secondary schools carried out in partnership with universities, government organizations and industry led to increased enrolments and higher student retention rates in university agricultural science courses. Programmes included school visits to senior secondary school science classes by selected academic staff, industry placement scholarships for short internship-type activities, professional development activities for senior secondary school science teachers, and production of teaching packages to provide teachers with appropriate resource material. The outcomes observed by Russell et al. (2006) are supported by Stone et al. (2005), who suggested development of specialized information programs in partnership with industry and government agencies to promote horticulture to high school students.

Encouraging better partnerships between universities and high school career counsellors and securing greater involvement in student recruitment by alumni and industry have been cited in a Canadian Labour Market Information on Recruitment and Retention in Primary Agriculture Report (2009) as important factors that attract young people into horticulture. Bell and Biddulph (2009) also found that previous experience in, or connections with agriculture, an interest in science and learning, the possibilities of real-world employment, and the potential lifestyle benefits were important in studying agriculture at university. Further, they concluded that to inspire young people to take up a career in agricultural science required a strong commitment from the industry to secure the levels of remuneration comparable with other equally qualified professions, to improve the career path by providing for job security and continuity, and to promote the wider social issues that are associated with these work opportunities. Partnerships with industry, educational institutions, and/or government agencies can supply these experiences, making it more likely that students will enter into horticulture programs at the college or university level.

Successful programmes that have inspired young people towards a career in horticulture and horticultural science range from industry websites that promote the benefits of careers in horticulture (Cooper 2009; McEvilly and Aldous 2009; Aldous and McEvilly 2011) and short videos that showcase the diverse career options within horticulture and horticultural science (Chapman 2006) to an alumni online mentoring program (Lineberger 2009a, b, c). Coutts et al. (2004) suggested a whole-of-horticulture employment service that not only engaged the employer with potential employees through a website, but also listed vacancies for interested persons to lodge their resume.

Most Government Departments, University and VET providers and the wider horticultural industry in Australia have initiatives that communicate information about postgraduate, undergraduate and vocational and technical education courses. Successful institutional initiatives in Australia range from programs like field days and expo activities to career markets. Many universities are involved with programs such as "Science in School", "Science in the Bush" and "Science in the Suburbs". Other universities provide for Year 11 agriculture camps, Supermarket Botany road shows and a Giant Science program to interest primary school students. A successful event has been the "Degree in a Day Talented Students" programmes offered for Year 11 and 12 students as well as field days in local rural locations where they offer small science-based activities. The successful video production of "Pick of the Crop" features New Zealand horticulture and promotes the diverse range of positions and vocations.

Many USA universities have similar programmes, including Summer Science Enrichment programmes for high school students, agricultural career expos, alumni recruitment activities and K-12 programmes that provide resources to enhance agriculture curriculum in elementary and secondary schools. "Agriculture in the Classroom" is a programme coordinated by the United States Department of Agriculture (USDA) to provide a network that seeks to improve the awareness, knowledge, and appreciation of agriculture among K-12 teachers and their students (Anon 2012e). The programme is carried out in each state by representatives from industry, education, and government. There are also less formal opportunities to expose precollege students to agriculture in the form of youth enrichment programmes such as 4-H and the National Future Farmers of America.

Many USA universities have articulation agreements with state community colleges, where any graduate of a state-approved public community college is eligible for admission to a state university if the student has completed the prerequisites for their intended major. For example, the College of Agricultural and Life Sciences at the University of Florida has a long tradition of working with community college students and advisors to aid in transferring to the University, particularly in applied agriculture majors such as horticultural sciences (Anon 2012e). Community college or vocational education programmes that foster progression of students into university programmes offer substantial opportunities for students who may have been unsuccessful earlier in their lives.

The Report "Transforming Agricultural Education for a Changing World" (Anon 2009) recommended that academic institutions with undergraduate programmes in agriculture develop a range of steps with industry and government that better meet the needs of students, employers, and the broader community. Similarly the International Society for Horticultural Science (ISHS) on-line "Forum: Future of Horticultural Science within Academia" discussed the opportunities associated with the research, teaching and extension of horticultural science (Bogers 2006). In the United Kingdom, the "Future of UK Horticulture" report (Anon 2006) recommended, *inter alia*, that a radical change was required in terms of the use of higher levels of technology, better wages, and improved career prospects in the education/training in horticulture for improved horticultural production performance.

Smarter partnering should include changing the perception of horticultural science to a more dynamic, rewarding career. Such careers should demonstrate the positive impact that horticulture has on the economy and quality of life through better nutrition and health, improved environmental and social sustainability, as well as enhanced community amenities for sport and recreation.

Although partnering is an important strategy for addressing the problem of declining student enrolment in horticulture, this strategy will be unsuccessful unless colleges and universities develop a more innovative and flexible horticulture curriculum. Reganold et al. (2011) concluded that a "transformative approach" that builds on the understanding that agriculture is a "complex socioecological system" is required to improve USA agriculture and that whole system redesign—not single technological improvements—is needed. So, too, must the curriculum undergo continuous assessment and "whole system redesign" in order to meet the current and future challenges facing horticulture and horticultural science? Evidence suggests that many students are interested in the science, business, environmental, and social issues related to food production, but "educators have not helped students…make the connection between those issues and a degree in agriculture" (Anon 2009).

Academic institutions must maintain relevant curricula in the face of the current focus on multi-disciplinary scientific approaches to solving horticultural problems, rising concerns for environmental issues, increasing consumer demand for organic and/or local food, and the changing demographics of students. New horticultural majors, ranging from Organic to Sustainable Cropping Systems to Food Systems to Plant Molecular/Cellular Biology, have been developed in several land-grant universities in the US to address these issues. But curriculum must remain flexible, and faculty must remain committed to assessment and revision, so that future issues in horticulture continue to be addressed.

Opportunities outside of formal classroom settings must also be available to students in order to increase retention. Research experiences for undergraduate students have been shown to help recruit and retain students in their major (Anon 2009) and the precedent for this has been established in basic science, through the National Science Foundation, National Institutes of Health, and the Howard Hughes Medical Institute (HHMI). The Horticultural Sciences Department at the University of Florida has an established history of recruiting students into their Plant Molecular/Cellular Biology option within the Horticultural Science degree through the HHMI program. Study abroad programs are another avenue that attracts students into horticulture, particularly those interested in international agriculture experience.

Employers are seeking not only students knowledgeable in scientific content, but also those skilled in "transferable competences, including interpersonal communication skills, critical thinking skills, writing skills, and computing skills" (Anon 2009). Team-building, leadership, and conflict resolution skills are also being sought, as is the ability to work in diverse, sometimes global, environments. Thus, horticultural curricula and/or extracurricular activities should encompass aspects of all these skills.

## Distance Education and Electronic Delivery

Opportunities for studying at a distance from formal institutional educational facilities and at a pace controlled by the individual student have many academic and practical attractions. This process may be defined as structured learning where the student and teacher are separated by space and possibly time. This is the fastest growing form of domestic and international education across most disciplines (Gunawardena and McIsaac 2004). Distance learning is not new, as from the late 1800s the University of Chicago offered correspondence courses. Australia developed its schools of the air as an offshoot of the Radio Flying Doctor Service from 1946 onwards. Currently, this still serves the needs of people who inhabit 1.5 million $km^2$ of Australian Outback. Even in a country as confined geographically as Great Britain school radio broadcasts were important supplements to formal education in the years immediately after World War II. This was followed in the 1970s with the establishment of the Open University, and in 1986 by Wisconsin University's innovative development of distance learning. Opportunities offered by electronic technology were identified as ready for exploitation by Kerry and Isaksson (2000) in their report to the US Congress "The Power of the Internet for Learning".

Technology-mediated instruction is now a term applied collectively to a variety of methods that use digital technology in its many forms to facilitate the acquisition of knowledge by students. Increasing opportunities for distance learning have many academic and practical attractions and is an important recruitment tool especially for place-bound students. Distance education is the fastest growing form of domestic and international education across most disciplines (Gunawardena and McIsaac 2004). Technology-mediated instruction is now a term applied collectively to a variety of methods that use digital technology in its many forms to facilitate the acquisition of knowledge by students. Early adopters used the World Wide Web (www) for archiving and delivering instructional materials such as class notes, slide sets, and assignments. Concomitant with the development of advanced web browsers and wider availability of broadband connections, a wide array of interactive tools, including streaming video delivery and desktop conferencing, have enabled instructors to create a virtual classroom environment that is independent of distance and time. Horticultural teaching and learning have grasped the opportunities offered by these facilities (Dixon 2005b; Lineberger 2009a, b, c). Electronic technology may be used for simple short courses, individual assessments of teaching and learning as described by Mason (2005), or for extended provision of complete courses. Excellent courses in specialty topics such as arboriculture and turfgrass science have been developed for undergraduates who wish to pursue their studies on a part-time, home-based basis as developed by Myerscough College, Lancashire, England (www.myerscough.ac.uk). Similar developments in this specific sub-discipline are reported from the USA. Access to specialized course content such as Introductory Turfgrass Science (Bigelow 2009) indicates that this online version of the course is a suitable substitute for those students unable to take the traditional lecture version but wishing to gain fundamental knowledge related to turfgrass science.

In the US, regional teaching collaboration uses web technology to design, develop, and deliver academic coursework (Lineberger 2009a, b, c). The growing importance of providing continuing education and "re-education" of former students who are extending and changing careers means that traditional semester-long, on-campus approaches to degree programmes are insufficient. Currently, students who are place-bound by work or family obligations require distance-delivered degree programmes. Furthermore, the rate of change of horticultural production technology has increased such that all practitioners must constantly update their knowledge and skills.

Some of the problems inherent in electronic teaching and learning as used at the von Humboldt University, Berlin, are described by Tietze and Schmidt (2008). Specialized technical and methodological skills for handling information and using communication technology are required from both students and lecturers. Teachers are required to revise their teaching approaches. It is not sufficient simply to attempt to transfer existing face-to-face lecture materials straight into an e-format. All material including assessments and examinations requires to be changed in order to suit the new leaning environment. Institutional managers must not be allowed to think that electronic teaching and learning is simply a cheaper alternative to face-to-face provision. When used effectively electronic teaching and learning requires high grade pedagogic knowledge and skills of no lesser quality and quantity to other systems. Communication gaps, a lack of human resources, and the inertia of students are problems which will emerge once these systems are put in place. But once students and teachers embrace this technology it can provide a very effective means of knowledge exchange and there is some unquantified evidence that the students collaborate and communicate with each other more than in conventional education. Communication gaps, a lack of human resources, and the inertia of students are problems which will emerge once these systems are in place. Such systems are appropriate for horticultural teaching and learning at all levels. For example, computer simulation software has been used to model the influence of climate/environment on the phenology and growth of horticultural crops (Kobayashi 2003). The software enabled students to run multiple rapid and easy simulations with attractive graphs and tables. This made active learning possible by providing hands-on experience for students. Similarly, an understanding of crop genotype x environment interactions using simpler modelling is described by McKay et al. (2005) and McKay and Fisher

(2005). Their systems illustrated relationships between crop density and yield using carrots and maize, respectively.

Manual skills training may also be provided by distance learning, as described by Hennigan and Mudge (2004) and Mudge and Hennigan (2001). Their web/CD-based course, "The How, When and Why of Grafting", was used to teach the hands-on horticultural skill of grafting to non-traditional students in collaboration with Cornell Cooperative Extension. *Hibiscus* plants, grafting knives, and other grafting supplies were provided to the students in advance for the grafting laboratory exercises. Students and instructors independently evaluated student grafts after four weeks based on pre-defined criteria of grafting success. Student self-evaluations indicated that it is possible to learn grafting at a distance in the absence of face-to-face instruction. Similarly Cornell University's "The Nature of Plants" programme is aimed at acquainting students with nature through experiential learning (Bauerle and Park 2012).

The Extension Master Gardener (EMG) volunteers are a key means for disseminating horticultural information to the U.S. public. Electronic techniques that increase the capacity and effectiveness of EMG volunteers have been identified. Regional web-based, advanced training was used to expand volunteers' knowledge, increase their effectiveness, and reduce the costs normally associated with face-to-face training (Anon 2012c). This type of approach can be of immense value in countries such as India and China with huge rural populations who need access to education. Ramachandraiah (2000) describes course content and mode of instruction of a Certificate Program in Mushroom Cultivation offered in Open University format at Hyderabad University, Andhra Pradesh, India. Extensive use of electronic technology as described by Dixon et al. (2003) and Bauske et al. (2011) is valuable in practice for extension and advisory services. Web delivered services included information on crop monitoring and protection, weather forecasting, yield evaluation and accountancy and taxation services. Future technologies will be of increasing benefit in the teaching, research and extension of horticulture and horticultural science.

A further strategy to attract and retain students is to develop advocacy agencies for horticultural education. Bogers (2006) reported that, from the European perspective, stronger advocacy measures of the horticultural sector are needed and the industry should take responsibility for making its educational needs clear to government and other funding bodies. Although many agencies advocate and lobby with respect to horticulture production, few advocate in respect of the need for, and improvements in, agricultural and horticultural education and training. One such Australian body is the Australian Council of Deans of Agriculture (2009) that plays an important ambassadorial role in horticultural and agricultural education and has engaged effectively with industry, institutions and government for the common good (Pratley 2012a, b).

## Conclusions

This paper has endeavoured to canvass the issue of the need for better education and training in horticulture and its provision. Such a review has generated the following outcomes:

- to provide for undergraduate and postgraduate education in horticulture and horticultural science there need to be effective strategies and partnerships put in place by the institution, the government and most importantly the industry;
- horticulture must regain its position as a subject recognised for its intellectual rigour and its key role in solving the major social and economic problems facing mankind resulting from climate change, the erosion of natural biodiversity, over population and diminishing food supplies;
- successful programmes and case studies need to be developed to provide better information to high school career counsellors, improve the image of horticulture and encourage greater involvement from alumni and the industry in recruitment such that attraction and retention of young people into horticulture and horticultural science are enhanced;
- the providers of horticultural education must ensure that courses are offered using the latest forms of electronic technology since the coming generations of students will expect nothing less;
- initiatives to improve career aspirations, promote greater awareness and recognition of the profession in the wider community, ensure improved levels of re muneration, promote the social features of the profession, and provide for job security and continuity need to be developed;
- educational institutions, government, and industry need to work together to improve communication about horticulture and horticultural science, in order to attract enrolments of more and talented students to ensure the future prosperity and sustainability of the industry; and,
- horticulture curricula must be continuously assessed and revised so that they remain relevant to future challenges facing the industry.

**Acknowledgements** The authors wish to acknowledge Ms. Jane Wilson, volunteer archivist, University of Melbourne, 3000 Victoria Australia in assisting with images from the University of Melbourne Burnley Archives, as well as Dr Ashley Elle (née Basinger), Texas Tech University College, Lubbock Texas 79409-2122 USA for the abbreviated table of horticultural competencies.

## References

Aldous DE (1990) Horticultural education and training in Australia. Prof Horticulture 4:36–40
Aldous DE, McEvilly G (2011) Horticultural education and training futures in Australia and New Zealand. Acta Hort 920:63–69
Aldous DE, Offord CA, Silk JP (2011) The origin of horticulture in Australia: the early European colony in Sydney 1788–1850. Chronica Horticulturae 51(4):9–13

Anon (2006) Future of UK Horticulture A case study analysis and overview of the UK horticultural production industry and its future over the next 10–20 years. National Horticultural Forum and Promar International. http://www.hortforum.net/uploads/7/2/9/5/7295387/nhf_future_of_hort_pdf3.pdf. Accessed 11 June 2012

Anon (2009) Transforming agricultural education for a changing world. http://dels-old.nas.edu/ag_education/report.shtml. Accessed 9 July 2012

Anon (2012) Agriculture in the Classroom. National organisation agriculture in the classroom. Washington, DC. http://www.agclassroom.org. Accessed 1st Oct 2012

Anon (2012c) Extension Master Gardener. http://www.extension.org/mastergardener. Accessed 1 October 2012

Anon (2012d) Horticulture, Wikipedia. http://en.wikipedia.org/wiki/Horticulture. Accessed 26 Oct 2012

Anon (2012e) Horticulture courses. Universities Clearing and Admissions Service (UCAS). http://search.ucas.com. Accessed 18 Sept 2012

Australian Council of Deans of Agriculture (2009) http://www.csu.edu.au/special/acda/. Accessed 15 June, 2012

Basinger AR, McKenney CB, Auld D (2009) Competencies for a United States horticulture undergraduate major: a national Delphi study. HortTechnology 19(2):452–458

Bauerie TL, Park TD (2012) Experiential learning enhances student knowledge retention in the plant sciences. HortTechnology 22(5):715–718

Bauske EM, Kelly L, Smith K et al (2011) Increasing effectiveness of cooperative extension's master gardener volunteers. HortTechnology 21(2):150–154

Bell L, Biddulph B (2009) For love not money: insights on the career choice of early career agricultural scientists. Agr Sci 21(2):22–29

Bigelow CA (2009) Comparing student performance in an online versus a face to face introductory turfgrass science course—a case study. N Am Coll Teach Agr (NACTA) J 53:2 unpaginated

Bogers RP (2006) The future of horticultural science and education: a European perspective. Chronica Horticulturae 47(2):4–6

Bokelmann W (2007) The German approach of horticultural education. Acta Hort 762:401–406

Canadian Labour Market Information on Recruitment and Retention in Primary Agriculture (2009) http://www.cahrc-ccrha.ca/sites/default/files/files/publications/LMI-Recruitment-Retention/Labour%20Market%20Information%20Final%20Report.pdf. Accessed 15 July 2012

Cecchettini CL, Sommer R, Leising JG (1992) Australian students' perceptions of agricultural careers. J Agric Educ 33:30–35

Chapman JC (2006) Submission to the enquiry into rural skills training and research. Australian Society of Horticultural Science Inc. http://www.aushs.org.au. Accessed 9 April 2010

Collins RJ, Dunne AJ (2009) Can dual degrees help to arrest the decline in tertiary enrolments in horticulture: a case study from the University of Queensland, Australia. Acta Hort 832:65–70

Coons JM (ed) (2001) The fate of horticulture in merged departments. HortTechnology 11(3):398

Cooper B (2009) Horticultural careers—your guide to courses in Australia. http://horticulture.realviewtechnologies.com/?cdn=0&xml=Courses_and_Careers_in_Horticulture. Accessed 15 June 2012

Coutts J, Stone G, Casey M et al (2004) Strategy to attract young people to horticulture. Final Report March Horticulture Australia Limited Project Number AH2021

Darnell RL (2006) The future of horticultural science within academia. Chronica Horticulturae 46(2):8–9

Darnell RL, Cheek JG (2005) Plant science graduate students: demographics, research areas, and recruitment issues. HortTechnology 15:677–681

Delwasse L (2008) The Apocalypse Tapestry of Angers. Éditions du Patrimone, Centre des Monuments Nationaux. Paris

Dixon GR (1991) Horticulture education and training. Chronica Horticulturae 49(4):51–52

Dixon GR (1993) Opportunities for education in horticulture in the United Kingdom through course articulation and student progression. Acta Hort 350:7–18

Dixon GR (2001) Institute of horticulture position paper: education provision for horticulture. Horticulturist 10(4) autumn:13–16
Dixon GR (2005a) A review of horticulture as an evolving scholarship and the implications for educational provision. Acta Hort 672:25–34
Dixon GR (2005b) The challenge of distance learning in horticulture. Chronica Horticulturae 45(4):5–8
Dixon GR, Hardiman L, Payne V et al (2003) HorTIPS—a new concept in horticultural knowledge transfer. Acta Hort 641:125–130
Falvey JL (1998) Are faculties of agriculture still necessary? Australian Academy of Technological Sciences and Engineering No. 103 July/August. www.atse.org.au. Accessed 1st Oct 2012
Gerwin V (2010) Cultivating new talent. Nature 464:128–130
Gill M (2007) Review of the provision for land-based studies. Higher Education Funding Council for England (HEFCE) Swindon by JM Consulting and SQW Ltd. http://www.hefce.ac.uk/whatwedo/kes/sis/land-basedstudies/. Accessed 1 Oct 2012
Guisard Y, Kent J (2009) Optimisation of undergraduate horticulture course design at Charles Sturt University (Australia): a structure for the future. Acta Hort 832:87–94
Gunawardena CN, McIsaac MS (2004) Distance education. In: Jonassen DH (ed) Handbook of research on educational communications and technology. Chapter 14 Lawrence Erlbau. http://www.help.senate.gov/imo/media/for_profit_report/Contents.pdf. Accessed 15 Nov 2012
Hendrick UP (1950) A history of horticulture in America to 1860. Oxford University, New York
Hennigan K, Mudge KW (2004) Effect of interactivity and learning style on developing hands-on horticultural skills via distance learning. Acta Hort 641:85–92
Ingram D (2012) Calling all horticultural faculty at non-land grant institutions. Reflections. ASHS Newsletter 28(4):3, 10
Jin X (1994) Mother of gardens strives for a bright future with flowers: a report on the development of ornamental horticulture in China. J Korean Soc for Hort Sci 35(Suppl):155–160
Jung W (2012) Philosophy of science and education. Sci Educ 21(8):1055–1083
Kerry B, Isaksson J (2000) The power of the internet for learning. Report of the Web-Based Education Commission to the President and the Congress of the United States. http://www2.ed.gov/offices/AC/WBEC/FinalReport/WBECReport.pdf. Assessed 5th Dec 2012
Klose E, Whitehouse D (2004) The Niagara parks commission school of horticulture. Acta Hort 641:145–146
Kobayashi K (2003) Using computer simulation software to enhance student learning. Hortscience 38(5):695
LaWell C (2011) The state of horticulture departments. Lawn & Landscape. GIE Media Richfield US, Sept pp 82–89
Li B, Li XX, Hofmann-Souki S (2011) Beyond lectures and exercises—establishing mentorship to complement professional development of students in Asian horticultural and agricultural study programmes. Acta Hort 920:55–62
Lineberger RD (2009a) Technology-mediated instruction: shifting the paradigm of horticultural education. Acta Hort 832:107–112
Lineberger RD (2009b) Evolution of web-based collaborative learning environments in horticulture. Acta Hort 832:13–18
Lineberger RD (2009c) IT evolution of web-based collaborative learning environments in horticulture. Acta Hort 832:3–18
Looney NE (2004) Future of horticultural science within academia. Acta Hort 44:13–13
Mason J (2005) Applying technology to horticultural education. Acta Hort 672:47–55
Matthews B, Falvey JL (1999) Year 10 students' perceptions of agricultural careers: Victoria (Australia). J Inter Agricult Extens Educ 6:55–67
McEvilly G, Aldous DE (2009) Guiding young people to horticulture. Chronica Horticulturae 50:16–18
McGrath S (2012) Vocational education and training for development: a policy in need of a theory? Inter J Educ Dev 32(5):623–631

McKay BR, Fisher PR (2005) Interactive studies on the internet: the Ramosus maize tool. Acta Hort 672:217–225
McKay BR, Reid J, Love R (2005) Virtual carrots: an online tool for teaching yield x density relationships. Acta Hort 672:227–231
McSweeney PK, Raynor K, Rayner J et al (2009) Developments in Australian horticultural vocational education. Acta Hort 832:121–130
McVittie J, (2008) National qualifications: a short history. Policy and new products research report No. 3. Scottish Qualifications Agency (SQA), Edinburg, pp 21
Mechan-Schmidt F (2012) German palates slow to adapt to Bogna's quick-cook recipe. Times Higher Education Supplement no 2067 issue of 13–19 Sept 20–21
Mudge K, Jannick J, Scofield S (2009) The history of grafting. In: Janick J (ed) Horticultural reviews vol 35. Wiley, pp 437–493
Mudge KW, Hennigan K (2001) The how, when and why of grafting, a distance learning approach, emphasizing computer-mediated hands-on learning. Hortscience 36(3):432
NCVER (2007) National Centre for Vocational Education. Industry and Training. http://www.ncver.edu.au/publications/1806.html. Accessed 11 Nov 2012
Parker LH (2005) Investing in students' learning at three levels: national, institutional and individual support the scholarship of teaching. Acta Hort 672:17–24
Powell JJ, Bernhard N, Graf L (2012) The emergent European model in skill formation: comparing higher education and vocational training in the Bologna and Copenhagen processes. Sociol Educ 85(3):240–258
Pratley JE (2008) Workforce planning in agriculture: agricultural education and capacity building at the crossroads. Farm Policy 5(3):27–41. (August Quarter)
Pratley JE (2012a) Professional agriculture—a case of supply and demand. Paper 1. http://www.csu.edu.au/special/acda/papers.html. Accessed 11 Nov 2012
Pratley JE (2012b) The workforce challenge in horticulture. Agric Sci 24(1):26–29
Pratley JE, Copeland K (2008) Graduate completions in agriculture and related degrees from Australian universities-2001–2006. Farm Policy 5(3):1–10
Pridham B, O'Mallon S, Prain V (2012) Insights into vocational learning from an applied learning perspective. Vocations Learning 5(2):77–97
Puflett D (2009) Horticulture industry based training: does it really work? Acta Hort 832:177–184
Ramachandraiah M (2000) Teaching mushroom cultivation through distance education. In: Griensven LJLD van (ed) Proceedings of the 15th International Congress on the Science and Cultivation of Edible Fungi. Maastricht, Netherlands
Rayner J, McSweeney P, Rayner K et al (2009) Where to now for horticultural education in Australia? Acta Hort 832:185–194
Reganold JP, Jackson-Smith D, Bitie SS (2011) Transforming U.S. Agriculture. Science 332(6030):670–671
Rindermann H (2012) Intellectual classes, technological progress and economic development: the rise of cognitive capitalism. Person Individ Differ 53(2):108–113
Russell D, Hawke C, Stone G (2004) A model promoting excellence in science teaching. Acta Hort 672:319–324
Russell D, Stone G, Green S (2006) Scoping study: the national primary industry centre for science education (PICSE). School of agricultural science, University of Tasmania and Department of education, science and training, Australian government. http://www.picse.net/HUB/docs/scopingStudy.pdf. Accessed 15th Dec 2014
Sansavini S (2010) Master of science in horticulture: new approaches in Europe. Chronica Horticulturae 50(3):10–15
Scott PR, Dixon GR (2004) Knowledge management for science—based decision making. Acta Hort 642:115–118
Spiertz JHJ, Kropff ML (2011) Adaptation of knowledge systems to changes in agriculture and society: the case of the Netherlands. Wageningen J Life Sci 58(1–2 June):1–10
St. Hilaire R, Thompson JM (2005) Integrating a university and community college course in landscape construction. HortTechnology 15:181–184

Stone G, Casey M, Coutts J (2005) Strategy to attract young people to horticulture. Acta Hort 672:339–346

Strawbridge CL, Emmett MR, Ashton I (2011) Development of vocational training programmes with active links to current research. Acta Hort 920:71–75

Thompson JC, Russell EB (1993) Beliefs and intentions of counsellors, parents, and students regarding agriculture as a career choice. J Agric Educ 34(4):55–63

Tietze J, Schmidt U (2008) E-Learning within horticultural sciences. Acta Hort 801:687–691

Trip G, Maijers W, Lossonczy T (2004) Educating competent professionals for the horticultural job market; analysis of the new model for higher education in the Netherlands. Acta Hort 655:451–460

Winzenried AP (1991) Green Grows Our Garden: the Centenary History of Horticultural Education at Burnley. (VCAH-Burnley Richmond) 199 pp

# Chapter 33
# Extension Approaches for Horticultural Innovation

**Peter F. McSweeney, Chris C. Williams, Ruth A. Nettle, John P. Rayner and Robin G. Brumfield**

**Abstract** The focus of this chapter is towards the changing extension climate surrounding the horticultural industry and the implications for horticultural extension now and into the future. Extension as a function and a practice is being redefined in many countries alongside changes in the institutional arrangements for extension, changing funding models and varying degrees of involvement of the private sector. The chapter analyses:

- industry/sector changes and implications for extension
- traditional and more recent interpretation surrounding extension definitions and delivery models
- the evolving enabling environment, resource constraints and institutional roles surrounding extension service delivery
- the extension practitioner (their skills, competencies, roles)
- elements of a model suited to support industry needs with high, ongoing innovation requirements.

P. F. McSweeney (✉) · R. A. Nettle
Department of Agriculture and Food Systems,
Melbourne School of Land and Environment, The University of Melbourne,
Parkville Campus, Melbourne, VIC 3010, Australia
e-mail: peterm1@unimelb.edu.au

R. A. Nettle
e-mail: ranettle@unimelb.edu.au

C. C. Williams · J. P. Rayner
Department of Resource Management and Geography,
Melbourne School of Land and Environment, University of Melbourne,
Burnley Campus, Richmond, VIC 3121, Australia
e-mail: chriscw@unimelb.edu.au

J. P. Rayner
e-mail: jrayner@unimelb.edu.au

R. G. Brumfield
Department of Agricultural, Food and Resource Economics,
Rutgers, The State University of New Jersey, 55 Dudley Road,
New Brunswick, NJ 08901–8520, USA
e-mail: brumfield@AESOP.Rutgers.edu

G. R. Dixon, D. E. Aldous (eds.), *Horticulture: Plants for People and Places, Volume 3*,
DOI 10.1007/978-94-017-8560-0_33, © Springer Science+Business Media Dordrecht 2014

**Keywords** Rural advisory services · Extension models · Agricultural extension · Urban · Amenity · Environmental · Consumer horticulture · Public horticulture · Production horticulture · Green infrastructure

## Introduction

Much has been written regarding the many dimensions surrounding extension for agriculture and horticulture. 'Extension' is a construct which has gained meaning through its praxis. It is linked to concepts such as outreach, knowledge or technology transfer, innovation diffusion, change management, capacity building, empowerment, 'suasion' and, to some extent, business incubation. The term 'rural advisory services' (RAS) has also gained recent currency as an alternative to the perceived outdated term of 'extension' (Adolph 2011). Implicit in all of these terms is the assumption of a pool of referent knowledge in specialist discipline areas, much of this based on research and development built up over many years.

Roling (1988) refers to extension science as the development of the body of knowledge (from extension research) into extension practice. Early agricultural extension involved the linked functions of diagnosis of farmer situations and opportunities, message transfer, gathering feedback and developing linkages between industry participants (Farrington 1995). On an international scale, approaches tend to differ between and within countries, and are shaped by the history and culture of extension, the governance of extension at a national, state and industry level, the degree of involvement of the public and private sector and the maturity and varying needs of sectors. While the provision of the extension function has had its roots in public service administration, models for extension provision are continually evolving in response to the growing sphere and need for extension services particularly within the challenge of fewer public resources. In some contexts, the outlook for future extension is a pessimistic one. Hunt et al. (2012) take the view that following periods of maturation and growth, the traditional extension models are "unravelling" which is directly related to government policy shifts, which in the Australian context at least, are moving away from public sector extension.

The focus of this chapter is towards the changing extension climate surrounding the horticultural industry and the implications for horticultural extension now and into the future. Initially we define the term 'horticulture', expanding on the historical context surrounding public and consumer horticulture in particular. We consider trends and recent tensions and challenges surrounding extension more broadly with particular reference to institutional changes affecting extension, funding arrangements and the relative involvement of the public and private sector. We review changes in extension delivery models and extension roles and review the key skills and competencies of extension practice. We then return to the horticultural sector, separating out some selected industry segments with respect to their extension needs, including mature areas of the horticulture industry (e.g. environmental or ornamental horticulture), pioneering and growth areas (e.g. urban green infrastructure), with particular reference to the Australian context. We consider some of the models or paradigms that might assist the industry to develop the innovative thinking and capacity required across the development of green infrastructure, an emerging knowledge area.

## Defining Horticulture

Horticulture as an industry and discipline has always been difficult to define. Two broad industry-based definitions have historically been used – 'amenity or ornamental horticulture' and 'production horticulture'. In recent times these definitions have been challenged through more specific industry sectors and/or groupings, and also through economic, social and technological changes driving industry development, diversification and employment. Amenity horticulture has included arboriculture, interior plant hire, landscape design and construction, nursery production and retailing, parks and gardens and turf production and maintenance. Amenity horticulture is also referred to as 'environmental horticulture', 'landscape horticulture', 'lifestyle horticulture' and 'urban horticulture'. The latter definitions also encompass newer design and community-based applications, particularly in urban and peri-urban settings, such as green infrastructure technologies, urban agriculture, therapeutic horticulture, community gardens and related areas. Production horticulture as it suggests is based around the more agricultural and productive food and resource industries and encompasses fruits, vegetables, flowers, mushrooms, nuts and some specialised crops. Some definitions also include nursery and turf production.

Urban horticulture can be usefully divided into two related sectors: private, residential landscapes, gardens and amateur gardeners, called consumer horticulture here; and then there is public landscape and vegetation management, comprising advanced vocational and professional skills, referred to here as public horticulture. Whilst the two sectors have historically been closely linked, since the 1980's there has been increasing divergence in their respective needs for information, practice and management. In public horticulture this has been driven by the increasing complexity of open space, landscape and vegetation management and the rise of relevant degree and post-graduate qualifications. In consumer horticulture differences have been driven by the growth of residential gardens and landscapes for leisure and recreation and the development of service and landscape businesses in support of these outcomes. There are however fundamental differences between the two, which also reflect the provision of extension services.

## Consumer Horticulture

Consumer horticulture, or amateur and residential gardening, is largely a product of the rapid development of suburban landscapes in many countries over the last century, but particularly in the USA, Australia, United Kingdom, New Zealand and

Canada. A strong gardening culture emerged as middle classes grew more wealthy and leisured, aided by an increased focus on suburban development and sizable domestic gardens (Pollan 1996). It was further assisted by improved access to home garden machinery, equipment and materials and new plant introductions, supported by the plant nursery industry, garden-based media and publications, garden clubs and interest groups. Consumer horticulture places a high priority on specimen values and seasonal display, largely achieved by intensive garden practices and resource inputs to optimize soils, growing environments and maintenance.

The needs in consumer horticulture are largely derived from gardening culture translated freely into the horticultural practices found in public parks and gardens and vice versa. Indeed there was an expectation that skilful horticultural display in parks provided a model for private gardens. The nineteenth century vision of the urban parks movement was to create public spaces that used landscape design and horticulture as a civilising influence on the population as a whole. At the same time, private gardening was also seen as an important part of nation building and fostering civic pride. As secular, compulsory and free education expanded many governments created gardening classes for children (Whitehead 2001).

The focus on specimens and display has driven extension practice in consumer horticulture. Much of this has been delivered as garden cultivation information around plant groups (e.g. dahlias, roses, conifers, etc.), life forms (e.g. small trees, bedding plants, etc.) or garden practices (e.g. pruning, composting, pest and disease control, etc.). As private gardening developed, all the perceived inputs needed to optimise plant performance created a lucrative market for sales of horticultural products. This market grew rapidly after the Second World War, with extension increasingly becoming focused on translating agricultural research outcomes, aimed at commercial farmers, into advice pitched to the consumer horticulture sector. Home gardeners were encouraged to use pesticides, fertilisers and soil modifiers which had been developed for broadacre industrial agriculture, including chemicals that have subsequently been banned such as DDT (Anon 1965).

Since the advent of the environmental movement following the publication of Rachel Carson's *Silent Spring* in 1962 (Carson 1962), there has been increasing availability of 'organic' and environmentally friendly products. Like their mainstream counterparts, these are sometimes not tested for amenity horticulture applications and there is often minimal independent scientific evaluation of their effectiveness. In other words, the alternative products for home use, like those derived from agricultural research are not really subjected to dispassionate, objective research and then integrated into extension programs. A key question for responsible extension in this context is establishing whether recommended agricultural inputs are really necessary for optimal plant growth in the amateur gardening context? For example a common reason given for using gypsum in home gardens and promoted as such by many sources of gardening information is that will make it easier to dig into compacted soils. The reality is that gypsum, as tested for agriculture and production horticulture, is recommended for frequently cultivated sodic clay soils, cropped annually and not for soils that are simply "hard to dig" and then planted with perennial shrubs and trees (Chalker-Scott 2010).

In North America the tradition of extension to home gardeners has been largely maintained through the Cooperative Extension System delivered by state land-grant universities. Since the 1970's this has also encompassed Master Gardener programs, particularly in the USA. Delivered through universities using rigorous, scientifically-based curricula (Chalker-Scott and Collman 2006), Master Gardeners act as volunteers in training and education programs, provide residential horticultural advice (often at county offices), work in community gardens, schools and undertake a variety of other garden and horticultural outreach projects. An evaluation of the Washington State University Master Garden program over 30 years (1973–2004) identified 4,015 active volunteers, servicing>350,000 clients with an estimated dollar value (of hours) at $US 4,058,796 (Chalker-Scott and Collman 2006). Recently land grant universities have responded to public interest by delivering programs on organics, home food growing and nutrition (Weisenhorn 2012), increasing their relevancy and currency to consumers. One of the more interesting recent examples of extension practice in action for home gardeners is the work of plant physiologist Linda Chalker-Scott at Washington State University in the USA. Chalker-Scott's blog and books set out to explain or de-bunk horticultural myths and to separate gardening folk-lore from scientific fact. Chalker-Scott examines topics as diverse as compost tea, water crystals, tree-staking and xeriscaping set within the context of how particular products or practices are heavily promoted through popular magazines or websites. She then uses peer-reviewed academic literature and her own experience as a scientist to examine the validity of claims made for various consumer horticulture products or practices (Chalker-Scott 2008, 2010).

In England extension services for consumer horticulture are provided through a registered charity, the Royal Horticultural Society (RHS). The RHS claims horticulture to be the biggest employer across the United Kingdom with 18.5 million gardeners, spending more than £ 2,000 million a year on plants and gardening products (www.rhs.org.uk). RHS services include its impressive information-rich and interactive website, public gardens, horticultural publications, garden shows and events, training and education programs and an increasing involvement in school and community garden programs. In 2012 the RHS had 383,046 members, an increase of 70,197 from the previous year, and recorded 1.455 million visitors to its four flagship gardens at Wisley, Rosemoor, Harlow Carr and Hyde Hall (Royal Horticultural Society 2012).

In Australia, agency based extension to residential gardeners has largely been provided through garden advisory services. Most of these state-based services were terminated in the 1990s after government cut backs, but many of the publications remain on-line and cover a range of topics as diverse as composting, lawns, indoor plants, fruit tree pruning and mulches (Garden Advisory Service 1989). While these agencies no longer exist, governments have more recently had to respond to the challenge of severe droughts and water restrictions, particularly the impacts on residential gardening. New agencies have been funded to assist extension in this area, including savewater!® (www.savewater.com.au). Formed through an alliance of eastern Australian water retailers, savewater!® provides detailed advice on water conservation for residential landscapes. This includes design, plant selection, irrigation and garden practice information and participation in garden shows, including demonstration gardens.

## Public Horticulture

While gardening as a trade is well established, public horticulture as an advanced vocation and profession is more recent. Built around urban landscapes that comprise diverse vegetation elements and systems, extensive community engagement, comprehensive resources and assets and related issues of use and safety, over the last 30 years public horticulture has been transformed from parks and gardens maintenance to the management of designed and complex spaces with multi-purpose outcomes and needs (Cobham 1986). This has also occurred in the wider context of greatly reduced budgets for public landscapes, while at the same time more professions and disciplines have begun analysing and promoting the wider social, economic and social benefits of urban green space (Ward-Thompson and Travlou 2007). This has increased the need for more scientifically informed management and more sophisticated approaches to urban landscape management (Hitchmough 1994).

Public horticulture is largely derived from the extensive growth of cities over the past 150 years and the consequent creation of urban parks and gardens. Iconic urban green spaces such as Central Park in New York City have provided models for parks across the western world from the latter half of the nineteenth century. The philosophy behind the urban parks movement is well-documented and is essentially one which values the ideal of *rus in urbe* ('country in the city') as a way to provide beauty and health benefits to the wider population. In order to create the experience of the 'country in the city' the most influential designers and landscapers of the Victorian era combined elements of the picturesque and the gardenesque at the same time. In other words, public open space at its best sought to combine the experience of being in larger, free space, enjoyed through walking and taking in views, along with admiration for the display of carefully tendered individual plant specimens and planting beds. With varying degrees of success, but with great overall commitment over several generations, municipal parks and gardens became an essential part of the 'green infrastructure' of urban life.

By the middle of the twentieth century, the quality of urban green space began to decline in the West even while public open space, its need enshrined in local planning law, was still being created. This was the period of "middling Modernist" municipal park creation (Rabinow 1989) that saw many cookie-cutter spaces created, using simple combinations of mown turf and scattered trees and shrubs with a standard set of play equipment inserted for children and mothers. This urban "non-descript" was thus essentially "non-designed" in the sense that a formal design process was absent for the space as a whole but also in terms of where and why a particular planting regime would be used (Clouston 1986). A standard list of trees and shrubs used over many decades acted as a 'plant selection process'. In this context, extension for amenity horticulture was as largely informal and anecdotal as it was for private gardening. Information was largely relayed through the gardening apprenticeship system which prevailed in municipal government during this time.

The era of neo-classical economics, exemplified by the elections of Margaret Thatcher and Ronald Reagan in 1979 and 1980 respectively, marks the beginning of privatisation of many government services or their complete closure. For amenity horticulture this meant the end of the era for in-house horticultural crews at the municipal level. It can be argued that urban green space had been in decline for several decades despite over-employment of parks gardeners. However, it can equally be argued that compulsory competitive tendering for routine maintenance in municipal green space saw this decline only exacerbated. Whatever the complexity of this situation, it also seems the case that this process of outsourcing traditional in-house vegetation management focused the minds of professionals on a couple of long neglected issues: firstly, on how to do more for less i.e. defining and achieving quality cost effectively; and, secondly, on setting objectives and outcomes to work towards and in ways that were more responsive to what clients/customers (tax payers) actually wanted from urban green space. What emerges is some kind of rudimentary value chain for the kinds of services to be derived from professional amenity horticulture.

Several prevailing trends have led to this situation. One of these has been the rise of landscape architecture as a profession, especially since the 1970s and 1980s. As a design profession, landscape architecture has provided a much needed critique of the old horticultural formula of mown turf, scattered trees and some play equipment as the basis of park implementation. Interest in quality of life through good design has put the spotlight back on the greater aspirations for public open space and its central role in creating stimulating and sustainable landscapes. Ecology too has taken a greater interest in urban areas since the 1980s, helping to make the protection of nature and the restoration of ecosystems a standard component of managing city environments. Projects to bring nature back to cities frequently need on-ground implementation by staff with horticultural training. Arising out of broader ecological and sustainability concerns the concept of green infrastructure has gathered sometimes disparate uses of urban vegetation under one umbrella. This has been especially influential for our understanding of the ecosystem services delivered by trees in parks and streets, for example, in terms of energy efficiency. Ecology has also given horticulture new intellectual tools with which to manage urban vegetation.

Public health advocates and policy makers have begun looking at public open space as places to facilitate respite from stressful urban living and to provide physical exercise. This is essentially the same agenda that created the urban parks movement in the nineteenth century now framed within discourses about epidemics of mental illness, diabetes and obesity. For the first time since the allotment gardens of the Second World War, a range of advocates, both agency-based and non-governmental organisations (NGOs) are also calling for more food to be grown in cities and sometimes by the community at the municipal level in public open space. Essentially the confluence of design, ecology, urban health and financial constraint concerns have heightened awareness of the serious information gaps and associated extension needs in public horticulture.

## Changing the Understanding of Extension

In examining the extension needs and challenges for the horticultural industry segments, a broader perspective on extension is also appropriate. Coutts and Roberts (2011) segregated recent extension history and influences into phases associated with linear or technology transfer (1960s), the influence of farmer discussion groups on farming systems research (1970s), systems thinking (1980s), pluralist approaches incorporating participatory methods (1990s) and then the capacity building and community engagement movement of the 2000s.

Early extension, based around top-down service delivery, and employing linear, unidirectional information flows, was frequently criticised. Such an approach did not make use of the multiple sources of "new agricultural inputs, ideas and practice" (Farrington 1995). The 'training and visit' (T and V) system, sometimes disparagingly referred to as the 'touch and vanish' system, was also criticised in part, at least in development agriculture, for its failure to achieve cost-effective change (Howell 1982). The 'train the trainer' model, also linked to extension, was also dependent upon people being adequately trained in a range of core discipline areas. The disconnect occurs with this approach when there is a mismatch between the knowledge and capacity of extension staff and the specialist programs required (Ward et al. 2011).

Over time, extension has applied different lenses in responding to changing needs of client groups. The traditional foundations for much extension work lies in agricultural science and horticultural science (i.e. building agronomic or horticultural expertise) and agricultural economics (i.e. enterprise decision making for profit). The Food and Agriculture Organization of the United Nations (FAO) encourages 'Good Agricultural Practice' (GAP) as a suitable basis for "local development of optimal good agricultural practice" (Poisot et al. 2007). This extension lens builds upon the three pillars of economic viability, environmental sustainability and social acceptance.

The agribusiness model also provides a framework to analyse enterprise performance within the supply/value chain. According to Ward et al. (2011, p. 135) agribusiness extension is directed to "improving management skills, decision making, and strategic thinking within value chain development". The process of value chain thinking also tends to sharpen managerial focus on responding to customer needs as opposed to a single production lens.

The role of extension has also been examined within a broader social context. Macadam et al. (2004) discussed extension programs as being complementary to capacity building. In this sense, agricultural extension overlaps with rural extension which takes on a broader rural development perspective. Taking the Rutgers New Jersey Agricultural Experiment Station (NJAES) Cooperative Extension as a typical example, its mission is to help "the diverse population of New Jersey adapt to a rapidly changing society and improve their lives and communities through an educational process that uses science based knowledge. Through science-based educational programs, Rutgers Cooperative Extension truly enhances the quality of life for residents of New Jersey and brings the wealth of knowledge of the state university to local communities." (http://njaes.rutgers.edu/extension/)

The breadth of contemporary extension and its directions far beyond strictly agrarian needs can be captured within the ambit of co-operative extension within the USA. Robson (undated) provides an overview of extension in the USA as part of the Land Grant System which has the three pillars of teaching, research and extension. In 1862, during the Civil War in the USA, the Morrill Act (1862) passed which provided grants of federal lands to create public institutions for teaching agriculture and mechanical arts. The subsequent Hatch Act (1887) established federal agricultural experiment stations to conduct research at land-grant institutions. In 1914 the Smith-Lever Act passed, established the Agricultural Extension Service at each land-grant institution with the intent of extending non-biased, scientific-based research findings to the citizens. The term 'cooperative' comes from the partnership of federal, state, and local government who share the funding and established Cooperative Extension Offices in every county in the USA to serve the citizenship.

The capacity of communities to deal with change in this context aligns with contemporary definitions of extension. For example, Coutts et al. (2005, p. 7) states that "extension is the process of engaging with individuals, groups and communities so that people are more able to deal with issues affecting them and opportunities open to them". A group of Australian state extension leaders (State Extension Leaders' Network)(Anon 2006) defined extension as "the process of enabling change in individuals, communities and industries involved in the primary industry sector and with natural resource management". A change response is also implicit in the definition by Leeuwis and van den Ban (2004) which states that "[e]xtension [is] a series of embedded communicative interventions that are meant, among others, to develop and/or induce innovations which supposedly help to resolve (usually multi-actor) problematic situations."

As the delivery approaches, lenses and capacities of extension adapt and adjust to the changing operating environment, it is increasingly suggested that the term 'extension' itself is losing currency. In a broad ranging synthesis report Adolph (2011) suggests the term rural advisory services (RAS) better represents "...the different activities that provide the information and services needed and demanded by farmers and other actors in rural settings to assist them in developing their own technical, organisational and management skills and practices, so as to improve their livelihoods and well- being" than the term 'extension'.

## Changing Institutional Roles to Extension

Extension approaches adopted by individual countries are driven by many factors. As Coutts et al. (2005) state, "[e]xtension approaches do not develop in a vacuum. The structural, social, political, economic and philosophical contexts of the time all contribute to the types of projects that are developed, proposed and funded". So given the many variables at play within any one country and across countries, it is natural to find considerable diversity in approaches to extension. Extension practitioners and analysts, in seeking strategic, coherent mixes of extension solutions to

**Market Reforms**
**Funding**

| | | Public | Private |
|---|---|---|---|
| Delivery | Public | Revision of public sector extension via downsizing and some cost recovery<br><br>(Canada, Israel, USA) | Cost recovery (fee-based) systems<br><br>(OECD countries, previously in Mexico) |
| | Private | Pluralism, partnerships, power sharing<br><br>(Chile, Estonia, Hungary, Venezuela, South Korea, and Taiwan) | Privatization (total) Commercialization<br><br>(The Netherlands, New Zealand, England and Wales) |

**Non-Market Reforms**

| | | |
|---|---|---|
| Political Fiscal Administrative Issues | Decentralization of lower tiers of government<br><br>(Colombia, Indonesia, Mexico, The Philippines, Uganda and others) | Transfer (delegation of responsibility) to other entities<br><br>(Bolivia, to farmer organizations; Ecuador, mixed with farmer-led NGO programmes; Peru, extension devolved to NGOs) |

**Fig. 33.1** Public sector extension reform strategies. (Source: Food and Agricultural Organization of the United Nations, 2001, Rivera, W.M., Agricultural and rural extension worldwide; options for institutional reform in developing countries. Extension, Education and Communication Service. ftp://ftp.fao.org/docrep/fao/004/y2709e/y2709e.pdf)

their own problems, will logically explore and learn from the approaches used in contexts similar to their own in addressing the common challenges.

Extension is also frequently claimed to be a strategic policy instrument (Anon 2006), and there is evidence supporting the effectiveness of systematic change through considered policy initiatives.

In the context of agricultural development, Rivera (2001) argues that "no single approach best suits extension development in all circumstances, just as there is no single approach that best suits development. Otherwise the problems of extension and, for that matter, of development, would have been solved long ago." Fig. 33.1 highlights the diversity of approaches adopted across the globe with different in-country institutional arrangements and responses of NGOs many of which are highly idiosyncratic. In some of these countries, policy surrounding extension is clearly articulated. In others, the direction and sphere of extension efforts is much less clear. This is particularly the case in countries using pluralist approaches combining private and government service delivery, especially where government support operates through multiple tiers. Australia is one example of a country applying pluralist approaches to both funding and delivery of extension.

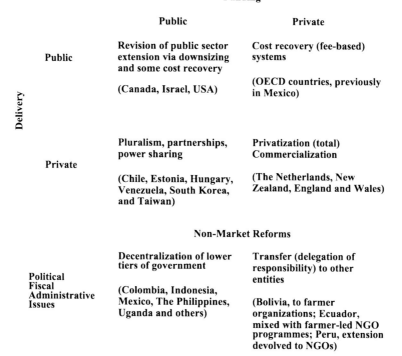

Funding or financing extension services or programs has been a particularly vexed issue. Any recent dissection of extension reforms and practices has focused on the divergent responses to funding shortages and to the consequences for extension efforts. This has applied to developed, developing and least-developed countries alike. Farrington (1995) discusses the fiscal crisis impacting on public sector services in least-developed countries and describes the "picture [as] one of resources spread too thinly to be effective, inflexibility and inability to respond to the changing infrastructural and institutional contexts" (p. 540). According to Kidd et al. (2000) "[a] particular challenge will be to find a strategy with a coherent mix of mechanisms for financing and providing extension that can help rigid public extension systems to evolve gradually in a flexible manner".

While scarcity of resources invariably focuses attention on privatisation and commercialisation of extension services, such pathways can be problematic in terms of the ideological tensions they may create, or, counter-productive impacts such as reduced knowledge sharing and learning stifling productivity and innovation. For example, Rivera (2008) argues that there is "a significant divergence between privatisation measures". There is an expectation privatisation brings "demand-led extension", "farmer participation in extension decision-making" and "farmer empowerment". This has proven not to be the case in all circumstances with potential risks of farmers becoming increasingly "beholden to commercial forces" (Rivera 2008).

A review of some of the studies into private and public investment in extension demonstrates some of the issues. Developed economies have adopted different mixes of public and private extension. The Netherlands, New Zealand, England and Wales have commercialised/privatised agricultural and related extension (Rivera 2001) with New Zealand "at the forefront of privatisation of government services since 1983…." (Botha et al. 2008, p. 125). The transformation of New Zealand's government extension services over the last 20 years has been particularly contentious. Whilst it has been argued that the transition to private consulting services has led to improvements in the quality, relevance and timeliness of advice provided at an individual grower level, this has come at the cost of a lack information sharing, more regionally-focused service delivery, poor staff recruitment and a decline in horticultural statistics compilation, biosecurity capability and the provision of leadership and advocacy (Warrington et al. 2004).

Countries like Australia are clearly at the crossroads in the extension space in terms of 'who does what and for whose benefit'. Marsh and Pannell's (2000) view on Australia was that "rapid change [was] occurring at the federal level and in all states …" (p. 605) and that "[a]gricultural information [was] increasingly perceived by policy-makers to have 'private-good' characteristics". In drawing on the lessons from neighbouring New Zealand's experience, Stantiall and Paine (2000) observed that "[while] it is legitimate to devolve consultancy to the private sector, it is crucial to retain a critical mass of extension capability to achieve public-good goals". Overall, Australia's extension approach sees the work of industry extension distributed among layers of government and private service providers. The country also employs research and development corporations (RDCs) which are designed to integrate the industry linkages and efforts in research and development initiatives. The RDC model works on a matching funding system where government matches industry funds raised through industry levies. A recent Australian Productivity Commission report into Rural RDCs (Anon 2011) acknowledged the 'systems integrating' role of the RDCs yet questioned the return to the community on the sizable government investment. This focus is significant for horticulture in the sense that Horticulture Australia Limited (HAL) is one of the largest of the rural RDCs. The Productivity Commission identified many initiatives that could have been funded directly through farmer levies or stakeholder support. It should be noted that in a study of farmer levy-funded research, development and extension arrangements in the Netherlands, Klerkx and Leeuwis (2008) found that although end-users had the opportunity to raise issues, other groups in the research and development (R&D) planning process influenced the process so that farmers' innovation needs were not adequately reflected in the outcomes.

The USA is similarly challenged in a resourcing sense. In discussing the Californian extension system, Murray (2005) identifies the challenges as "[being] declining public support monies, changing societal needs or situations, the inability to behave as 'change agents' we so often claim to be, and competition from the public sector for many of the services we have traditionally offered". The "Great Recession" of 2008 has increased these challenges. As public funding continues to shrink, extension is turning to 'eXtension', webinars and other technologies to 'do more with less' and provide information to its clients in a more cost-effective way. Much of this mirrors, or is linked, to the use of web-based technologies, communications and instruction methodologies in horticultural education and training (Lineberger 2009a, b, c).

## Evolving Enabling Environments

Overlapping with the above discussion regarding the changing institutional landscape and 'private/public good' debate in extension are trends within what may be described as the enabling environment surrounding extension. This includes some analysis of agricultural, horticultural and related education and the development of extension (or rural advisory services) capacity.

Commenting on agricultural education to start with, worldwide, there is considerable diversity in the relationship between agricultural and related education and research, advisory services, networks and stakeholder relationships (Adolph 2011). For example, in countries where agriculture remains the main economic and employment driver, the provision of agricultural and related education at secondary, vocational and higher education levels seems largely intact (Adolph 2011). On the other hand, the experience with some of the developed economies is quite different. Again, using Australia as the example with which we are most familiar there has been a steady decline in the uptake and provision of higher education (HE) agricultural and related programs since the 1990s (McSweeney and Rayner 2011). The reality is that more students are gravitating to areas of increased interest e.g. environmental sciences, biofuels, food science etc. and away from traditional major areas of study. One consequence of this lack of interest is the adverse effect on the professional services and extension capacity necessary for a vibrant rural sector (Falvey 1998; Malcolm 2010; Pratley 2008).

Closely allied to the changes in traditional areas of agricultural education, enrolments in most areas of Australian horticultural education at HE levels have fallen significantly (Collins and Dunne 2009; Dunne 2010; McSweeney and Rayner 2011; Pratley 2012a and b). While enrolments at vocational education and training (VET) levels seem somewhat more stable, they remain concentrated at the lower curriculum levels (Rayner et al. 2009). Of concern for the development in professional capacity in horticulture are the lack of pathways between VET and HE systems and, not surprisingly, low levels of upward student progression. McEvilly and Aldous (2010) also highlight the difficulty of communicating and guiding young people into areas such as horticulture.

Following on from the trends surrounding public and consumer horticulture discussed earlier in this chapter we attempt to synthesise in Fig. 33.2 the current levels of support for horticulture in formal training (VET and HE), research and extension services, again for the Australian context. The four categories chosen are not mutually exclusive (e.g. public horticulture overlaps with green infrastructure), yet they serve to highlight some of the changes in terms of enabling support for the sector. Of particular interest with Fig. 33.2 are:

- the declining HE presence and increasing VET significance in traditional areas of consumer, public and production horticulture,
- the declining government role in extension in these segments and the growing reliance on industry-funded support for horticulture,
- the significant transformation away from the traditional advisory service models for consumer horticulture,
- how growth areas e.g. green infrastructure, are to be supported in light of trends in education, and
- to what extent niche curriculum areas will be developed and sustained in light of the growth areas.

It is also important to recognise that horticulture comprises segments that are at different stages of maturity which are supported by different levels of industry-driven extension facilitation. For example, both public and production horticulture segments typify mature segments in one sense in that they have progressed from cottage-type industries to those where larger scale commercial operations are more prevalent. In terms of the enabling environment for public and production horticulture, there are isolated pockets of higher education (HE) delivery into some traditional areas (nursery, turfgrass, cut flowers, parks and landscapes), yet most support comes from the vocational education and training (VET) sector through generic training packages or industry specific training initiatives.

In horticultural production and service areas, many enterprises are self-sufficient in technical expertise and employ in-house training; many remain family-owned and managed yet are operating on a larger scale; and many can access and afford agronomic and other technical advice as required. The sector is generally well supported through relevant industry associations. In Australia's case, the

| | | | |
|---|---|---|---|
| **1. Consumer horticulture** | | | |
| **Research / knowledge base** | **Higher Education (HE)** | **Vocational Education and Training (VET)** | **Extension provision** |
| **Long established principles**<br><br>**Many untested myths**<br><br>**Research exploration of plant functionality / molecular biosciences and biotechnology** | **Minimal undergraduate footprint in traditional areas**<br><br>**Growth in some specialist program areas e.g. sustainable horticulture, permaculture, organic.**<br><br>**Variable pathways from VET** | **Trend toward generic vocational education and training via training packages**<br><br>**Little or no integration into HE programs** | **Traditional advisory services discontinued**<br><br>**Growth in online media** |
| **2. Public horticulture** | | | |
| **Traditional links to horticultural science**<br><br>**Supported through the research and development corporations**<br><br>**Static since rapid growth phase of the 1980 – 90s**<br><br>**Loss of core institutional leadership / core players** | **Historically tied to older style colleges**<br><br>**Variable HE presence but few undergraduate / postgraduate programs**<br><br>**Linkages to some related discipline areas** | **As above** | **Declining areas of specialist expertise in government**<br><br>**Targeted industry development**<br><br>**Industry association initiatives**<br><br>**Decline in professional development activity** |

**Fig. 33.2** Enabling environment elements supporting areas of Australian horticultural extension

aforementioned Horticulture Australia Limited (HAL) (Anon 2012a) is the peak industry body funded mainly by a combination of statutory industry levies, voluntary contributions, and Commonwealth government matching funds. Such funds are applied strategically to research and development and extension and provided primarily through industry development officers. One horticultural sector organisation that operates under the auspices of HAL, Nursery and Garden Industry Australia (NGIA), provides a useful illustration. In recent years the NGIA has shifted

| 3. Production horticulture | | | |
|---|---|---|---|
| Traditional links to agricultural science<br><br>Interest in sustainability initiatives / community responses | Highly variable presence from State to State | As above | Key roles of industry associations bodies and peak bodies, Horticulture Australia Limited<br><br>Agronomic advice through rural services<br><br>Public service provision rotates focus and its regional strengths |
| **4. Green infrastructure** | | | |
| Rapid response innovations being sought<br><br>Inter-disciplinary research activity e.g. plants, water, design, climate, engineering, social | Emerging curriculum area across disciplines | Minimal coverage and at low skill level | Conference-based networking and knowledge diffusion<br><br>Emerging associations (e.g. Green Roofs Australasia) |

**Fig. 33.2** (continued)

its focus more towards urban landscapes, green infrastructure, urban forestry and lifestyle horticulture activities. To drive industry development and investment, this focus, together with outcomes in improved communications and enhancing skills, knowledge and practice, form part of the NGIA's strategic investment plan over coming years (Anon 2012b).

One could argue that the traditional industry research and education foundations systems have supported horticulture well. Significant innovations in horticulture and horticultural science have taken place through plant breeding, plant biotechnology, production system innovations, environmental management, improvements in media and fertilisers, irrigation design and protected cropping, plant health, integrated pest management, postharvest protocols and improved market access, to name a few. In more recent years, production horticulture has been exposed to some level of automation, mechanisation or sophisticated application of greenhouse technology.

Moving away from a mature segment, such as production horticulture, the growth of 'green infrastructure' in cities has brought about a new and pioneering approach to extension within horticulture, where new and different extension skills and knowledge are required in communicating. Green infrastructure has been defined as "all natural, semi-natural and artificial networks of multifunctional ecological systems within, around and between urban areas" (Tzoulas et al. 2007). Operating at a range of scales and spaces, it includes traditional urban greening, such as trees, turfgrass, parks and gardens, etc., newer technologies, such as green roofs and green walls, and water-sensitive urban design strategies, such as rain gardens and swales. Green infrastructure is often hampered by the need to incorporate a range of different disciplines and skills, particularly in the design phase. Horticulture is ideally placed to ensure that functional outcomes from green infrastructure are achieved as it requires integrating design principles with a knowledge of plant materials, soils, plant husbandry, engineering, irrigation and climate systems. The challenge is to ensure this integration occurs in urban horticulture through research, education and teaching, and then relates this effectively and appropriately to extension.

To what extent and how extension provision in its broadest sense will be impacted in the long term by such trends in the underpinning education systems is unclear. The reality is that in the short to medium term at least, government agencies, industry associations, educators and private firms will be faced with declining output of graduates from traditional sources. This will be more critical to some segments than others. While at face value this is problematic it may be that human resource capacity to meet industry needs and the extension and advisory service provision into the future may be sourced and developed in a different way.

## Evolving Extension Practice in Horticulture

In recent years, as a result of the sharply declining institutional extension capacity, the growth in horticultural information needs and rapidly changing models for information sharing, consumer horticulture, in particular, has experienced an explosion of on-line gardening information. User-generated content on the web, largely by private gardeners, provides information about plants and horticulture that far outweighs the available extension information provided by universities and colleges. Nevertheless, for those who seek less anecdotal information on plants, practices and products, science-based advice, partly sponsored by governments, still exists in some jurisdictions.

In public horticulture, while the information needs are potentially great, the sources of quality information are either lacking or belong to diverse and often informal networks. We can, however, conceive of an enabling environment for access to better quality information and therefore to different forms of extension for this sector. This enabling environment is derived from the resurgent interest in public open space and vegetation to achieve positive social, environmental outcomes for urban communities. Internationally, there are few examples of well developed extension services that support public horticulture. The United Kingdom, with its focus on gardening as a major leisure and recreation activity, also has a large number of representative industry organisations that support members engaged in public horticulture. This includes the Horticultural Trade Association (HTA), Institute of

Horticulture (IOH), Professional Gardeners' Guild, GreenSpace, British Association of Landscape Industries (BALI), The National Trust UK, The National Trust for Scotland, English Heritage, Botanic Gardens Conservation International (BGCI), The Arboricultural Association (AA), and the Institute of Groundsmanship (IOG). The range of extension services provided by these groups is impressive and includes business and technical information, guidance and publications, visual media and training resources, professional development through specific conferences, seminars and training and participation in shows, events and related outreach activities. There is also a weekly trade magazine, "Horticulture Week" (www.hortweek.com) which has a focus on key issues across all areas of horticulture and specifically on careers.

In the American context, extension for urban horticulture is seeking contemporary relevance in the face of new policy concerns. This requires extensive interdisciplinary collaboration to meet new extension challenges. Programs that tackle childhood obesity, diabetes, and crime in urban areas often use gardening and horticulture projects as part of their package of strategies e.g. 4-H programs on nutrition and anti-gang education for youth. At the same time, the resurgence of farmers markets and growing food generally has created opportunities for horticultural extension to provide good quality advice to home gardeners and small-scale commercial producers. Niche growers now grow and sell food, plants, and flowers to local specialty markets and upscale restaurants and need advice on a range of novel edible species that are relatively new to mainstream horticulture (Carlson 2012). Seeking commercial opportunities for extension services in these niche areas has become especially important in the post-GFC American economy when funding streams from federal, state, and county governments have declined or have been cut altogether.

Despite the many changes and challenges for extension and the continued questioning of the relevance of the term itself to the current needs of agriculture and horticulture, the recognised practices of extension appear to be increasingly in demand. Knowledge of farming or horticultural systems, an ability to understand and work with the goals and aspirations of producers and land managers (Nettle and Lamb 2010), an ability to understand, interpret and translate science to practice and policy (Nettle and Paine 2009), facilitation, communication and networking skills are referred to as essential capacities for agricultural industries and rural development more broadly (Roberts et al. 2004). The need for social and organisational capability, not just technical knowledge is considered part of an ability to support farm business systems in their community and environmental context (Nettle 2003, p. 4). Further, emerging challenges for agricultural and horticultural industries represent new professional situations requiring rural advisory services to engage in workplace learning alongside and with their clients, rather than fulfill information delivery functions (Nettle and Paine 2009). Finally, as agricultural and horticultural innovation systems emerge as an alternative conceptual framework in contrast to agricultural or horticultural 'RD&E', it is interesting to note that the features ascribed to the central role of "innovation intermediaries and brokers" (Klerkx and Leeuwis 2009) for successful innovation mirror the extension capacities described above. Despite this, as already discussed, the training ground for extension skills and capacities (via public sector agencies and in Universities) has declined in recent years alongside broader trends.

## Extension Models to Support High Innovation Requirements

The growing demand for the skills underpinning innovation intermediation/broking is best illustrated in the green infrastructure industry segment. The development of knowledge and capacity underpinning the expansion of green infrastructure, particularly new technologies such as green roofs and green walls, for urban domestic and commercial applications provides an excellent case for considering the future of extension direction in Australian horticulture. Fig. 33.3 captures the traditional elements of a seemingly straightforward supply chain (marketing, sales, design, procurement etc) for any green infrastructure installation, yet more importantly it emphasises the knowledge requirements essential for the area (knowledge generation, adaptation, dissemination, application). Such knowledge roles are performed by multiple actors, institutions and intermediaries. For a horticultural segment in its relative infancy, the linear unidirectional information flows associated with traditional extension, as discussed early in this Chapter, are by necessity being replaced by many interactions. While many of these interactions collectively form the basis for 'small step' or incremental improvements; others contribute to 'big step' innovation.

The green infrastructure segment is but one challenge faced by agriculture and horticulture alike to consider the traditional agricultural and horticultural RD&E 'pipeline' and its replacement with 'innovation systems' thinking. This call flows from an increasing recognition that there are many sources of innovation (not just science) and innovation is a process of co-production of new knowledge, products and processes. The purpose of innovation is to provide benefits in society and requires technological, social, economic and institutional change (Hekkert et al. 2007). In this process, research and extension are part of a broader network of actors (Knickel et al. 2009) including in the case of green infrastructure, designers, ecologists, consumers, urban health specialists, and financiers. The environment for innovation to proceed requires institutional frameworks that support innovation as a co-production process and effective governance arrangements that allow people to work across organisational boundaries, work adaptively and create pathways to the desired benefits. Extension in an innovation context becomes an important practice for stimulating networks, translating different disciplinary knowledge, helping the piloting of new approaches and facilitating rapid learning (Howells 2006; Klerkx and Leeuwis 2008, 2009; Klerkx and Nettle 2013). In an Australian context, a program-team model drawing on innovation systems ideas has been described as one mechanism emerging within the current extension climate in Australia (Nettle et al. 2013).

## Conclusion

Contemporary horticultural extension has become a diverse, multi-dimensional enterprise that has moved largely away from facilitating the adoption of new techniques or products by growers and farmers. Extension still embraces the need to provide objective, science and technology-based advice to producers, users and

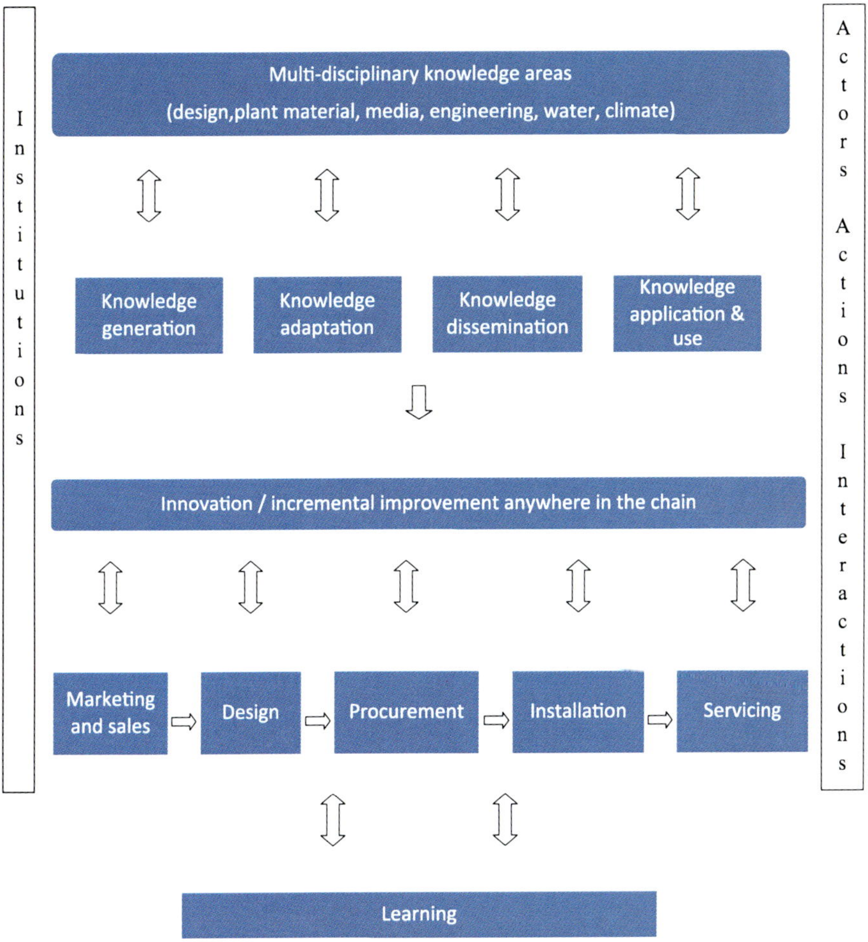


**Fig. 33.3** Integration of value chain and innovation for green infrastructure. (Source: Adapted from Anandajayasekeram and Gebremedhin 2009)

managers of vegetation, but the overall scope has greatly expanded to include issues such as succession planning, natural resources management, rural development and emerging greening technologies. However, even within the broader rubric, extension for public horticulture has long been less formal and focused than extension to agriculture and production horticulture. Arguably, this is simply because the sector has always had fewer commercial and economic imperatives to drive the perceived need for practice change in ways comparable to agriculture. Significant components of horticulture have effectively by-passed the top-down phase of science-driven extension, despite the fact that there is a pressing contemporary need for this in many sectors, particularly in emerging, multi-disciplinary areas, such as green infrastructure. The broad challenge for horticultural extension in the future is to create an overall enabling environment that sustains and builds capacity in order to foster and support innovation and change toward industry development and growth.

## References

Adolph B (2011) Rural advisory services worldwide: a synthesis of actors and issues. Revised edn. Global Forum for Rural Advisory Services. Lindau, Switzerland

Anandajayasekeram P, Gebremedhin B (2009) Integrating innovation systems perspective and value chain analysis in agricultural research for development: implications and challenges. Improving Productivity and Market Success (IPMS) of Ethiopian Farmers Project Working Paper 16. International Livestock Research Institute, Nairobi, Kenya

Anon (1965) Yates garden guide for Australian home gardener. Angus & Robertson, Sydney

Anon (2006) Enabling change in rural and regional Australia: the role of extension in achieving sustainable and productive futures—a discussion document. State Extension Leaders Network http://www.seln.org.au/attachments/uploads/061205SELN_Enabling_change_12pp.pdf. Accessed 5 Oct 2012

Anon (2011) Australian Productivity Commission Report in Rural Research and Development Corporations. http://www.pc.gov.au/projects/inquiry/rural-research/report. Accessed 30 Nov 2012

Anon (2012a) Horticulture Australia limited strategic plan 2012–2015 http://www.horticulture.com.au/about_hal/strategic_plan.asp. Accessed 4 Jan 2013

Anon (2012b) Nursery and Garden Industry Strategic Investment Plan 2012–2016. www.ngia.com.au/Category?Action=View&Category_id=139. Accessed 30 Dec 2012

Botha N, Coutts J, Roth H (2008) The role of agricultural consultants in New Zealand in environmental extension. J Agric Educ Ext 14(2):125–138

Carson R (1962) Silent spring. Houghton Mifflin, Boston. (reprint. 1987)

Carlson S (2012) The new extension service: uban, ubane. The Chronicle of Higher Education, 30 Nov http://chronicle.com/article/The-New-Extension-Service-/135912/. Accessed 5 Jan 2013

Chalker-Scott L (2008) The informed gardener. University of Washington Press, Seattle

Chalker-Scott L(2010) The informed gardener blooms again. University of Washington Press, Seattle

Chalker-Scott L, Collman SJ (2006) Washington State's master gardener program: 30 years of leadership in university-sponsored, volunteer-coordinated, sustainable community horticulture. J Clean Prod 14:988–993

Clouston JB (1986) Landscape design: user requirements. In: Bradshaw AD, Goode PA, Thorp E (eds) Ecology and design in landscape. Blackwell Scientific Publications, Oxford

Cobham RO (1986) Professional integration in place of ineptitude. In: Bradshaw AD, Goode DA, Thorp EHP (eds) Ecology and design in landscape. Blackwell Scientific Publications, Oxford

Collins RJ, Dunne AJ (2009) Can dual degrees help to arrest the decline in tertiary enrolments in horticulture? A case study from the University of Queensland, Australia. Acta Hortic 832:65–70

Coutts J, Roberts K (2011) Theories and approaches of extension: review of extension in capacity building. In: Jennings J, Packham R, Woodside D (eds) Shaping change: natural resource management, agriculture and the role of extension. ISBN:978-0-9577030-7-0

Coutts J, Roberts K, Frost F, Coutts A (2005) Extension for capacity building: a review of extension in Australia in 2001–2003 and its implications for developing capacity into the future. Rural Industries Research and Development Corporation (publication No. 05/094). RIRDC, Kingston ACT

Dunne AJ (2010) Contemporary issues in the provision of tertiary agriculture programmes: a case study of The University of Queensland. Australa Agribus Perspect 18(82):1–12

Falvey JL (1998) Are faculties of agriculture still necessary? Australian Academy of Technological Sciences and Engineering, Focus 103:July/August

Farrington J (1995) The changing public role in agricultural extension. Food Policy 20(6):537–544

Garden Advisory Service (1989) The gardener's guide. Department of Agriculture and Rural Affairs: Victoria. http://trove.nla.gov.au/work/8204451?q=Garden+Advisory+Service&c=book&sort=holdings+desc&_=1357518692215&versionId=40294945. Accessed 7 Jan 2013

Hekkert MP, Suurs RAA, Negro SO, Kuhlmann S, Smits REHM (2007) Functions of innovation systems: a new approach for analysing technological change. Technol Forecast Soc Change 74:413–432

Hitchmough JD (1994) Urban landscape management. Inkata/Butterworths, Sydney
Howell J (1982) Managing agricultural extension: the T and V system in practice. Agric Adm 11:273–284
Howells J (2006) Intermediation and the role of intermediaries in innovation. Res Policy 35:715–728
Hunt W, Birch C, Coutts J, Vanclay F (2012) The many turnings of agricultural extension in Australia. J Agric Educ Ext 18(1):9–26
Kidd AD, Lamers JPA, Ficarelli PP, Hoffman V (2000) Privatizing agricultural extension: caveat emptor. J Rural Stud 16:95–102
Klerkx L, Nettle R (2013) Achievements and challenges of innovation co-production support initiatives in the Australian and Dutch dairy sectors: a comparative study. Food Policy 40:74–89
Klerkx L, Leeuwis C (2008) Institutionalizing end-user demand steering in agricultural R & D: farmer levy funding of R & D in The Netherlands. Res Policy 37:460–472
Klerkx L, Leeuwis C (2009) Establishment and embedding of innovation brokers and different innovation system levels: insights from the Dutch agricultural sector. Technol Forecast Soc Change 76(6):849–860
Knickel K, Brunori G, Rand S, Proost S (2009) Towards a better conceptual framework for innovation processes in agriculture and rural development: from linear models to systemic approaches. In: 8th European International Farming Systems Association Symposium, Clermont-Ferrand (France), 6–10 July
Lineberger RD (2009a) Evolution of web-based collaborative learning environments in horticulture. Acta Hortic 832:13–18
Lineberger RD (2009b) Technology-mediated instruction: shifting the paradigm of horticultural education. Acta Hortic 832:107–112
Lineberger RD (2009c) IT Evolution of web-based collaborative learning environments in horticulture. Acta Hortic 832:13–18
Leeuwis C, van den Ban A (2004) Communication for rural innovation: rethinking agricultural extension. Blackwell, London
Macadam R, Drinan J, Inall N, McKenzie B (2004) Growing the capital of rural Australia—the task of capacity building. Rural Industries Research and Development Corporation (publication 04/034). RIRDC, Kingston
McEvilly G, Aldous DE (2010) Guiding young people to horticulture. Chron Hortic 50(3):16–18
McSweeney P, Rayner J (2011) Developments in Australian agricultural and related education. J Higher Educ Policy Manag 33(4):415–425
Malcolm B (2010) Agriculture and agricultural science: where have all the young people gone? Agric Sci 22(3):35–39
Marsh SP, Pannell DJ (2000) The new environment for agriculture: fostering the relationship between public and private extension (RIRDC Publication No. 00/149). Rural Industries Research and Development Corporation, Canberra
Murray M (2005) The late, great California extension system: what went wrong? Acta Hortic 672:277–284
Nettle R (2003) The development of a National Dairy Extension Strategy—a literature review. Institute of Land and Food Resources, Innovation and Change Management Group, The University of Melbourne
Nettle R, Brightling P, Hope A (2013) How programme teams progress agricultural innovation in the Australian dairy industry. J Agric Educ Ext 19(3):271–290
Nettle R, Lamb G (2010) Water security: how can extension work with farming worldviews? Ext Farming Syst J 6(1):11–22
Nettle R, Paine M (2009) Water security and farming systems: implications for advisory practice and policy-making. J Agric Educ Ext 15(2):147–160
Poisot A, Speedy A, Kueneman E (2007) Good Agricultural Practices—a working concept. Background paper for the FAO internal workshop on good agricultural practices, Rome, October 2004. Food and Agriculture Organization of the United Nations
Pollan M, (1996) Second nature. Bloomsbury, London

Pratley J (2008) Workforce planning in agriculture; agricultural education and capacity building at the crossroads. Farm Policy J 5(3):27–41
Pratley J (2012a) Professional agriculture—a case of supply and demand. The Australian Council of Deans of Agriculture, Paper 1 http://www.csu.edu.au/special/acda/papers.html. Accessed 4 Jan 2013
Pratley J (2012b) The workforce challenge in horticulture. Agric Sci 24(1):26–29
Rabinow P (1989) French modern: norms and forms of the social environment. MIT, Cambridge
Rayner J, McSweeney P, Raynor K, Aldous DE (2009) Where to now for horticultural higher education in Australia? Acta Hortic 832:185–194
Rivera WM (2001) Agricultural and rural extension worldwide; options for institutional reform in developing countries. Extension, Education and Communication Service, FAO, Rome. ftp://ftp.fao.org/docrep/fao/004/y2709e/y2709e.pdf. Accessed 4 Jan 2013
Rivera WM (2008) Pathways and tensions in the family of reform. J Agric Educ Ext 14(2):101–109
Roberts K, Paine MS, Nettle RA, Ho E (2004) Mapping of rural industries service providers. Co-operative venture for capacity building. RIRDC, Canberra
Robson MG (undated) Rutgers New Jersey Agricultural Research Station Newark. http://njaes.rutgers.edu/about/NJAES-presentation-Newark.pdf. Accessed Jan 3 2013
Roling N (1988) Extension science: information systems in agricultural development. Cambridge University, Cambridge
Royal Horticultural Society Annual Review (2011–2012) Royal Horticultural Society, London. www.rhs.org.uk. Accessed 8 Dec, 2012
Stantiall J, Paine M (2000) Agricultural extension in New Zealand- implications for Australia. Paper presented at the Australasia Pacific Extension Network national forum Creating a Climate for Change; Extension in Australasia, Melbourne. http://www.regional.org.au/au/apen/2000/4/stantiall.htm. Accessed 4 Jan 2013
Tzoulas K, Korpela K, Venn S et al (2007) Promoting ecosystem and human health in urban areas using green infrastructure: a literature review. Landsc Urban Plan 81:167–178
Ward RA, Woods T, Wysocki A (2011) Agribusiness extension: the past, present and future? Int Food Agribus Rev 14(5):125–140
Ward-Thompson C, Travlou P (eds) (2007) Open space, people space. Taylor and Francis, London
Warrington IJ, Wallace BD, Scarrow S (2004) International perspectives on horticultural extension: a New Zealand viewpoint. HortTechnology 14(1):20–23
Weisenhorn J (2012) University of Minnesota Extension website. http://blog.lib.umn.edu/efans/mgdirector/. Accessed 7 Jan 2013
Whitehead A (2001) Trade, trade liberalization and rural poverty in low-income Africa: a gendered account. Background paper for the Least Developed Countries Report 2002. United Nations Conference on Trade and Development (UNCTAD), Geneva

# Chapter 34
# Increasing the Economic Role for Smallholder Farmers in the World Market for Horticultural Food

**Roy Murray-Prior, Peter Batt, Luis Hualda, Sylvia Concepcion and Maria Fay Rola-Rubzen**

**Abstract** Smallholder farmers will be critical to meeting the growing demand for food in the next 40 years. However, currently they face many challenges in meeting the changing demands of modern markets, including the effects of climate change, deficiencies in their enabling environment, resources, capacities and institutional models for change and development. In this chapter we set the context by defining these deficiencies and their implications for development of the smallholder horticultural sector. We present a dualistic agribusiness systems framework that helps focus analysis on the interactions in the system and the complexity of the problems. This framework helps highlight the need to develop new institutional approaches to link smallholder farmers to markets and to improve their productivity. We then review some options for linking them to markets and conclude that a range of solutions will be required, but that contract farming and traditional cooperatives will only be relevant to a limited range of contexts. We suggest that cluster marketing arrangements will be another important solution, because they are suited better to smallholder resources and capacities. They can also be used as a means to develop a horticultural innovation system that meets the needs of smallholder farmers rather than just the needs of larger enterprises.

R. Murray-Prior (✉) · P. Batt · L. Hualda
School of Management, Curtin University, GPO Box U1987, Perth 6845, WA, Australia
e-mail: roy@agribizrde.com

P. Batt
e-mail: p.batt@curtin.edu.au

L. Hualda
e-mail: luis.hualda@postgrad.curtin.edu.au

S. Concepcion
School of Management, University of the Philippines Mindanao, Mintal, Davao, The Philippines
e-mail: sbconcepcion@yahoo.com

M. F. Rola-Rubzen
CBS—Research & Development, Curtin University, GPO Box U1987,
Perth 6845, WA, Australia
e-mail: f.rola-rubzen@curtin.edu.au

G. R. Dixon, D. E. Aldous (eds.), *Horticulture: Plants for People and Places, Volume 3*,
DOI 10.1007/978-94-017-8560-0_34, 

**Keywords** Agribusiness · Development · Cooperatives · Clusters · Value chains · Enabling environment · Smallholder farmers

## "Business as Usual" is not an Option

'"Business as usual" is not an option' is the striking heading at the beginning of the *World Economic and Social Survey 2011* (DESA 2011 p. v). The statement refers to the need for a transformation of the models of economic growth and development because the current paradigms will lead to the depletion of the world's resources and the pollution of the natural environment. It also acknowledges that economic progress must improve if the populations of developing countries are to have a decent standard of living. This coincides with an increasing awareness by world organisations such as the World Bank (WB), the Food and Agriculture Organization of the United Nations (FAO) and the Organisation for Economic Co-operation and Development (OECD), non-government organisations (NGOs such as OXFAM and the International Food Policy Research Institute (IFPRI)) and national governments that investment in agricultural development and innovation has not kept pace with the rising demand for food (Viatte et al. 2009; Nelson et al. 2010; OECD/FAO 2012). Underinvestment in agricultural productivity, population growth and the effects of severe climatic disturbances have caused more volatile and higher world food prices. The undernourished population worldwide had been declining from around 20 % in 1990 (Nelson et al. 2010), but this trend began to reverse at the turn of the century, with numbers increasing to over one billion following the food price spikes of 2008, 2011 and 2012 (DESA 2011). Food production will need to increase to meet the increased demand, with much of this increase to come from smallholder farmers, many of whom produce horticultural crops (DESA 2011).

To enable smallholder horticultural farmers to improve their productivity so that they can become part of the solution to the emerging food security problems, a transformation of the horticultural sector is required in developing countries that involves the whole agribusiness system and the food chains that smallholder farmers currently supply and will supply into the future. The focus of this chapter is therefore on the social, economic and environmental justifications and the approaches to including smallholder farmers in horticulture food markets, including the modern institutional markets. It begins by discussing the context for increasing the role of smallholder farmers in a discussion of constraints in the enabling environment that currently limits their ability to compete with larger farms. It outlines a framework for analysing the agribusiness system incorporating the enabling environment, the actors in the chain and other elements required to develop solutions to the complex range of issues to be addressed if smallholder farmers are to be involved in modern markets. Finally, it discusses some models for linking smallholder farmers to these markets and suggests that small cluster marketing groups will be an important solution to this issue as well as providing a means to integrate them into research, development and extension programs to meet the needs of smallholder farmers.

## The Context for Increasing the Role of Smallholder Farmers

### *Need for Increased Food Production and Productivity*

At the High-Level Expert Forum on How to Feed the World to 2050, attention was drawn to the greatest challenge humankind is facing—how to feed a growing population in the face of declining growth in agricultural productivity, climate change and a fast changing consumer demand (FAO 2009). According to FAO, by 2050, the world's population would have grown to over 9 billion, with the majority of this increase occurring in developing nations, where most of the poor live and where most of the smallholder farmers reside. Failure to meet the food requirement will lead to food insecurity with consequences for the entire world including hunger, malnutrition and conflict.

Undernourishment is a key indicator of food insecurity, with increased food insecurity for the latter part of the last decade causing rioting in some countries and resulting in changes to governments and political systems. This has the potential to become a continuing issue that will have consequences for all countries because of the projected increase in the world's population (DESA 2011). The projected increase in population, along with increased living standards in some countries, will lead to increased demand for food, requiring food production to increase by at least 60 % by 2050 (DESA 2011; OECD/FAO 2012).

While the so-called 'green revolution' of the 1960s and 1970s increased food productivity and food production, it did not lead to sustainable management of resources (DESA 2011). The increases necessary to meet the demand for food by the current world population have led to adverse environmental outcomes including land degradation, loss in biodiversity, climate change, reduction in forests, and pollution of water and marine ecosystems. The exponential increase in some of these adverse outcomes will have serious consequences if we do not adopt more sustainable production systems at the same time as we are increasing food production to meet the growing world demand (Nelson et al. 2010; DESA 2011; OECD/FAO 2012). At the same time, we face additional problems due to increased $CO_2$ in the atmosphere (leading to global warming and climate change), increasing prices of traditional energy sources, and little additional arable land available for development.

Climate change is leading to drying climates in some parts of the world and could be the reason for the increased volatility in world grain prices between 2007 and 2012 due to droughts in the grain belts of Russia, Ukraine, USA and Australia. In other places it is leading to increased flooding and other violent weather events. It appears that the effect of climate change is to increase the probability of extreme weather events (DESA 2011), which accentuate yield volatility and hence price volatility. Food inventories may need to be increased in many countries to provide a safety net for the poor.

Increased fossil fuel prices are a problem for the developed country production systems in particular because their key inputs such as diesel, fertiliser, pesticides and transport are linked closely to energy prices. Many of the green revolution innovations in developing countries also relied on these inputs (DESA 2011) and they are a key component of grain and some export crop production in these countries. Increased energy prices and concerns about $CO_2$ emissions have led to increased demand for biofuels, which compete with food crops for land and inputs and drive up the price of food. The pricing of carbon to tackle $CO_2$ emissions and climate change will increase the price of fossil fuel-based inputs further and drive research and development to find alternatives.

Competition for land and water resources will also increase from alternative uses such as the growth of cities and recreational and environmental uses. While OECD/FAO (2012) predicts a slight increase in arable land used for agriculture by 2050 (less than 5 %), this will be due to an increase in some developing countries offsetting a decline in developed countries. However, they also predict a decline in water availability for agriculture by 2050, with uneven distribution being a key constraint. Consequently, the required increases in food production will have to come from improvements in land and water productivity rather than from increased land area.

Despite the need for agricultural production to grow and meet demand, OECD/FAO (2012) predicted agricultural production is to slow from 2 % in recent decades to 1.7 % in the next decade. Consequently, the 'key issue facing global agriculture is how to increase productivity in a more sustainable way to meet the rising demand for food, feed, fuel and fibre' (OECD-FAO 2012, p. 15). While productivity has been increasing, the increases have not been consistent across regions (DESA 2011; OECD/FAO 2012), with the largest increases being in developed countries. However, there is evidence that productivity has begun to slow in developed countries and developing countries (OECD/FAO 2012), which has been partly linked to pressure on resources, but also because the easier options of improved seeds, fertilisers and other inputs have been adopted in many cases (Hazell et al. 2006). Accordingly, the largest increases in production and productivity in the next 40 years are projected to come from developing countries. This is due to a greater availability of land for agriculture and greater potential to increase productivity through reducing the gap between actual and potential crop yields and efficiencies. Therefore, farmers in developing countries will have a major role in providing the increase in food required for humankind.

## *Importance of Smallholder Agriculture in Meeting the Challenges*

The increases in food production resulting from the 'Green Revolution' of the 1960s and 1970s and the influence of free market thinking on investment priorities, led to a level of complacency about food supplies and resulted in a decline of investments in agricultural productivity by developing country governments and international donor agencies (Oxfam 2008; FAO 2010; Heady and Fan 2010; Nelson et al. 2010; DESA 2011; Islam 2011). This also coincided with a declining importance of the agricultural sector as a proportion of GDP and in some countries expanding mineral and manufacturing sectors that are perceived to provide a better return. However, some governments, donor agencies and international institutions have recognised that this has gone too far and it has begun to be reversed (FAO 2010; Heady and Fan 2010).

Most food (particularly fresh fruit and vegetables) is locally produced and consumed, much of it by smallholder farmers (DESA 2011). In the developing world, 3 billion people live in rural areas, of which 2 billion live on small farms (less than 2 ha) (Hazell et al. 2006). These people include half the world's undernourished people and a majority of the people living in absolute poverty. Consequently, smallholder farmers are at the heart of the food insecurity challenge (DESA 2011).

Historically, economic development has led to a migration of people from rural areas to higher paying jobs in urban areas and an increase in farm size, reducing the disparity in incomes between the rural and urban populations (Davis and Goldberg 1957; Hazell et al. 2006; DESA 2011). This development has mostly been combined with investment to improve agricultural productivity, which has been both an engine for economic growth and a major contributor to decreasing poverty. However, the scale and speed of the changes necessary to meet the rising food demand in developing countries over the next 40 years, means that these changes may not occur quickly enough (Hazell et al. 2006). Conversely, when governments have invested in improved agricultural productivity on small farms, poverty and undernourishment in rural areas have been dramatically reduced (Diao et al. 2010). In fact, investment in agriculture and particularly in staple foods leads to much greater reductions in poverty than investments in other areas (Oxfam 2008). It also leads to broad-based growth and through reducing food prices, leads to improvements in the local economies. Evidence from the Philippines is that it leads to increased local employment and spending (Rola-Rubzen et al. 2012). When this investment does not occur, the rural poor remain so and this often leads to degradation in the ecosystem (DESA 2011) and resultant political unrest. Therefore, investment in improvements to smallholder agricultural productivity is required to meet the rising food demand, but also as a driver of economic development and declining poverty levels.

Another key to improving the diet of poor people is to increase their consumption of vegetables and fruits. Much of this product is perishable and therefore has to be grown close to the point of consumption if it is to be affordable (Moustier 2012). Therefore, smallholder horticultural farmers will have an especially important role in these markets. The increasing role of global trade and the emergence of global food manufacturers, food service chains and global retailers present both an opportunity, but also a problem for smallholder farmers.

## *Changing Global Agrifood Industry*

With economic development, significant changes become evident in food supply chains. In the first instance, greater urbanisation means a larger proportion of the population is disconnected from food production and reliant on the food distribution system. Rising incomes lead to a substantial reduction in the consumption of cereals, roots and tubers and an increased demand for meat, dairy, oil and fresh fruit and vegetables (Gehlhar and Regmi 2005). In parallel, there is a marked increase in the consumption of food away from home and with more busy lifestyles, an increasing demand for more convenient ready-to-eat foods. With the greater ownership of motor vehicles, refrigerators and microwave ovens, consumers not only shop less regularly, but they are more inclined to purchase from modern retail outlets (Shepherd 2005).

Not only is there a greater demand for a greater variety of food, but consumers are showing a greater interest in the holistic attributes of the food that they consume. Consumers want to know who produced the food, where and how. There is a growing demand for food that is more healthy, that contains less fat, less salt, less sugar, fewer additives and fewer preservatives (Batt et al. 2006). The desire for better health and greater nutritional value leads to the development of more functional foods with added vitamins, minerals and fibre. The growing awareness of the impact of food miles on greenhouse gas emissions has led to a growing desire for local food that has been produced in a more sustainable manner.

However, few producers sell directly to consumers; most sell through one or more market intermediaries. In this respect, the increasingly globalised nature of the food processing, retailing and food service sectors is having a profound effect on producers. In the first instance, the amount of fresh produce traded internationally has increased rapidly (Humphrey 2006). Not only has the composition of exports changed dramatically, there has been a marked increase in the number of alternative suppliers, intensifying the competition in the market. In parallel, in response to saturation in their home markets and new opportunities arising from economic growth, population growth and a progressive easing of the restrictions on foreign direct investment, aggregation and concentration in food processing, manufacturing, retailing and the food service sector have intensified (Batt 2006).

For these large institutional buyers, purchasing on the spot market is no longer a viable option. The inherent variability in the quantity, quality and range of products makes it impossible to adequately price the product or to engage in any generic promotion or product merchandising (Batt 2006). To overcome these impediments, food retailers, processors and manufacturers have developed alternative purchasing strategies including centralised procurement, specialised or dedicated wholesalers, preferred supplier systems and concessionaires (Shepherd 2005).

Preferred suppliers are able to offer a regular and reliable supply of a range of good quality products at a predetermined price. While quality is a physical description of the product in terms of its size, shape, colour, freedom from pests and diseases, purity (in terms of its freedom from chemical contaminants, pathogenic organisms and genetically modified plants), maturity or freshness, it also describes the manner in which the product has been packed and the way a supplier goes about delivering the product to the customer (Batt 2006). Though this means being able to deliver the product when the customer wants it, by implication it also involves many inter-related activities such as production scheduling, post harvest storage and warehousing, logistics, ordering and invoicing.

With the need to differentiate products in saturated markets, Codron et al. (2005) extends the quality concept to include the sensory attributes, health attributes, process attributes and convenience. The sensory attributes refer to the classical aspects of food quality: taste, appearance and smell. Health, as a choice criterion, is primarily about communicating both the short-term and long-term benefits arising from the consumption of various foods. The process attributes relate to the consumers interest in the processes used in food production, even though such processes may have no tangible impact on the final product. Nevertheless, greater numbers of consumers are demonstrating that they are willing to pay significant price premiums for natural or organic products, Fairtrade products (products registered as coming from smallholder producers) and those that minimise the impact on the environment. The convenience attributes are those aspects of a food product that reduce the amount of time household members typically spend on shopping, food storage, food preparation, eating and food disposal. Convenience is also associated with 'eating on the run', where consumers chose those products that can be eaten in one hand without making a mess (Martech Consulting 2005).

However, implicit in any definition of food quality is the underlying assumption that the food is safe to eat. Regrettably, with the increasing reliance on convenience foods, the greater consumption of food away from home, the increased volume of trade in fresh and processed food products, and the increasing desire for fresh and natural food products, there has been a marked increase in the number of food safety incidents (Kaferstein 2003). For fresh produce, the major health concerns appear to relate to the presence of chemical residues (Shepherd and Galvez 2007). With limited knowledge, illiteracy, inappropriate labelling and persuasive sales representatives, the overuse of chemicals is frequent among smallholder farmers (Ketelaar 2007). Many smallholder farmers apply pesticides too often, at rates often much greater than label recommendations and too close to harvest (Shepherd and Tam 2008). Other farmers apply chemicals immediately prior to harvest to improve the physical appearance (Davies et al. 2006; Shepherd and Tam 2008).

More recently, the microbiological contamination of fresh produce has become a major issue, with some of the most recent and serious food safety incidents involving spinach in the US and organic bean sprouts in Germany. Biological contamination may arise from the irrigation or washing of fresh produce in water contaminated by both human and animal waste, poor personal hygiene and the frequent use of poorly composted animal manures (Shepherd and Tam 2008). In some instances, the reuse of fertiliser bags or bags used for the transport of animal manures is a common practice.

In order to protect both consumers and the integrity of their brands, most retailers and food manufacturers have implemented one or more quality assurance programs to identify those critical steps in the chain that are most likely to lead to contamination. Rather than rely on end-point inspection, the preferred strategy for minimising the risk of contamination is the Hazard Analysis Critical Control Point methodology (HACCP), which focuses on prevention (Baines et al. 2006).

In addition to requiring suppliers to meet food safety standards, several global retailers and food manufacturers have also specified product quality criteria. Not only does this enable buyers to specify how products should be grown, harvested, transported, processed, packaged and stored, it has provided them with the power to impose their requirements on other actors in the value chain and to reward compliance (Humphrey 2006). The majority of these standards are based on Good Agricultural Practices (GAP), which not only provide an assurance of food safety, but also focus greater attention on the adoption of sustainable farming practices. These endeavour to ensure that farmers are adopting and following prescribed crop rotations, minimising the application of fertilisers, pesticides and herbicides to reduce environmental contamination, protect ecological diversity and minimising the eutrophication and pollution of waterways from excessive run-off. Good practices extend to the appropriate use of chemical storage facilities and protective equipment to protect against accidental poisoning, occupational health and safety, and animal welfare (Akkaya et al. 2006).

Recent shifts in the regulatory environment have exacerbated the widely held view that smallholder farmers may be marginalised—if not excluded completely—from participating in modern supply chains (Humphrey 2006). Many buyers believe that smallholder farmers, even when they are organised into collaborative marketing groups, are unable to supply a sufficient quantity of good quality product. Even with appropriate training, as the farmers effectively pool their produce, the risk of non-conformance to prescribed standards is multiplied greatly, thereby demanding more frequent monitoring and inspection. Furthermore, the majority of smallholder farmers are unable to comply with many of the requirements such as concrete floors, foot operated hand basins, or to provide a reliable source of potable water (Shepherd 2005).

## *Constraints in the Enabling Environment for Smallholder Agriculture*

De Oliveira (2007, p. 57) defined the enabling environment as 'all the factors that are external to the agribusiness itself but which affect the way businesses operate and impinge on the development of the private sector'. This is based on the context that the government should lay out policies that are conducive for the business of the private sector. Rottger and Da Silva (2007, p. 5) identified that the factors for creating an enabling environment included 'macroeconomic and political stability, efficient land markets and tenure systems, consistent open trade policies, rural and agricultural financial service delivery, availability of human resources, well-functioning public-private partnerships (PPP), good governance, and the availability of improved technologies'. Conditions in the enabling environment were categorised by Christy et al. (2009) as including essential, important and useful enablers. Essential enablers include land tenure and property rights, infrastructure and trade policies. Important enablers are financial services, standards and regulations, and research, development and extension services. Useful enablers include business development services, business linkages and ease-of-doing business. In the Philippines, impediments to investments in agriculture are summarised by Habito and Briones (2005) as including access to public and private land, inadequate infrastructure, poor local governance, limited access to long-term finance, limited access to technology, limited access to raw materials, lack of global market access, unstable peace and order, widespread corruption and weak enforcement of contracts.

## Land Tenure and Policy

Land is a necessary resource in agriculture production and a limiting factor, because only so much can be grown on a given area. In the Philippines, average land sizes of farms were reduced to 2 ha in 2002 from an average of 3.2 ha in the 1960s (Canlas et al. 2011). The reason given for this reduction in average farm size is that land is divided among children for their inheritance.

Aside from the reduction of farm sizes, there is also the threat of competition for agricultural land from foreign companies (Polack 2012). Foreign companies in biofuel production and plantation crops such as bananas are investing in land. There is also the conversion of land use from agricultural to industrial, residential and commercial (Kelly 2003). Agricultural land located in the periphery of urban areas is converted to other uses as population increases and urban areas expand. Both of these concerns occur because of weak policies (Kelly 2003; Polack 2012).

The agrarian reform program in the Philippines aimed to provide access for landless farmers to land resources. It was able to provide positive impacts to beneficiaries with evidence of higher income and reduced poverty incidence (Reyes 2002). Access to land can be of equal importance to ownership when it comes to land concerns. Marginalised farmers have lesser land to cultivate which limits their productivity. However, if access to land is granted through institutional mechanisms such as policies that provide disincentives for idle lands, then smallholder producers can be granted opportunities to expand their production (Smit and Nasr 1992).

Access to land for crop production does not necessarily require providing ownership to smallholder producers. Rather, what is necessary is that there are policies and tenure instruments that provide access and security in the use of land. It is a matter of implementing policies and creating innovative solutions to provide smallholder producers access to a necessary resource. It is also necessary for the government to set priorities for the allocation of land between uses and users.

## Infrastructure

National government projects such as the Strong Republic Nautical Highway (SRNH) in the Philippines provided benefits to agricultural producers by reducing transportation costs by as much as 33 %, and providing access to new markets (Basilio 2008). Teruel and Kuroda (2005) concluded that the reduced investment in infrastructure is one of the causes of the decrease in the productivity of agriculture in the Philippines. The presence of national level infrastructure that connects communities helps in utilising comparative advantages in agricultural/horticultural production, which can enhance efficiency in the use of national resources. This can also create better access to commodities for consumers.

Communication facilities improve coordination among smallholder producers. It is also used in expanding the coverage of extension services (Olchondra 2010). Communication services are limited in that these are usually owned by the private sector and investments are made where it will be most profitable. Schrekenberg and Mitchell (2011) suggested that governments can provide policies and incentives to the private sector to expand coverage of communications services. Public-Private Partnerships (PPP) is an avenue that can be used in improving infrastructure facilities in rural areas, but requires projects to be profitable (Warner et al. 2008). In this regard, delivering conditions that will assist smallholder producers should also involve providing conditions to the private sector that can assist smallholder producers.

## Research, Development and Extension

Research, development and extension remain important to smallholder producers because of changing consumer demands, changing structures and climate change. Smallholder producers need to be able to adapt to consumer demands with respect to variety. Increasing demand will also mean that productivity must be enhanced. As agricultural productivity has plateaued, it is critical that research and extension continue to spur growth in the agricultural sector (FAO 2009). According to Beintema and Elliott (2009), declining rate of growth in agricultural/horticultural R&D investment has been associated with a decline in the growth of agricultural/horticultural productivity, which has been statistically linked to the change in the composition of research away from productivity-enhancement at the farm or grower level.

Of equal importance are extension services that bring research outputs to rural communities. In the Philippines, weaknesses in the extension services can be traced to the decentralisation of agriculture services from the national to the local level (Prantilla 2011). Prantilla (2011) found that the performance of agriculture extension services is dependent on the support given to agriculture by the local chief executive. Coordination between the levels of government was reduced because decentralisation gave local governments autonomy in deciding priorities and developing and implementing agriculture programs.

## Financial Services

Interest rates given to larger firms by financial institutions are 7% per annum while smaller firms are charged 10-12% per annum (Canlas et al. 2011). However, the higher interest rates can be explained by higher risks and the operational expenses involved when dealing with smallholder producers (Armendariz and Morduch 2010). Microfinance institutions (MFIs) are committed to serving their clients by ensuring that they are able to reach them even with high operational costs (Mendoza and Vick 2010).

Microfinance in the Philippines has been deregulated, giving MFIs the freedom to develop their own strategies (Quinones and Seibel 2000). MFIs tend to favour clients who have better capacity to repay their loans, which results in mission drift (Aubert et al. 2009). Mission drift can put smallholder producers at a further disadvantage due to limited access to financial services.

High interest rates charged by MFIs to smallholder producers is a constraint that cannot be solved easily because this is tied to operational effectiveness of firms. Without higher interest rates, the viability of the operations of MFIs can be compromised, leaving smallholder producers with less access to financial services.

### Business Development Services

As farming enters the century of globalisation, smallholder farming increasingly has to operate in a complex business environment. Unfortunately farmers are often ill-equipped to deal with players in the modern chain, and access to business development services can facilitate their participation in modern markets.

Business services are delivered by public sector providers, private providers, NGO providers and cooperatives or membership-based organisations (Kahan 2007). Business development services can be a function performed by non-government organisations that operate on a specific timeframe and project budget. Upon the completion of the project, business development services can cease to exist except if they build in sustainability of operations. UMFI adopted a model whereby they provide paid services to the communities they serve. This means that they can generate funds to sustain their operations.

The Pecuaria Development Cooperative Incorporated (PDCI) was supported by the Upland Marketing Foundation Incorporated (UMFI) (Concepcion et al. 2011) to find the right marketing channel for their organic rice. Services also included developing the brand of the organic rice including labelling and development of packaging. It was able to provide higher income for smallholder producers as it realised the potential value of organic rice.

### Issues at Different Levels

Constraints in the enabling environment emerge at the national, meso and micro levels, although most of the constraints may emanate from national level decisions because of conflicting policies. For example, the decentralisation of agricultural services in the Philippines was designed to enhance the provision of services based on the rationale that it is the local government that knows the local situation better, and thus it is in the best position to provide services. The policy inadvertently weakened agricultural services because not all local government units had supportive local chief executives.

Institutions at the micro level are in a better position to provide favourable conditions to smallholder producers primarily because they have greater control of the situation, and macro level influences are also beyond their control (Bryant 1989). Even though there are macro-environment factors that are exerted at the micro level, institutions and stakeholders at the local level can make the necessary adjustments.

Smallholder producers find it difficult to comply with modern market requirements in terms of quality, variety, volume and consistency. However, research by the Regoverning Markets Programme found that modern markets do not necessarily exclude smallholder producers (Vorley 2011). Examples in the Philippines like PDCI and NorminVeggies show how smallholder producers are able to supply supermarkets and institutional buyers with the help of non-government organisations (Concepcion et al. 2007a, b).

Pinstrup-Andersen and Watson (2011) found that the role of the government in food policy passed through several phases. These phases included leaving markets alone, followed by heavy interventions, then the "government as a problem" phase, and "getting the institutions right". Heavy interventions resulted in governments creating more problems rather than setting things right. The recognition that harm has been done by heavy government interventions needs to be corrected by getting the institutions right. This highlights the recognition of the importance that public-private partnerships can contribute in promoting inclusive growth.

The fourth phase of getting institutions right recognises that each stakeholder has its role to fulfil in the agribusiness system. Addressing constraints in the enabling environment involves participation from different stakeholders of the agribusiness system. It is also acknowledged that these stakeholders have their own objectives to meet, and implies that compromises should be made. The role of the government is to support policies that will allow the private sector to support the inclusion of smallholder producers.

## *Summary of the Context for Increasing the Role of Smallholder Farmers*

In this section, we have shown that the world has serious and complex problems to address if it is to continue to feed its growing population, while maintaining a liveable quality environment. The 'success' of the 'Green Revolution', which resulted in lower food prices and declines in the proportion of the world's population who were undernourished led to complacency about food security and underinvestment in agricultural research and development. The recent food price spikes have led to increased focus on food insecurity in recognition of the rapidly increasing population, combined with rising energy prices, global warming and climate change, and depleting natural resources. Because of their intimate involvement with the production of food and the large numbers of farm households living impoverished and undernourished lives, they will need to be actively involved in meeting these challenges. Apart from their obvious lack of human and produced economic capital, they face additional issues in increasing their productivity including difficulties in accessing the expanding value chains, which in turn are constrained by deficiencies in their enabling environment. Nevertheless, solutions to these challenges are required, so the rest of this chapter will outline some ideas about how smallholders can become part of the solution rather than part of the problem.

## A Framework for Undertaking a 'Sustainable Agribusiness Transformation' in the Horticulture Sector of Developing Economies

Because 'Business as usual' is not an option, we must craft new ways to address the complex issues involved in developing innovative smallholder farming systems able to adapt fast enough to meet food production targets. We argue that a 'Sustainable Agribusiness Transformation' is required rather than simply another 'Green Revolution' to meet the economic, social and environmental requirements for future generations. This section outlines a framework for this to occur. This requires a holistic approach to analyse and understand the issues involved.

### *Defining the Elements of an Agribusiness System*

John Davis (1955) was the first to publicly use the term agribusiness in a presentation to the Boston Conference on Distribution in 1955. In 1957 Davis and Goldberg (1957, p. 2) defined agribusiness as:

> the sum total of all operations involved in the manufacture and distribution of farm supplies; production operations on the farm; and the storage, processing, and distribution of farm commodities and items made from them.

In other words, agribusiness is the set of interacting organisations that jointly provide food and fibre products for consumers. It includes all organisations that produce process or distribute food and fibre products and those organisations that provide inputs to those organisations. Davis and Goldberg (1957, p. 74) went on to say 'the problems of commercial agriculture ... need to be approached as agribusiness issues because both their cause and their solution encompass the off-farm functions of supply manufacturing and processing-distribution as well as on-farm production. *The point is that the approach to solutions must be as comprehensive as is the bases of the problems themselves*' (original italics). Essentially, they were arguing that the food system needs to be treated as a holistic holistic system and that attempts to address problems in the food system will fail if they attempted to address only portions or segments of the system.

Their arguments can be extended by drawing on general systems theory to consider the food system as an open system—an open agribusiness system. General systems theory stresses the need to understand the system as a whole and the arrangements of the components and relationships in the system (von Bertalanffy 1968; Checkland 1981; Kaine and Cowan 2011). Open systems interact with their environment, and the state of the environment influences its behaviour and changes in the state of the environment will change the behaviour of the system.

An agribusiness system can therefore be considered as those organisations that produce, process and deliver food and fibre products to consumers and it exists within a socio-economic and political or enabling environment that provides a framework of institutions and norms and values that define and constrain the operations of the system (Murray-Prior and Ncukana 2000). Such a representation is consistent with the call in DESA (2011, p. 83) that 'all actors, institutions and processes, within the whole food chain must be part of the policy innovation framework'.

For example, the vegetable agribusiness system for smallholder farmers in Mindanao in the southern Philippines can be represented as a supply chain with its associated components and actors and the associated environmental suprasystem (Fig. 34.1). In this diagram the systems boundary is defined for the actors and functions for a particular industry, although, it is possible to consider a broader system, incorporating all elements of the food system within a country (Murray-Prior et al. 2004). The suprasystem consists of the socio-cultural and political environment (which incorporates the enabling environment) and the agro-climatic-ecological environment (which incorporates the natural capital and effects of environmental changes). The issues affecting the suprasystem have been discussed in the global context in the previous section. Traditionally, smallholder vegetable farmers in Mindanao supply the wet markets through a system of traders that deliver variable quality product to the poorer consumer segments. The product passes through many hands and its quality is affected by the inefficient and substandard production, packaging, handling, logistic and marketing systems. While the markets are generally competitive (although not always for farmers), relationships are often adversarial. Information flows from the market back along the chain are mostly non-existent and most elements of the chain have a poor understanding of market requirements and demand. Consequently there is no mechanism in the traditional system for smallholder farmers to receive a higher price if they produce a premium product. In fact they have little understanding of the market requirements outside their immediate locations.

The model in Fig. 34.1 also shows products being delivered to supermarkets. While much of the vegetable product in supermarkets in the Philippines comes from traditional supply chains, mostly from wholesalers in the wet markets, increasing amounts of product are being supplied by sophisticated value chains, either sourced from large corporate farms or imported from overseas. The quality of much of this product meets the requirements of value chains for reliable, consistent quality that incorporates food quality, safety and traceability systems. Conceptually, these chains can be considered as a dual system to the traditional system.

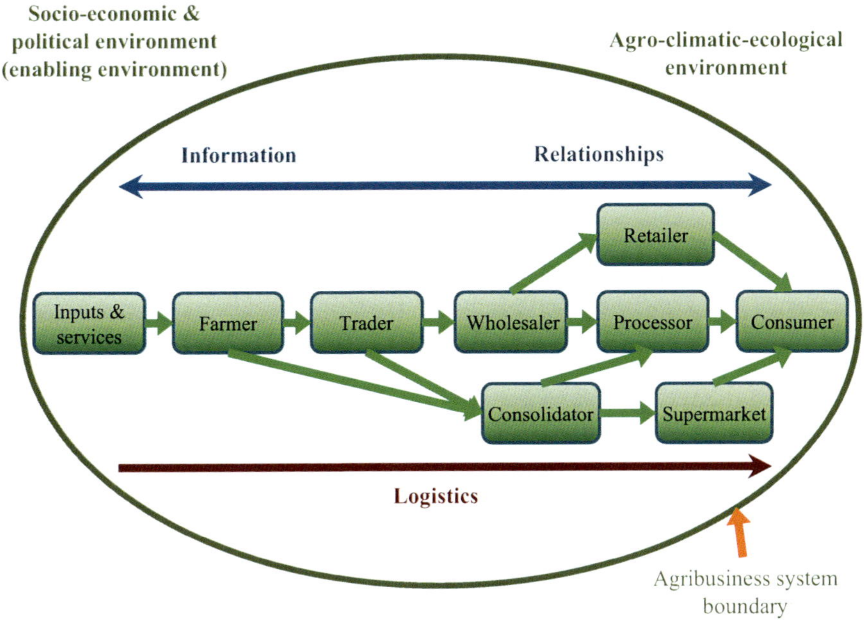


**Fig. 34.1** Simple representation of the agribusiness system for smallholder vegetable farmers in Mindanao, southern Philippines. (Adapted from Murray-Prior et al. 2004)

## *A Model to Analyse the Dual Agribusiness Systems for Horticultural Industries in Many Developing Economies*

The model of a dual agribusiness system was developed by Murray-Prior and Ncukana (2000) in order to conceptualise the issues facing agricultural development in South Africa. It provides a framework to analyse some of the issues that arise in attempting to outline a way forward for people in the 'resource poor' chains that supply traditional markets. Theoretical models of dual economies began with Lewis (1954) and were based around the ideas of two sectors in the economy, one an 'advanced' capitalist sector and the other a 'backward' predominantly rural sector. Singer (1970) outlines four key elements of dualistic economies:

1. The dual systems exist in a given space or country. The coexistence of these systems is often based on a dependency relationship.
2. Coexistence is persistent and will not necessarily disappear over time.
3. There may even be a tendency for the discrepancies between the two systems to increase rather than decrease.
4. There is no 'trickle' down effect from the 'advanced' sector to the 'backward' sector, in that the former does not pull up the latter and may even keep it down.

In many postcolonial, developing countries a 'resource rich' agribusiness sector coexists with a 'resource poor' agribusiness sector, with the former drawing on la-

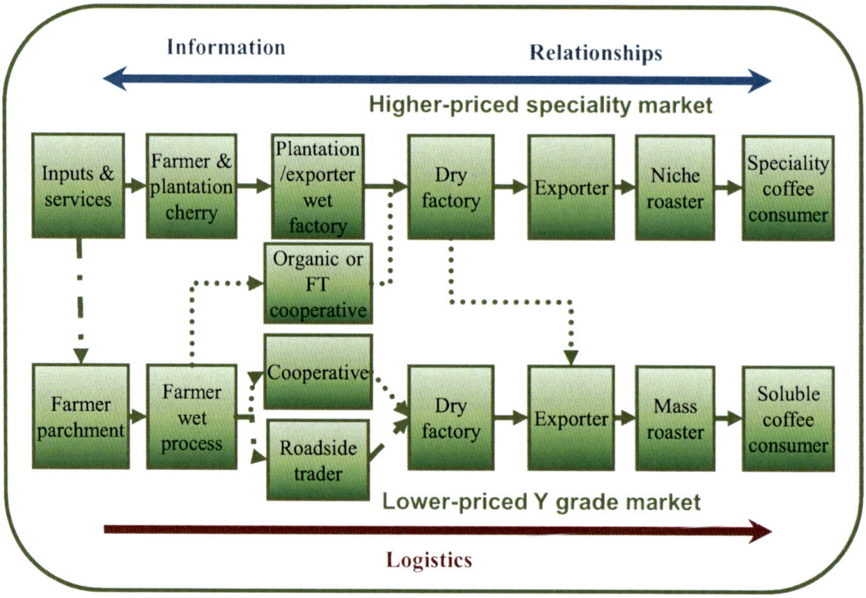


**Fig. 34.2** Model of the dualistic agribusiness system for Arabica coffee in Papua New Guinea. (Adapted from: Murray-Prior et al. 2008)

bour and some product from the latter, but providing little of value to it. In fact the existence of a highly developed, 'resource rich' agribusiness sector with its economies of size, complex systems of management and high standards of quality control make the markets they supply extremely difficult for 'resource poor' chains to penetrate. The latter are also at a disadvantage when competing for resources, whether the resources are physical, financial or human.

A dualistic agribusiness system can therefore be characterized as having two types of chains: a 'resource-rich' value chain that supplies high quality product to a higher priced market and a 'resource-poor' supply chain that supplies poorer quality product to a lower priced market. A good example of this can be found in Papua New Guinea's Arabica coffee industry in the highlands (Fig. 34.2). It has a plantation-based sector that produces Arabica green bean that is mostly sold to the speciality coffee markets and ends up in coffee houses such as Starbucks (Murray-Prior and Batt 2007). In this chain coffee cherry is processed in large wet factories that are run commercially and have exacting quality control standards. Green bean from these chains is sold at a premium to the Other Mild Arabicas contract on the New York ICE Futures market. On the other hand, green bean from the smallholder sector is sold to coffee roasters and is blended to produce soluble instant coffee. Smallholder coffee cherry is processed using village smallholder processing techniques that lead to variable quality and other defects in taste and presentation. Consequently it is sold at a discount to the Other Mild Arabicas contract. However, smallholder coffee cherry when processed through commercial wet factories (as is shown in Fig. 34.2) is sold to the speciality market. The use of such models provides a framework for analysing and finding solutions to the problems faced by smallholder farmers in these dualistic chains (e.g. Murray Prior et al. 2006, 2008). These solutions involve changes to the system and also to the enabling environment.

More broadly a dualistic agribusiness systems model has been incorporated within a pluralistic research framework, based on Checkland's soft systems methodology, to conduct research and development with agribusiness supply chains (Murray-Prior et al. 2004, 2007a). It enabled the research teams to identify which issues needed to be researched, what methodologies were appropriate for that research and to integrate research conducted by a multidisciplinary team of researchers.

## *Summary of a Transformation Framework*

In this section we have expanded the concept of agribusiness first expounded by Davis (1955) and Davis and Goldberg (1957) into a dualistic agribusiness systems framework that helps an analyst take a holistic view of the policy and research challenges facing the development of horticultural industries in the developing world. It also helps identify the components and the relationships critical to the functioning of a horticultural system, particularly those along smallholder supply and value chains and the constraints in the enabling environment that limit its ability to adapt and respond to the challenges it faces. This requires a focus along the value chains. One of the key issues identified in the context section and through the use of the agribusiness systems framework is the need to find ways to integrate smallholder farmers from developing countries into the changing modern markets. By doing so they will be able to help meet the increasing demand for food as well as improve their incomes to help move them out of poverty which the DESA report suggests is required.

## *Some Models for Linking Smallholder Farmers to Modern Markets*

The small volume produced by smallholder farmers as individuals means that an arrangement is required whereby their product can be consolidated to achieve the volumes required by modern retail markets. There has been a range of reviews of models for involving smallholder farmers in value chains (Batt 2007; Singh 2007; Vorley et al. 2009; Vermeulen and Cotula 2010; Moustier 2012). These models can be conceived as following two broad approaches: 'top down' approaches that involve a company structuring its arrangements with smallholder farmers in order to capture value and 'bottom up' approaches that involve smallholder farmers organising themselves in order to supply institutional markets. Of course, these are the two ends of the spectrum and there are examples of partnership models in the middle of the spectrum. There is insufficient space in this chapter to undertake a detailed examination of all the models, so our approach is to define the elements of the different approaches briefly while focussing on the 'bottom up' and partnership approaches.

## *'Top Down' or Buyer Driven Models*

Two drivers in modern value chains have led to the interest in buyer driven models that involve smallholder farmers. These are the move by supermarkets to source from preferred suppliers rather than wholesale markets (Vorley et al. 2009) and the large-scale acquisition of land by investors to supply institutional value chains (Vermeulen and Cotula 2010). Partly because of backlashes to some of these changes, but also because of the recognition that they are leading to the marginalisation of smallholder farmers and the concerns of some consumers about social and environmental issues, some businesses have begun to develop models for dealing with smallholder farmers.

Buyer driven models are normally organised through contracts between retailers or processors and farmers and are commonly known as contract farming arrangements. It is a form of vertical integration in which retailers or processors try to gain a competitive advantage over their competitors through creating efficiencies in the chain or improved product quality. They do this by establishing greater control over production processes and therefore improving reliability, consistency and quality of the final product (Vorley et al. 2009; Prowse 2012). It can also enable them to implement quality assurance procedures, which has risk management advantages. While other models are possible, such as management and lease contracts and joint ventures, these are not necessarily associated with linking smallholder farmers to market and are not considered in this discussion.

Contract farming arrangements involve advance contracts between farmers to deliver a specified quantity and quality of a product to a buyer at a specified time, place and price (Singh 2007; Vermeulen and Cotula 2010). Singh (2007) divides the contracts into three main types: procurement or marketing contracts, which are only about obtaining access to the product from the farmer; partial contracts, which involve a marketing contract and the provision of some inputs to the farmer; and a total contract, which involve a marketing contract and the provision of all the inputs and the management systems, with the farmer mainly supplying the land, labour and day-to-day monitoring services. Variations of contract farming include: a centralised model, involving contracts between a firm and a large number of independent farmers; a nucleus-estate model, involving a plantation that obtains extra product from independent farmers; a tripartite model, involving a joint venture of a public entity and private firm that contracts with farmers; an informal model, involving smaller firms organising annual agreements with a limited number of farmers; and an intermediary models, involving a firm sub-contracting to an intermediary who obtains product from farmers (Singh 2007; Prowse 2012). The types of contract for these variations will differ depending on the structure of the model.

Contract farming can have benefits for farmers including: access to markets, improvements in financial approval, improved prices, technical assistance, specialised inputs and new technologies; reduced price variation and risk; all of which can increase income and help with rural development (Key and Runsten 1999; Singh 2007; Prowse 2012). It can also be an alternative to corporate farming that makes smallholder farming competitive, while allowing firms to have improved product quality and lowering transaction costs for firms as well as farmers. However, many contract farming arrangements favour larger farmers, particularly in dualistic agrarian economies, which can exclude smallholder farmers and exacerbate income and asset inequalities. The total contract arrangements can lead to a loss of control with farmers becoming 'serfs with two-way radios' (Singh 2007) and be patchy in capacity building, with the emphasis being on technical competency rather than managerial competency (Vorley et al. 2009). Other disadvantages of contract farming include the loss of flexibility in choosing crops or enterprises, increased market power of agribusiness firms and in some cases, manipulation of quotas such that not all farm production is purchased by the company leaving farmers to shoulder production losses (Sofranko et al. 2000; Eaton and Shepherd 2001; Singh 2007). On the other side of the coin, there have been studies showing that some farmers do not honour the contracts either deliberately or due to misinterpretation or differences in interpretation of the contract (Glover and Kusterer 1990). If a firm has market power as a monopsony buyer (the only buyer), it can influence markets. They can demand exclusivity of supply or reduce the availability of market signals from spot markets, which reduce the ability of farmers to determine realistic prices for their product (Singh 2007; Vorley et al. 2009).

The success or otherwise of various contract farming initiatives seems to be highly dependent on context, including such issues as culture, policy, land tenure systems, asymmetry of information, differential access to information, differences in negotiating power and other characteristics of the enabling environment (Singh 2007; Vermeulen and Cotula 2010; Prowse 2012). These issues have implications for the design and implementation of buyer-driven models for linking smallholder farmers to modern markets. Most contract farming models struggle in a competitive environment, particularly where there are economies of scale, unless they have a comparative advantage. Prowse (2012) reviewed 44 cases and concluded that economies of scale, variations in quality, perishability and price per kilogram were linked to success of models. Some fruits, vegetables and tree crops were suited to contract farming, but Singh (2007) suggests that in some cases these schemes may wither when the reasons for their comparative advantage are removed.

Vermeulen and Cotula (2010) suggest a common set of principles that sum to '*systemic competiveness*', based on collective efficiencies rather than individual actor efficiencies are apparent in successful models. This means that the business must focus on a new approach to corporate social responsibility that incorporates an inclusive model and facilitates collaborative problem solving rather than on supplier codes and compliance. The key is to build a chain model 'that balances risk, responsibilities and benefits along the chain while not undermining competiveness' (p. 217).

## *'Bottom Up' or Farmer-Driven Collaborative Marketing Models*

Farmer-driven models for linking farmers to markets have a long history and are normally associated with cooperative models. Cooperatives have provided farmers access to inputs, access to credits, encouraged sharing off agricultural knowledge, fostered new technologies and innovations, facilitated transport, storage and processing and linked farmers to markets (Trewin 2004; Bacon 2005; FFTC 2006; Bakucs et al. 2007; Bernard and Spielman 2009). Often farmers form groups to increase their bargaining power, pursue a common enterprise or interest including accessing government or other external programs that require group membership (Trewin 2004).

We prefer to use the term collaborative marketing models or collaborative marketing groups (CMGs) as an all-encompassing term to describe 'a group of farmers who have organised to collectively market their produce' (Murray-Prior 2007b, p. 2). This definition includes structures such as cooperatives, growers associations, cluster marketing groups and bargaining cooperatives. Historically farmers have formed cooperatives for three main reasons: increasing bargaining power (often with processors), in response to government programs and policies, and to take advantage of entrepreneurial opportunities.

The outcomes from cooperatives and CMGs organised by or for smallholder farmers in developing countries have been mixed, with many examples of their failure (Lele 1981; Murray-Prior 2007b; Vorley et al. 2009). Most of the literature on the deficiencies of traditional cooperative forms have been for developed countries (e.g. Cook and Chaddad 2004; Nilsson et al. 2012), but there is a growing literature (Chibanda et al. 2009; Batt and Murray-Prior 2011; Thomas and Hangula 2011) assessing CMG models that are appropriate for developing countries, which will be the focus of this section. Although Reardon and Huang (2008) found membership of producer organisations was correlated with participation in modern markets for 4 out of 8 developing countries, we believe smallholder CMGs will be a key part of the solution to meeting the increasing demand for food over the next 40 years and in coping with changes in the agribusiness sector and in the climate. The key will be finding contexts and models that are suited to improving smallholder access and profitability.

However, if CMGs are to be successful, they must have a comparative advantage over alternative marketing structures, within the environment of smallholder farmers in developing countries, and they must be able to deal with the issues of trust and member commitment (Murray-Prior 2007b). As with contract farming, there are many structures and models for cooperatives and CMGs, but in this section we will briefly discuss two forms: cooperatives, mainly linked to Fair Trade and organic markets and cluster marketing groups (Cluster MGs).

### Cooperatives

Most successful supply or marketing cooperatives have been in a few developed countries and many have been associated with processing activities. However, in developing countries the chances of success for these types of cooperatives if they are managed by smallholder farmers are more problematic. For a start, they have to compete with internationally competitive processing companies because of the more globalised environment for most agricultural products. This means it is very difficult for them to achieve the economies of size, raise the capital to build the plants and to acquire the management expertise to run the operations (Murray-Prior et al. 2009). Without substantial and long-term outside assistance, their chances of setting up and succeeding are very small.

If a marketing cooperative is to have a comparative advantage for smallholder farmers in a competitive environment, then it has to find a market where this form of organisation has an advantage. Fairtrade (FT) and to a lesser extent organic markets can be niche markets for smallholder farmers in some countries and in some industries. FT is most prevalent for non-perishable products such as coffee, cocoa, cotton and rice (Fairtrade International 2012). While some fresh fruits are sold under the label, they have a relatively limited market penetration, so to a certain extent this option is only available for a limited number of farmers. Because FT consumers are in developed countries, smallholder producers in developing countries who produce perishable horticultural fruits and vegetables are mostly excluded by virtue of the distance, transport, packaging standards and associated constraints.

Particularly in the coffee industry, smallholder farmers can obtain organic and FT certifications that enable them to access these markets and potentially gain an advantage over large and corporate farms, which cannot obtain FT certification. Evidence of smallholder gains from participating in these markets is mixed. Becchetti et al. (2011) found participation in FT and organic certifications through cooperatives increased per capita income for rice farmers in Thailand. Two studies of FT-organic coffee production in Nicaragua suggest a slightly different picture, with Valkila (2009) finding that participation increased income for low intensity farmers but that this type of farming did not produce much coffee or income so that the farmers remained in poverty. However, Beuchelt and Zeller (2012) found no clear income affect and suggest that the business model of a cooperative was a more important factor in their success. In Tanzania, the benefits of involvement in FT cooperatives appear to be less than they are in Nicaragua, which is attributed to the large size of the Tanzanian cooperatives leading to a lack of member commitment to producing quality coffee (Pirotte et al. 2006).

One of the key problems for FT and organic certification is the requirement to have a functioning cooperative and to be able to meet the FT certification standards (Batt and Murray-Prior 2011). Considerable time and effort is required to establish and maintain such groups and this relies either on the smallholder farmers having the necessary human and social capital to achieve this or a private company or non government organisation (NGO) to provide the expertise and support. If a private company provides the necessary expertise and support and markets the farmers' product, it then has to overcome perceptions that it is acting fairly (Murray-Prior et al. 2009). This requires a mechanism or structure to maintain trust between farmers and management. One mechanism to achieve this is the involvement of third parties to act as arbiters and referees (Murray-Prior et al. 2009; Moustier 2012; Prowse 2012). The other issue is the time required to obtain organic and to a lesser extent FT certification, during which premiums are not available, but costs need to be borne to undertake the certification process. Therefore, while FT-organic markets are an option, they appear to provide limited opportunities for many smallholder horticultural farmers in developing countries.

## Cluster Marketing Groups

Because of the abundant evidence that collaborative marketing groups of smallholder farmers in developing countries are often not competitive in institutional markets, or even when competing with intermediaries in traditional markets (Murray-Prior 2008), models of collaboration are required that enable smallholder horticultural producers to be competitive in institutional markets and to be sustainable. One such approach that has shown promise is the clustering approach developed by the International Center for Tropical Agriculture (CIAT) and adapted by the Catholic Relief Services in the Philippines (CRS). CIAT developed the *Territorial Approach to Rural Agro enterprise Development* (Lundy et al. 2005) as a guide for service providers to facilitate collective marketing by smallholder producers and to strengthen the capacity for them and their chains to compete in their selected markets. CRS-Philippines (2007) have since adapted this process to organise 'small farmers into marketing clusters to enable them to equitably participate in the opportunities of evolving dynamic markets' (p. xv).

The CRS Clustering Approach to Agro-enterprise Development is referred to as the Eight Step Clustering Approach (CRS-Philippines 2007) (Fig. 34.3). In Step 1, the site is selected and partnerships are built with farmers and other stakeholders. Step 2 involves a process in which members of the farmer group identify the community's resources, products and production and marketing practices and decide what product or products they will focus on. In Step 3, farmers are trained to undertake a market chain study involving market visits to develop their understanding of the chains for their selected products. They also negotiate preliminary trading terms with potential buyers.

Step 4 involves formation of the cluster, selection of the leaders and agreement on a basic cluster arrangement and objectives. Normally, the number of farmers in a cluster ranges from 5 to 15, with most clusters kept below 20 in an attempt to ensure effective communication and maintain a trusting environment. In Step 5, a planting and harvest calendar for the products of the cluster and a test marketing plan are developed. The test marketing activities (Step 6) involve at least four trial product deliveries. After each delivery, the cluster assesses performance and adjusts the plans to improve performance. When the cluster and facilitators judge the test marketing activities to be successful they appraise the readiness of the cluster for scaling up and begin planning and conducting a scaling up process (Step 7). This involves producing more or additional products to supply existing or more diversified markets. Step 8 (cluster strengthening) comprises improving cluster maturity through expanding cluster capacity and networks with other clusters and businesses.

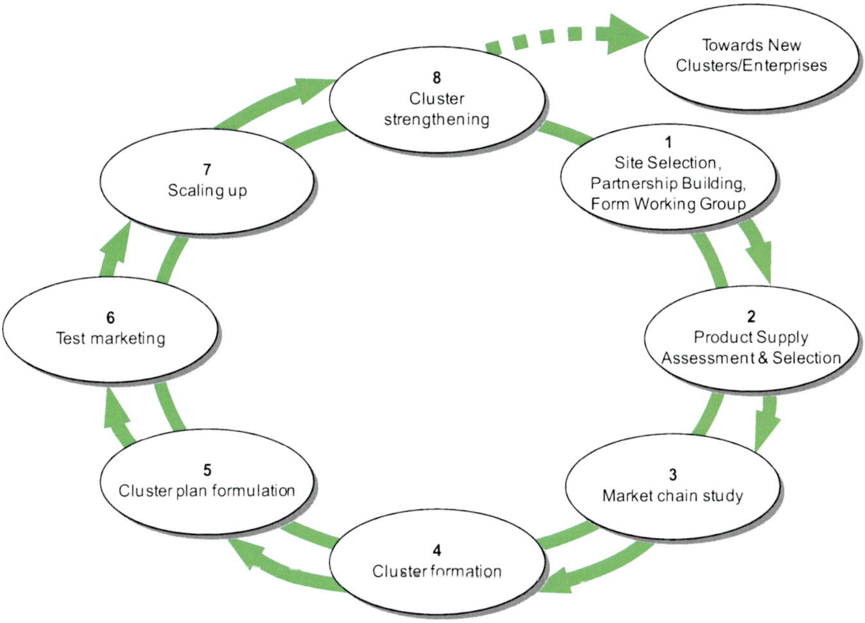


**Fig. 34.3** Eight-step process of the clustering approach to agro-enterprise development. (Source: CRS-Philippines 2007)

This clustering process was evaluated as part of an Australian Centre for International Agricultural Research project in Mindanao, southern Philippines that facilitated the establishment and development of 29 clusters that marketed vegetables and involved about 360 farmers (Rola-Rubzen et al. 2012). It found that clustering had a positive economic impact and increased the household income of cluster members over non-cluster members through increasing the range of vegetables produced and the volumes and prices of most vegetables. The process improved the production and marketing capacity of cluster members and in particular improved their negotiating skills, bargaining power, access to government, NGO and private sector services, and the quality and yields of their products.

These findings are consistent with Vorley et al. (2009) who analysed a range of business models and concluded that producer organisations can lead to improved negotiating skills and access to services. Furthermore, Moustier (2012) suggests they lead to reductions in transaction costs associated with training and quality inspections, two issues that are the focus of the clustering process. The size of the clusters also helps overcome the issues of trust and member commitment required to sustain successful CMGs (Murray-Prior 2007b). Clusters have a comparative advantage because they can combine products to achieve more marketable volumes, sort for quality, and improve packaging and transport that together enable access to higher priced markets.

Despite these successes, Murray-Prior et al. (2012) suggested the CRS Eight Step Clustering Approach should be adapted to incorporate processes to reduce some of the problems such as: input financing; risk of production failure; maintaining relationships with buyers; and building group resilience and independence. They also suggested the need for a formal exit strategy for the donor agencies. They suggested a Three-phase Clustering Framework incorporating: Phase 1—Establishment; Phase 2—Building Resilience; and Phase 3—Implement an Exit Strategy.

Markelova and Mwangi (2010); Vermeulen and Cotula (2010) suggest donor agencies develop clear milestones and exit strategies from the onset of a project to lessen dependency issues and to help increase the chances of the group being sustainable. The CRS clustering process already includes criteria for assessing cluster maturity (see CRS-Philippines 2007, p. 140), so the focus here is on how to incorporate these into a process for implementing an exit strategy for the donor agency. It should be made clear to the farmers from the beginning of the project that intensive support will be provided for a finite period and it is important to emphasise this reality to the cluster members and to the donor agency staff.

### *Summary of Methods of Linking Smallholder Horticultural Farmers to Modern Markets*

Improving the productivity of smallholder farmers and linking them to modern markets will be a key component of transforming the horticultural sector in developing economies. Our belief is that there is a need to focus on 'bottom up' and partnership approaches. Large scale acquisitions of smallholder land by investors are likely to lead to social unrest, while contract farming tends to favour large farmers and to be highly dependent on context. Some authors suggest a more collaborative approach is required. On the other hand cooperatives also have a variable track record and are less likely to be a solution for many smallholder horticultural farmers. Even cooperatives linked to FT and organic markets can struggle. New models of cluster marketing may help in other situations because they overcome some of the problems associated with larger cooperative models, but more research is required into the factors that improve the sustainability of these models without special donor support.

## Using Cluster Marketing Groups to Transform the Horticultural Innovation System

As we have argued above, a transformation of the horticulture sector requires development interventions along the smallholder chains, including the services and enabling environment to support the chains, a view that is supported by other authors (Anandajayasekeram and Gebremedhin 2009; Davis 2010; Christoplos et al. 2011; Hawkes and Ruel 2011). This needs to occur at multiple levels, the farmer-group level, the chain level and the industry and political level. Cluster marketing groups and their chains could form an important part of a horticultural innovation system if they are integrated into a multi-level, action-learning and action-research process (Murray-Prior 2011). The groups could help identify the binding constraints to development and when linked with research and development activities that work on clearly identified and relevant priorities, would develop appropriate solutions that could be scaled up and out with greater confidence and improved impact. Such a system would be more dynamic and could respond more quickly to emerging challenges. It would focus directly on developing solutions to smallholder opportunities and problems, rather than for medium and large-scale farmers, which is the norm with current research and development strategies.

## Conclusions

Smallholder horticultural farmers will be an important source of food to supply the growing world demand in the next 40 years, but if they are to achieve the improvements in productivity and effectiveness required, a transformation of the horticultural sector is essential. The complexity of the problems involved requires a holistic approach to transformation that needs to recognise the constraints in smallholder resources and enabling environments and involves addressing issues along the smallholder chains. An agribusiness systems framework is outlined, which for many developing economies is a dualistic agribusiness system that helps focus analysis on the critical constraints and opportunities for smallholder chains in supplying modern horticultural markets. A key issue is the need to adapt existing institutions and develop new institutions to help link smallholder farmers to modern markets. While contract farming works in some contexts, it tends to favour larger farmers and to only be appropriate in selected contexts. Cooperatives also have a role, particularly those linked to FT-organic markets, but once again depend on context and product and hence will also only be part of the solution. We argue that cluster marketing arrangements, because they are more suited to the resources and capacities of smallholder farmers will be an important component of the models for linking them to markets. They also provide an opportunity to identify research priorities, develop appropriate solutions to the relevant problems and opportunities, and test and scale these solutions out and up.

## References

Akkaya F, Yalcin R, Ozkan B (2006) Good agricultural practice (GAP) and its implementation in Turkey. Acta Hortic 699:47–52

Anandajayasekeram P, Gebremedhin B (2009) Integrating innovation systems perspective and value chain analysis in agricultural research for development: implications and challenges. Working Paper no. 16. International Livestock Research Institute, Nairobi

Armendariz B, Morduch J (2010) The economics of microfinance, 2nd edn. The MIT, Cambridge

Aubert C, de Janvry A, Sadoulet E (2009) Designing credit agent incentives to prevent mission drift in pro-poor microfinance institutions. J Dev Econ 90(1):153–162. doi:10.1016/j.jdeveco.2008.11.002

Bacon C (2005) Confronting the coffee crisis: can fair trade, organic and specialty coffee reduce small-scale farmer vulnerability in northern Nicaragua? World Dev 33(3):497–511

Baines RN, Davies WP, Batt PJ (2006) Benchmarking international food safety and quality systems towards a framework for fresh produce in the transitional economies. Acta Hortic 699:69–76

Bakucs LZ, Ferto I, Azabo GG (2007) Hungary Market Cooperative: a successful case of linking small farmers to markets for horticultural produce. IIED, London (Regoverning Markets Innovative Practice)

Basilio EL (2008) Linking the Philippine Islands through highways of the sea. Center for Research and Communication Foundation Inc., Pasig City. http://asiafoundation.org/publications/force-download.php?f=%2Fresources%2Fpdfs%2FRoRobookcomplete.pdf. Accessed 11 July 2012

Batt PJ (2006) Fulfilling customer needs in agribusiness supply chains. Acta Hortic 699:83–90

Batt PJ (2007) Alternative marketing strategies for smallholder farmers. Stewart Postharvest Rev 3(6):1–5. doi:org/10.2212/spr.2007.6.13

Batt PJ, Murray-Prior R (2011) Quality and ethical sourcing among smallholder coffee producers in Papua New Guinea. In: Jayachandran C, Seshadri S (eds) Proceedings of the 12th International conference of the society for global business and economic development: building capabilities for sustainable global business: balancing corporate success & social good, [CD]. SGBED, Singapore, pp 1488–1496, 21–23 July 2011

Batt PJ, Noonan JN, Kenyon P (2006) Global trends analysis of food safety and quality systems for the Australian food industry. Department of Agriculture Forests and Fisheries, Canberra

Becchetti L, Conzo P, Gianfreda G (2011) Market access, organic farming and productivity: the effects of Fair Trade affiliation on Thai farmer producer groups'. Australian J Agric Resour Econ 56:117–140

Beintema N, Elliott H (2009) Setting meaningful investment targets in agricultural research and development: challenges, opportunities and fiscal realities, how to feed the World in 2050. High-Level Expert Forum, Food and Agriculture Organisation of the United Nations, Rome, 12–13 October

Bernard T, Spielman DJ (2009) Reaching the rural poor through rural producer organizations? A study of agricultural marketing cooperatives in Ethiopia. Food Policy 34:60–69

Beuchelt TD, Zeller M (2012) The role of cooperative business models for the success of smallholder coffee certification in Nicaragua: a comparison of conventional, organic and Organic-Fairtrade certified cooperatives. Renew Agric Food Syst 1–17. doi:10.1017/S1742170512000087

Bryant CR (1989) Entrepreneurs in the rural environment. J Rural Stud 5(4):337–348. doi:10.1016/0743-0167(89)90060-0

Canlas DB, Khan ME, Zhuang J (2011) Critical constraints to growth and poverty reduction. In: Canlas DB, Khan ME, Zhuang J (eds) Diagnosing the philippine economy: toward inclusive growth. Anthem and Asian Development Bank (ADB), London, pp 33–97

Checkland P (1981) Systems thinking, systems practice. Wiley, Chichester

Chibanda M, Ortmann GF, Lyne MC (2009) Institutional and governance factors influencing the performance of selected smallholder agricultural cooperatives in KwaZulu-Natal. Agrekon 48(3):293–315. doi:10.1080/03031853.2009.9523828

Christoplos I, Sandison P, Chipeta S (2011) Guide to extension evaluation. Global Forum for Rural Advisory Services (GFRAS), Lindau

Christy R, Mabaya E, Wilson N, Mutambatsere E, Mhlanga N (2009) Enabling environments for competitive agro-industries. In: Da Silva CA, Baker D, Shepherd AW, Jenane C, Miranda-da-Cruz S (eds) Agro-industries for development. CAB International, UNIDO and FAO, Rome

Codron J-M, Grunert K, Giraud-Heraud E, Soler L-G, Regmi A (2005) Retail sector responses to changing consumer preferences: the European experience. In: Regmi A, Gehlhar M (eds) New directions in global food markets. USDA, Washington, pp 32–46 (Agriculture Information Bulletin no. 794)

Concepcion SB, Digal LN, Guarin R, Hualda LAT (2007a) Keys to inclusion of small-scale organic rice producers in supermarkets: the case of Upland Marketing Foundation Inc. Regoverning Markets Innovative Practice series. http://www.regoverningmarkets.org/en/filemanager/active?fid=810. Accessed 6 Sept 2012

Concepcion SB, Digal LN, Uy J (2007b) Keys to inclusion of small farmers in dynamic markets: the case of Norminveggies in the Philippines. Regoverning Markets Innovative Practice series. http://www.regoverningmarkets.org/en/filemanager/active?fid=810. Accessed 6 Sept 2012

Concepcion SB, Digal LN, Guarin R, Hualda LAT (2011) Small-scale organic rice producers sell to Philippine supermarkets: upland marketing foundation incorporated. In: Bienabe E, Berdegué J, Peppelenbous L, Belt J (eds) Reconnecting markets: innovative global practices in connecting small-scale producers with dynamic food markets. Gower Publishing Limited, Surrey, pp 77–93

Cook ML, Chaddad FR (2004) Redesigning cooperative boundaries: the emergence of new models. Am J Agric Econ 86(5):1249–1253

CRS-Philippines (2007) The clustering approach to agroenterprise development for small farmers: the CRS-Philippines experience—a guidebook for facilitators. Catholic Relief Services—USCCB, Philippine Program, Davao

Davies WP, Baines RN, Turner JC (2006) Red alert: food safety lessons from dye contamination in spice supply. Acta Hortic 699:143–150

Davis JH (1955, 17 Oct) Business responsibility and the market for farm products. Address to the Boston conference on Distribution. Retail Board, Boston

Davis JH, Goldberg RA (1957) A concept of agribusiness. Division of Research, Graduate School of Business Administration, Harvard University, Boston

Davis KE (2010) The what, who, and how of shaping change in African communities through extension. Ext Farm Syst J 6:83–90

De Oliveira W (2007) Agricultural risk management as a tool to improve the agribusiness environment in Ukraine. In: Tanic S (ed) Proceedings of the enabling environments for agribusiness and agro-industry development in Eastern Europe and Central Asia. Food and Agriculture Organization of the United Nations, Budapest, pp 57–71

DESA (UN Department of Economic and Social Affairs) (2011) World economic and social survey 2011: the great green technological transformation. Department of Economic and Social Affairs, United Nations Secretariat, New York

Diao X, Hazell P, Thurlow T (2010) The role of agriculture in African development. World Dev 38(10):1375–1383

Eaton C, Shepherd AW (2001) Contract farming: partnership for grow. Food and Agricultural Organization, Rome (Agricultural Services Bulletin 145)

Fairtrade International (2012) Products. http://www.fairtrade.net/products.html. Accessed 16 Aug 2012

FAO (2009) How to Feed the World in 2050. High-Level Expert Forum, Food and Agriculture Organisation of the United Nations, Rome, 12–13 Oct

FAO (2010) The state of food insecurity in the World: addressing food insecurity in protracted crises. Food and Agriculture Organization of the United Nations, Rome. www.fao.org. Accessed 3 July 2011

FFTC (2006) Agricultural cooperatives in Asia: innovations and opportunities for the 21st century. Food and Fertilizer Technology Centre, Annual Report, Taipei

Gehlhar M, Regmi A (2005) Factors shaping global food markets. In: Regmi A, Gehlhar M (eds) New directions in global food markets. USDA, Washington, DC, pp 5–17 (Agriculture Information Bulletin no. 794)

Glover D, Kusterer K (1990) Small farmers, big business: contract farming and rural development. Macmillan, London

Habito CF, Briones RM (2005) Philippine agriculture over the years: performance, policies and pitfalls. Paper presented to Asia-Europe Meeting (ASEM) Trust Fund, Asian Institute of Management Policy Center, Foreign Investment Advisory Services, Philippines Institute of Development Studies and the World Bank Conference "Policies to Strengthen Productivity in the Philippines", Makati City. 27 June 2005. http://siteresources.worldbank.org/INTPHILIPPINES/Resources/Habito-word.pdf. Accessed 27 July 2010

Hawkes C, Ruel MT (2011) Value chains for nutrition, Paper presented to IFPRI 2020 International Conference "Leveraging Agriculture for Improving Nutrition and Health", New Delhi 10–12 February. International Food Policy Research Institute, Washington, DC. http://2020conference.ifpri.info/. Accessed 4 July 2011

Hazell P, Poulton C, Wiggins S, Dorward A (2006) The future of small farms: synthesis paper. www.rimisp.org. Accessed 15 June 2012

Headey D, Fan S (2010) Reflections on the global food crisis: how did it happen? how has it hurt? and how can we prevent the next one? International Food Policy Research Institute, Washington, DC. www.ifpri.org. Accessed 3 July 2011

Humphrey J (2006) Horticulture: responding to the challenges of poverty reduction and global competition. Acta Hortic 699:19–41

Islam N (2011) Foreign aid to agriculture: review of facts and analysis. IFPRI Discussion Paper 01053. International Food Policy Research Institute, Washington, DC. http://www.ifpri.org. Accessed 3 July 2011

Kaferstein FK (2003) Actions to reverse the upward curve of foodborne illness. Food Control 14:101–109

Kahan DG (2007) Business services in support of farm enterprise development: case studies. FAO, Rome

Kaine G, Cowan L (2011) Using general systems theory to understand how farmers manage variability, systems research and behavioral science. doi:10.1002/sres.1073

Kelly PF (2003) Urbanization and the politics of land in the Manila Region. Ann Am Acad Polit Soc Sci 590:170–187

Ketelaar J (2007) GAP, market access, farmers and field realities: making the connection through better farmer education integrated production and pest management. In: Batt PJ, Cadilhon J-J (eds) Proceedings of the international symposium on fresh produce supply chain management. FAO, Rome, pp 345–349 (RAP 2007/21)

Key N, Runsten D (1999) Contract farming, smallholders, and rural development in Latin America: the organization of agroprocessing firms and the scale of outgrower production. World Dev 27(2):381–401

Lele U (1981) Co-operatives and the poor: a comparative perspective. World Dev 9:55–72

Lewis WA (1954) Economic development with unlimited supplies of labour. Manch School 22:139–191

Lundy M, Gottret MV, Best R, Ferris S (2005) A guide to developing partnerships, territorial analysis and planning together. Manual 1: territorial approach to rural agro-enterprise development, Cali. Rural Agro-enterprise Development Project, CIAT, International Center for Tropical Agriculture, Colombia

Markelova H, Mwangi E (2010) Collective action for smallholder market access: evidence and implications for Africa. Rev Policy Res 27(5):621–640 doi 10.1111/j.1541-1338.2010.00462.x

Martech Consulting. 2005. Trends that impact New Zealand's horticultural food exports. Growing futures cast study series. http://www.martech.co.nz/images/21trends.pdf

Mendoza RU, Vick BC (2010) From revolution to evolution: charting the main features of microfinance 2.0. Perspect Glob Dev Technol 9:545–580. doi:10.1163/156914910x499813

Moustier P (2012) Reengaging with customers: proximity is essential but not enough. Acta Hortic [in press]

Murray-Prior R (2007a) Methodological frameworks for improving linkages and the competiveness of supply chains. In: O'Reilly S, Keane M, Enright P (eds) Proceedings of the 16th international farm management association congress "a vibrant rural economy—the challenge for balance", University College Cork, Cork, 15–20 July. International Farm Management Association, vol Peer Reviewed Papers, vol I, pp 195–201. http://www.ifmaonline.org/pages/con_full_articles.php?abstract=416

Murray-Prior R (2007b) The role of grower collaborative marketing groups in developing countries. Stewart Postharvest Rev 3(6):1–10

Murray-Prior R (2008) Are farmers in transitional economies likely to benefit from forming collaborative marketing groups? Banwa Manag 8(2):10–21

Murray-Prior R (2011) A participatory market-driven approach to development & extension. Paper presented to innovations in extension & advisory services: Linking Knowledge to Policy & Action for Food & Livelihoods, Nairobi, 15–18 Nov 2011

Murray-Prior R, Batt PJ (2007) Emerging possibilities and constraints to PNG smallholder coffee producers entering the speciality coffee market. In: Batt PJ, Cadilhon J-J (eds) Proceedings of an international symposium on fresh produce supply chain management, Lotus Pang Suan Kaeo Hotel, Chiang Mai, Thailand, pp 372–387, 6–9 Dec. FAO, Rome

Murray-Prior R, Ncukana L (2000) Agricultural development in South Africa—a dualistic agribusiness systems perspective, paper presented to the 'African Studies Association of Australasia & the Pacific 23rd Annual & International Conference: African Identities. St Marks College, University of Adelaide, North Adelaide, 13–15 July 2000

Murray-Prior R, Concepcion S, Batt P, Rola-Rubzen MF, McGregor M, Rasco E, Digal L, Manalili N, Montiflor M, Hualda L, Migalbin L (2004) Analyzing supply chains with pluralistic and agribusiness systems frameworks. Asian J Agric Dev 1(2):45–56

Murray-Prior R, Batt PJ, Rola-Rubzen MF, McGregor MJ, Concepcion SB, Rasco ET, Digal LN, Montiflor MO, Hualda LT, Migalbin LR, Manalili NM (2006) Global value chains: a place for Mindanao producers? Acta Hortic 699:307–315

Murray-Prior R, Batt PJ, Dambui C, Kufinale K (2008) Improving quality in coffee chains in Papua New Guinea'. Acta Hortic 794:247–255

Murray-Prior R, Sengere R, Batt PJ (2009) Overcoming constraints to the establishment of collaborative marketing groups for coffee growers in the highlands of PNG. Acta Hortic 831:277–283

Murray-Prior RB, Batt PJ, Rola-Rubzen MF, Concepcion SB, Montiflor MO, Axalan JT, Real RR, Lamban RJG, Israel F, Apara DI, Bacus RH (2012) Theory and practice of participatory action research and learning with cluster marketing groups in Mindanao, Philippines. Acta Hortic [in press]

Nelson GC, Rosegrant MW, Palazzo A, Gray I, Ingersoll C, Robertson R, Tokgoz S, Zhu T, Sulser TB, Ringler C, Msangi S, You L (2010) Food security, farming, and climate change to 2050: scenarios, results, policy options. IFPRI Research Monograph. www.ifpri.org. Accessed 4 July 2011

Nilsson J, Svendsen GLH, Svendsen GT (2012) Are large and complex agricultural cooperatives losing their social capital? Agribusiness 28(2):187–204. doi 10.1002/agr.21285

OECD/FAO (2012) OECD-FAO Agricultural outlook 2012–2021. OECD Publishing and FAO. doi:10.1787/agr_outlook-2012-en

Olchondra RT (2010) Da, Irri, Globe Telecom to give rice farmers timely information. Philippine Daily Inquirer, 8 June 2012. http://business.inquirer.net/money/breakingnews/view/20100608-274592/DA-IRRI-Globe-Telecom-to-give-rice-farmers-timely-information. Accessed 31 May 2012

Oxfam (2008) Investing in poor farmers pays: rethinking how to invest in agriculture. www.oxfam.org. Accessed 15 Aug 2012 (Oxfam Briefing Paper No. 129)

Pinstrup-Andersen P, Watson D (2011) Food policy for developing countries. Cornell University, Ithaca

Pirotte G, Pleyers G, Poncelet M (2006) Fair-trade coffee in Nicaragua and Tanzania: a comparison. Dev Pract 16(5):441–451. doi 10.1080/09614520600792390

Polack E (2012) Agricultural land acquisitions: a lens on Southeast Asia. International Institute for Environment and Development, London. http://pubs.iied.org/pdfs/17123IIED.pdf. Accessed 3 May 2012

Prantilla EB (2011) Rapid field appraisal of decentralization: Davao Region. The Asia Foundation, Makati City

Prowse M (2012) Contract farming in developing countries—a review, a Savoir, vol 12, Paris: de Agence Francaise development. http://recherche.afd.fr. Accessed 11 June 2012
Quinones BR, Seibel HD (2000) Social capital in microfinance: case studies in the Philippines. Policy Sci 33(3):421–433
Reardon T, Huang J (2008) Patterns in and determinants and effects of farmers' marketing strategies in developing countries. Synthesis report: micro study. www.regoverningmarkets.org. Accessed 6 September 2012
Reyes CM (2002) Impact of agrarian reform on poverty. Philippine J Dev 29(2):63–131
Rola-Rubzen MF, Murray-Prior RB, Batt PJ, Concepcion SB, Real RR, Lamban RJG, Axalan JT, Montiflor MO, Israel F, Apara DI, Bacus RH (2012) Impact of clustering on vegetable farmers in the Philippines. In: Proceedings of smallholder hopes: horticulture, people and soil conference, Cebu Parklane International Hotel, Cebu, Philippines, 3 July 2012. ACT: Australian Centre for International Agricultural Research, Canberra [in press]
Rottger A, Da Silva CA (2007) Enabling environments for agribusiness and agro-industry development in Africa. Food and Agriculture Organization of the United Nations, Accra
Schrekenberg K, Mitchell J (2011) Who runs this place? The external enabling environment for value chain development. In: Mitchell J, Coles C (eds) Markets and rural poverty: upgrading in value chains. International Development Research Centre, Ottawa, pp 217–234
Shepherd AW (2005) The implications of supermarket development for horticultural farmers and traditional marketing systems in Asia. Agricultural Management, Marketing and Finance Service, FAO, Rome
Shepherd AW, Galvez E (2007) The response of traditional marketing channels to the growth of supermarkets and to the demand for safer and higher quality fruit and vegetables with particular reference to Asia. In: Batt PJ, Cadilhon J-J (eds) Proceedings of the international symposium on fresh produce supply chain management. FAO, Rome, pp 305–314 (RAP 2007/21)
Shepherd AW, Tam PTG (2008) Improving the safety of marketed horticulture produce in Asia with particular reference to VietNam. Acta Hortic 794:301–308
Singer H (1970) Dualism revisited: a new approach to the problems of dual-societies in developing countries. J Devel Stud 7(1):60–61
Singh S (2007) Contract farming: theory and practice in the 21st century. Stewart Postharvest Rev 3:1–6. www.stewartpostharvest.com. Accessed 13 August 2012
Smit J, Nasr J (1992) Urban agriculture for sustainable cities: Using wastes and idle land and water bodies as resources. Environ Urban 4(2):141–152. doi 10.1177/095624789200400214
Sofranko A, Frerichs R, Samy M, Swanson B (2000) Will farmers organize? structural change and loss of control over production. http://web.aces.uuiuc.edu/value/research/organize.htm. Accessed 1 May 2011
Teruel RG, Kuroda Y (2005) Public infrastructure and productivity growth in Philippine agriculture, 1974-2000. J Asian Econ 16(3):555–576. doi 10.1016/j.asieco.2005.04.011
Thomas B, Hangula MM (2011) Reviewing theory, practices and dynamics of agricultural cooperatives: understanding cooperatives' development in Namibia. J Dev Agric Econ 3(16):695–702. http://www.academicjournals.org/JDAE. Accessed 17 July 2012
Trewin R (2004) Cooperatives: issues and trends in developing countries. Report of workshop. ACIAR, Perth 24–25 March 2003
Valkila J (2009) Fair Trade organic coffee production in Nicaragua—Sustainable development or a poverty trap? Ecol Econ 68(12):3018-3025. http://www.sciencedirect.com/science/article/pii/S0921800909002742. Accessed 17 July 2012
Vermeulen S, Cotula L (2010) Making the most of agricultural investment: a survey of business models that provide opportunities for smallholders. IIED/FAO/IFAD/SDC, London
Viatte G, De Graaf J, Demeke M, Takahatake T, de Arce MR (2009) Responding to the food crisis: synthesis of medium-term measures proposed in inter-agency assessments. www.fao.org/. Accessed 4 July 2011
von Bertalanffy L (1968) General system theory: Foundations, development, applications, ringwood. Penguin Books, NSW

Vorley B (2011) Small farms and market modernisation. International Institute for Environment and Development, London. http://pubs.iied.org/pdfs/G03126.pdf. Accessed 29 July 2012

Vorley B, Lundy M, MacGregor J (2009) Business models that are inclusive of small farmers. In: da Silva CA, Baker D, Shepherd AW, Jenane C, Miranda-da-Cruz S (eds) Agro-industries for development. CAB International and FAO, Rome, pp 186–222

Warner M, Kahan D, Lehel S (2008) Market-oriented agricultural infrastructure: appraisal of public-private partnership. FAO, Rome

# Chapter 35
# International Plant Trade and Biosecurity

**Aaron Maxwell, Anna Maria Vettraino, René Eschen and Vera Andjic**

**Abstract** This chapter explores the current status of plant trade and international biosecurity regulatory mechanisms to safeguard economic, social and economic well being of nations, states and economic regions. We provide an account of the international biosecurity framework in a historical context. In doing so we outline some of the common approaches to managing and regulating biosecurity risks associated with the plant horticultural export trade. This exploration identifies many of the inconsistencies in the application of plant biosecurity measured internationally. The approaches for regulation of live plants are compared amongst regions and future improvements are identified.

**Keywords** Pant trade · International biosecurity · Biosecurity risks · Plant biosecurity measures

## Introduction

Humans have traded and transported horticultural products including live plants for millennia. The rate of trade accelerated at the end of the Middle Ages (1400s) and again at the beginning of the Industrial Revolution (1800s). The 21st century has seen another marked acceleration in the movement of plants and plant products around

---

A. Maxwell (✉)
School of Veterinary and Life Sciences, Murdoch University,
90 South St. Murdoch, WA 6150, Australia
e-mail: Aaron.Maxwell@daff.gov.au

V. Andjic
Department of Agriculture, 9 Fricker Rd,
Perth International Airport WA 6105, Perth, Australia
e-mail: vera.andjic@daff.gov.au

A. M. Vettraino
DIBAF, University of Tuscia-Viterbo, Viterbo, Italy
e-mail: vettrain@unitus.it

R. Eschen
CABI, Rue des Grillons 1, Delémont, Switzerland
e-mail: r.eschen@cabi.org

G. R. Dixon, D. E. Aldous (eds.), *Horticulture: Plants for People and Places, Volume 3*,
DOI 10.1007/978-94-017-8560-0_35, 

the world with the Era of Globalization (Hulme 2009). The 17th and 18th centuries in particular saw the beginnings of a boom in horticultural exploration, collecting and introducing plants into Europe from throughout the world. John Tradescants the Younger (1608–1662), Joseph Banks (1743–1820), William Bartram (1739–1823), Alexander von Humboldt (1769–1859) and Sir Joseph Dalton Hooker (1817–1911), amongst other explorer naturalists, collected both living and dead plant specimens. Banks collected vast numbers of mostly non-living specimens (3,000 species on the Transit of Venus voyage of 1768–1767) and a few live plants (Paterson 2000). As well as improvements in transport, the invention of the Wardian Case (Anon 2012) enabled the transport of live plants over the long distances and often several months' journey back to Europe. Even during this early era of plant collecting and transportation, founding taxonomist Carl Linnaeus (1701–1778) identified the need to practice some form of quarantine to reduce the spread of plant disease (MacLeod et al. 2010; Usinger 1964).

The benefits to human societies of the trade in plants and plant products has been immeasurable and has contributed strongly to increases in human population size and richness of experience in terms of the diversity of produce available to be grown and enjoyed throughout the world. Increasingly, food and other horticultural products have shifted from that which was available locally and seasonally to that which is available globally and all year round. In many parts of the world, when we enjoy a meal much of the produce we consume will have been grown in any number of different countries and continents. That which is sourced locally will almost certainly be derived from plant species or varieties that originated elsewhere. For example, most of Australia's agricultural and horticultural systems are based on introduced plants. The economic benefit to Australia from plant-based industries has been estimated at over $ AUD 50 billion annually and these industries provide unmeasured social and environmental benefits (Virtue et al. 2004). These benefits from world-wide trade, however, come at a cost and this is in the form of a rise in alien invasive species (AIS) worldwide. Biological invasions by alien species are currently recognised as the second cause of loss in biological diversity, behind the destruction of habitats, and have also wide economical consequences (Mack et al. 2000; Perrings et al. 2005; Pimentel et al. 2005; Vitousek et al. 1997; Wilcove et al. 1998). The annual monetary impact of alien invasions in Europe has been estimated as close to € 10 billion, a value that is thought to be an underestimate (Hulme et al. 2009).

In plants, the AIS that have impacts are pests that attack and cause emerging infectious diseases (EIDs). Here we refer the term IAS to a species, subspecies, race, or *forma specialis* which is introduced into a country where it was previously unknown, behaves as an agent of disease, and threatens the biological diversity of native or exotic forest trees and shrubs, as defined by Santini et al. (2013). This encompasses plant pathogens, invertebrates and weeds. Once introduced into a new ecosystem, AIS can become widespread and can have a remarkably broad range of economic, environmental and social impacts. Whether an ecosystem is prone to invasion is both due to biological traits of the invasive species and the environmental and community features in the new and original ranges (Goodwin et al. 1999; Mitchell and Power 2003; Alpert et al. 2000). Social and economic factors are crucial for species introduction (Sakai et al. 2001; Guo et al. 2012), whereas biogeographical and ecological factors are important for naturalization, with evolutionary forces being key mediators of invasiveness (Sax 2001). Countries with a wider range of environments, and higher human impact and international trade issues host more invasive forest pathogens (IFPs) (Santini et al. 2013) and invertebrates (Roques et al. 2009).

Well-documented impacts of plant pest introductions resulting from plant trade include the devastation of potatoes in Ireland during the period 1845–47 caused by potato blight (*Phytophthora infestans*) that led directly and indirectly to deaths of over 1 million people and ensued mass migration out of Ireland to the Americas and other destinations. This pathogen probably coevolved with wild potato (*Solanum*) species and initially emerged as a disease of the cultivated potato (*tuberosum)* when *P. infestans* was transported to Mexico from the South American Andes (Niederhauser 1991). It was introduced into the USA around 1840, and subsequently into Europe. From Europe, potato blight was introduced into Asia, Africa and South America, with further introductions from Mexico into the USA and Canada during the early 1990s (Goodwin et al. 1994, 1995). Around the world the disease causes around $ US 6 billion of damage to crops each year (Nowicki et al. 2012).

The reason for biosecurity legislation and agreements is to manage what is seen as an obvious link between the movement of people and goods, and the risk of introducing pests from the exporting country or region to the place of import. The formal regulation of plant trade to guard against risks posed by these associated pests (harmful, pathogens, invertebrates and weeds) is a relatively recent development. Today there are 180 signatory countries to the International Plant Protection Convention, the central regulatory mechanism for establishing phytosanitary measures to control the spread of pests of plants and plant products. The ever increasing growth in international trade (e.g. 33 % per decade over past 4 decades in the US) throws up ever increasing challenges in the management of biosecurity risks. There are myriad approaches to managing those risks across various plant protection organisations within and across national borders. Although the World Trade Organisation (WTO) works to harmonise national practices, there remains divergence between states depending on perceptions of risk and the setting of an appropriate level of protection (ALOP) by individual member states.

In this chapter we explore the current status of plant imports and international biosecurity regulatory mechanisms to safeguard economic, social and economic well-being of nations, states and economic regions. We provide an account of the international biosecurity framework in a historical context. In doing so we outline some of the common approaches to managing and regulating biosecurity risks associated with plant trade. This exploration identifies many of the inconsistencies in the application of plant biosecurity measures internationally. The approaches for regulation of live plants are compared amongst regions and future improvements are identified.

## The Link Between Trade and Alien Invasive Species

The introduction of alien pathogens or hosts that leads to disease emergence is the most important driver of plant EIDs (Anderson et al. 2004). Many authors have described a correlation between the level of trade, transport, travel and tourism,—the "four T's", and the global scale and patterns of alien invasive species (Perrings et al. 2005; Hulme 2009).

The industrial revolution ushered in a period of acceleration in the construction of canals, railways, roads and steam powered shipping. This served to increase the extent of transport networks as well as the speed and volume of commodity trade and movement of people. Rapid transportation and reduced delivery times increase the survival of pathogen propagules and their chances to establish in a new environment. Political, economic and the above-mentioned transport efficiencies also facilitated the emigration of over 50 million Europeans around the world at the end of the 19th and beginning of the 20th centuries (Findlay and O'Rourke 2007; Hulme 2009). These populations served as vectors of horticultural products and crops, many of which were accompanied by pests or were invasive in their transplanted habitat. The fall of the Iron Curtain started a period of important political changes, the increase of globalization and transport of goods, and consequently the faster rate of alien pest and pathogen introductions. In the USA, the rate of accumulation of pests has been relatively constant since 1860, however the wood-borers have increased faster than other insect guild since the 1980s. The number of alien pests introduced in Europe has increased exponentially in the last 200 years, with a boost after World War II (Fig. 35.1).

During the last 40 years, as the world economy has entered the era of globalisation, the trends in transport and trade have accelerated. The volume of sea cargo has risen by over 400% since the 1970s (Anon 2012a), with a current world fleet over 1.5 million vessels (Anon 2012a). Positive correlations between the amount of trade imports and numbers of invasive species in European countries have been documented for a range of taxa (Westphal et al. 2008), including alien fungi (Desprez-Loustau 2009), plants (Pyšek et al. 2009) and invertebrates (Roques et al. 2009). Other indirect measures of trade volume, such as GDP, are also shown to correlate with invasive pest introductions, for example plants into China (Liu et al. 2005).

The plant sector (plant products, germplasm, grafts and live plants) has been part of the general trend in increased trade. The importance of plant trade as an invasion pathway has been recognized by several authors. At least 43% exotic insect introductions into Switzerland and Austria and 38% of exotic arthropod introductions in Europe were the result of the horticultural and ornamental trade, including cut flowers and seeds (Kenis et al. 2007; Roques et al. 2009). The greatest risk for the introduction of plant pests is associated with live plants, including bare-rooted plants, bulbs, non-rooted cuttings destined for propagation and bonsai production. Smith et al. (2007) have estimated that 90% of exotic invertebrate introductions into the UK during the period 1970–2004 were associated with the live plant trade. In the USA almost 70% of damaging forest insects and pathogens established since 1860 were introduced via the live plant trade (Liebhold et al. 2012). Most (>75%)

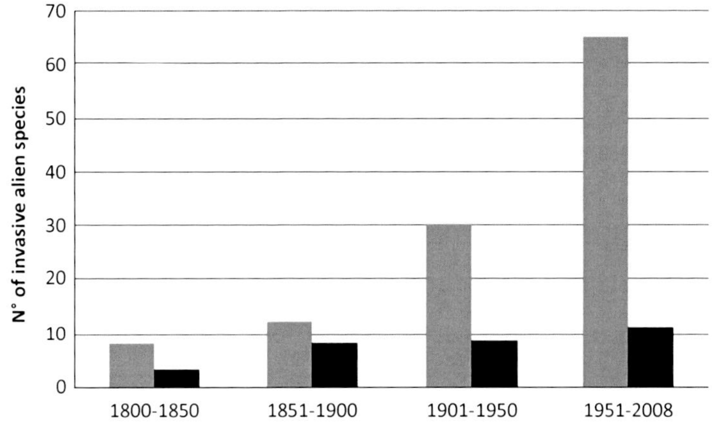


**Fig. 35.1** Total number of alien invertebrate (*grey bar*) and pathogen taxa (*black bar*) in Europe, since 1800, based on data adapted from Roques et al. (2009), Hulme et al. (2009) and Santini et al. (2013). (Graph used with permission from Elsevier.)

of the alien insect species in Europe came in as plant contaminants, whereas a few (< 10 %) travelled as hitchhikers. For example the horse chestnut leaf miner (*Cameraria ohridella*) and several ant species. Less than 2 % were the result of deliberate human activities such as biological control or leisure (Roques 2007).

Among the groups of pathogens, plant EIDs are caused mainly by viruses (47 %), fungi (30 %) and in lower proportion by bacteria, phytoplasmas and nematodes (Anderson et al. 2004). The rate of introduction of ascomycetes and oomycetes has increased dramatically since the beginning of 19th century. Of these IFP's 27 % are European species previously restricted to small areas of the continent, 22 % are aliens from temperate North America, and 14 % are from Asia. Alien terrestrial invertebrates, mostly insects (nearly 94 %), represent one of the most abundant groups of introduced organisms in Europe. A total of 1,296 alien species have established so far (Roques et al. 2009). The origin of invasive species has changed in the last 30 years, with a great increase in the number of pathogens from Asia and the appearance of new hybrid species (Brasier 2000; Roques 2007; Santini et al. 2013). The increment in the numbers of records of pests arrivals is associated with the increase of the "four T's", but can be also partially explained by the development of more effective diagnostics methods and greater attention given by the scientific community to the biological invasions. In comparison to Europe, Waage et al. (2008) found that Africa experienced an increase in reported invasive plant pests at the beginning of the 20th century which peaked mid-century and has since declined. The plateau in reports of new pests may reflect a possible improvement in the control of invasion pathways, or a decline in the emergence of new pathways through trade into Africa, that niches for invasive pests are increasingly occupied, or it may reflect a decrease in effort to detect pests.

## Impact of Recent Structural Change in Horticultural Production on Phytosanitary Risks

In addition to the growth in trade volumes, other trends that may shape plant pest management include the development of trade in South America, Asia and Africa, and the enlargement and economic integration of the European Union (Dehnen-Schmutz et al. 2010a). The small number of invasions in previous members of the Soviet Union may partly be the result of commercial isolation during the Soviet era. Future influences that carry biosecurity implications are climate change and the phasing out of chemicals such as methyl bromide for use in managing quarantine pests. Of current biosecurity impact have been some of the recent structural changes to the horticultural export industries (Dehnen-Schmutz et al. 2010a). Relevant changes include intensification of production, concentration of ownership in production and distribution, increased adoption of new technology such as micro-propagation, expansion of number of ornamental species traded internationally, proliferation of phytosanitary certification and accreditation schemes, and the adoption of more accurate and rapid diagnostic tools.

World-wide there has been an increasing concentration of horticultural production for both produce and ornamental trade into fewer entities. These entities have greater control over production values and distribution chains, which are also increasingly concentrated (Dehnen-Schmutz et al. 2010a). This trend may have both positive and negative implications for biosecurity pest management. The increasing role of supermarket chains and the overall concentration of horticultural enterprise is documented in parts of Africa (Neven and Reardon 2004), the Americas (Park and McLaughlin 2000), Europe, the Pacific and Asia (Dehnen-Schmutz et al. 2010a). This change has also been accompanied by increased penetration of distribution networks and the rapid deployment of produce across those networks. Some countries act as super-connected nodes for re-distribution of specific horticultural products. For example, the Netherlands acts as a super-connected node for cut flower and ornamental plant trade in that it dominates world production and export (O'Riordain 1999), as well as serving as a centre for redistribution from parts of Africa and Europe (Anon 1999; Lunati 2007), Asia and South America. Figure 35.2 shows the connectedness of European centres of horticultural trade and the central role played by the Netherlands.

The implications of the concentration of production and transport of the increasingly large number of live plants through a relatively small number of producers and transport hubs for pest distribution may be better managed through a more concentrated supply chain. However, on the negative side, when a new pest or disease does emerge and is not detected it may be dispersed much more rapidly and widely than through a diversified supply chain. Some models of disease distribution along networks suggest that highly interconnected nodes facilitate rapid pest spread (Jeger et al. 2007; Ercsey-Ravasz et al. 2012) and therefore more diverse supply chains, such as occur in Italy (Guzmán et al. 2009) and France (Anon 1999), may slow down the spread of emergent pests. A useful principle to apply from network theory is that control measures be targeted to critical (super-connected) nodes, as

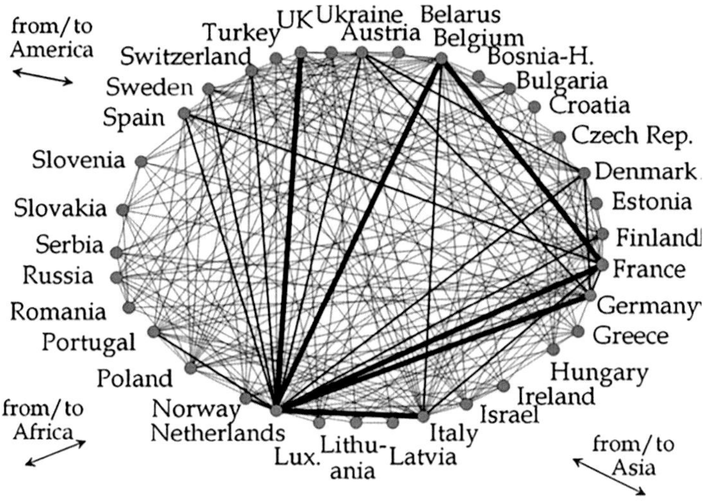


**Fig. 35.2** Network of trade interactions of European countries. Figure based on the sum of imports and exports of ornamental horticultural products (excluding seeds, 2003; from Anon (2004). Line thickness is proportional to trade volume. Albania, Andorra, Iceland, Kosovo, Liechtenstein, Macedonia and San Marino not shown for lack of data. Used with permission from (Dehnen-Schmutz et al. 2010a)

has been shown for network modelling of human disease pandemic (Hufnagel et al. 2004). However, pest management at the place of production, or more generally, in the country of origin to avoid entry of pests into transport may be the most effective prevention strategy.

Increased adoption of new technology such as soil-less production and micro-propagation has been taken up world-wide and benefits and limitations have been outlined, for example in the UK hardy nursery industry (Dixon 1987). Anticipation of new developments and recognition of their risks allows for strategies to be implemented that enable the benefits whilst minimising any associated risks (Dixon 2000). Micro-propagation facilitates a safer product when tissue culturing techniques are used. The inherent phytosanitary safety of this type of material is recognised in the Australian import conditions which allow for less stringent conditions on the import of certified medium risk genera as tissue culture plantlets than as bare-rooted plants (Anon 2013). New Zealand also has less stringent conditions for the import of tissue cultures than for the import of other live plants (Anon 2013a).

The inclusion of new ornamental plant species or varieties, or new origins of existing species in the international nursery trade may lead to new opportunities for movement of pests. Many pests were not known to be of harm in the region of origin before they became known as harmful AIS in the introduced range; or were previously undescribed species. Most countries around the world operate a so-called black-list approach and these countries regulate relatively few, known harmful pests. In practice this means that consignments of live plants or plant products must be free of the species listed by the importing country (see text of phytosanitary certificate, ISPM 12). In order to deal with new plant species and new origins, and to reduce the risk of introducing new pests on them, many countries require import permits and perform risk analysis on new commodities or pathways.

## Plant Biosecurity Regulation Past and Present: International and Regional Agreements

The increase in global horticultural trade in recent decades has been enabled by marked growth in production, and facilitated by the adoption of trade agreements and international and regional phytosanitary agreements. The current raft of international agreements can be understood in the context of the past evolution of these agreements.

Probably the earliest documented phytosanitary legislation was enacted in Rouen France in 1660 in a law that directed landholders to destroy barberry (*Berberis vulgaris*)` in order to protect wheat crops from black stem rust (*Puccinia graminis*) (Dehnen-Schmutz et al. 2010a). The first international phytosanitary agreement was the International Convention on Measures to be taken against Phylloxera (*Phylloxera vastatrix*) of 1878 and signed by seven European countries. This pest was introduced into Europe on imported infested US vines in 1862. The Convention outlined several measures, which still form the basis of many phytosanitary agreements today. These included providing written assurance of the *Phylloxera* free status of host material, providing powers to inspect and destroy infested material; and the requiring of signatories to establish an official body to administer and implement the measures. In the early part of the 20th century several nations enacted quarantine legislation, for example: the Australian Quarantine Act (1908), the Plant Quarantine Act in the USA (1912) and the Destructive Insects and Pests Act in Britain (1907).

The historic development of the scientific understanding of the role of microorganisms in causing plant disease was an important precursor in the evolution of quarantine procedures and subsequent legislation. Early work that demonstrated the role of microorganisms in causing plant disease can be traced back to Isaac Prevost (1755–1819) who in 1807 proved conclusively that wheat bunt was caused by a fungus and controlled by dipping in copper sulphate, and the work of Heinrich Anton deBarry (1831–1888) who in 1861 showed that *Phytophthora infestans* was the cause of potato blight (Prevost 1864; deBerry 1861). Prior to this, plant disease was thought to generate spontaneously (MacLeod et al. 2010). Without an understanding of the causes of plant disease and pest biology known more broadly the sophisticated phytosanitary measures of today could not have evolved. The potato blight, that was mentioned earlier, is a widely used example of a plant disease epidemic resulting from a new pest introduction (Schumann 1991). Contributing factors to the disease were that it was spread from related potato hosts in Mexico against which there was little natural resistance in the Lumper cultivar, a varietal white potato, grown as a monoculture in Ireland. Conditions such as the lack of genetic variation in Irish potatoes and the subsequent environmental conditions, were conducive to the disease at the time and there were no known fungicides with which to manage the disease. Although this event did not spawn the immediate introduction of quarantine measures, later potato pests did stimulate plant health legislation in Europe. In 1877 Britain passed the Destructive Insects Act which had the objective of preventing the introduction and establishment of the Colorado potato beetle (*Leptinotarsa decemlineata*) (Dehnen-Schmutz 2010b). As more pests were identified and their biology understood, governments began to regulate for a range of pests. For example, the British legislation was broadened with the Destructive Insects and Pests Act of 1907, to prevent entry of pathogenic fungi (Ebbels 1979) and later and viruses` and bacteria were included in the same Act of 1927 (Dehnen-Schmutz 2010b).

In June 1905, governments came together to form the International Institute of Agriculture (IIA) in Rome. Although the IIA did not deal specifically with phytosanitary measures, it published material relating to agricultural statistics, economics and legislation. Other events led to the 1929 International Convention for the Protection of Plants. Signatories to the Convention were required to establish an official inspection service to inspect and certify plants and control plant movements, establish a service to suppress dangerous plant diseases and to establish institutes for research to advise government. Countries could create their own actionable pest lists based on agreed criteria of pest impact, pest absence in importing country, and pest presence on import pathway (Dehnen-Schmutz et al. 2010a). Similar criteria are found in current phytosanitary agreements. However, the 1929 agreement was not particularly effective with only 12 states ratifying the agreement and the intervention of World War 11. In 1945, following World War II, the IIA was replaced by the Food and Agricultural Organization (FAO) which operates under the auspices of the United Nations (UN). Through the FAO, the International Plant Protection Convention (IPPC) was adopted in 1951 and remains the preeminent body for dealing with phytosanitary issues amongst a range of other current international agreements. As of 2013 there are currently 178 signatories to the Convention (Anon 2013b).

Amongst the almost 50 international agreements or guidelines that deal with invasive alien species, there are 4 that are key instruments for managing plant biosecurity and trade (MacLeod et al. 2010). These are the IPPC, the WTO Agreement on the Application of Sanitary and Phytosanitary measures (SPS Agreement), the Cartagena Protocol on Biosafety (CP) and the Convention on Biological Diversity (CBD). The interaction of these four agreements is summarised in Fig. 35.3. The SPS Agreement is under the auspice of the WTO which functions to reduce barriers to world trade and hence the SPS Agreement's purpose is to ensure that phytosanitary measures are based on science and are least restrictive to trade. The purpose of the CBD is to protect biodiversity and deals with the biodiversity threat posed by invasive pests. The IPPC and the CBD overlap in their influence. The Cartagena Protocol on Biosafety (CP) is an agreement under the CBD on the use of genetically modified organisms and may overlap with other agreements where those GMO's pose a threat as invasive pests or plants under a plant protection agreement.

The IPPC aims to prevent the introduction and spread of pests of plants and plant products, and to promote appropriate measures for their control internationally. Principles agreed to through the IPPC are similar to those of the IPPC of 1929:

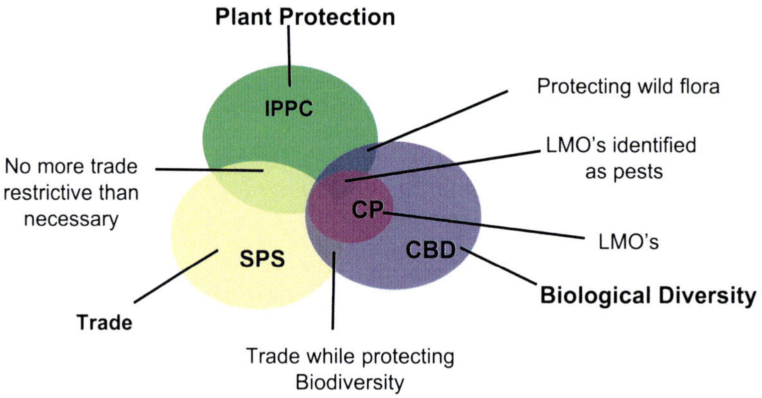


**Fig. 35.3** Links between international agreements protecting plant health. Key: IPPC = International Plant Protection Convention, SPS = WTO Agreement on the Application of Sanitary and Phytosanitary measures, CP = Cartagena Protocol on Biosafety, CBD = Convention on Biological Diversity, LMOs = Living Modified Organisms. From an original diagram by Lesley Cree, Canadian Food Inspection Agency Used with permission from MacLeod et al. (2010)

necessity, technically justified, transparent, non-discriminatory, and minimal interference with international trade (Anon 2002). The IPPC is governed by the Commission on Phytosanitary Measures (CPM) upon which National Plant Protection Organisation (NPPO) representatives may sit. Some of the roles of the IPPC are to set International Standards for Phytosanitary Measures (ISPMs), to share biosecurity pest information and to provide technical assistance, for example on the implementation of ISPM's. Examples of ISPM's include ISPM No. 1 (1993) Principles of plant quarantine as related to international trade, ISPM No. 2 (1995) Guidelines for pest risk analysis, ISPM No. 7 (1997) Export certification system, ISPM No. 23 (2005) Guidelines for inspection, and the recent ISPM No. 36 (2012) Integrated measures for plants for planting.

The national legislation of most countries is based on the principles of the IPPC, but the approach to biosecurity and the laws that regulate the implementation differ from country to country. However, there are several characteristics that are common to all countries and reflect the basic premises of the IPPC (and that go back to the International Convention on Measures against *Phylloxera vastatrix*). Each importing country collates a list of regulated organisms that it considers harmful to plant health and stipulates that imported plants should be free of those pests. Any consignment must be accompanied by a phytosanitary certificate on which the NPPO of the exporting country declares that, during pre-export inspection, the listed pests were not found. The importing country often requires additional measures to be taken in order to ensure the absence, or reduction in the prevalence of particular pests or the pest-free status of particular host plant species. Such measures vary, depending on the importing country and the targeted pest or host. Possibilities include a certified pest-free place of production, prescribed pesticide treatments, and a ban on the import of hosts from certain countries or regions. Additional declarations on the phytosanitary certificate can be required to ascertain that these additional measures have been undertaken. When the consignment is imported, the NPPO of the importing country carries out an inspection that is primarily aimed as a check for compliance with the required phytosanitary measures, including that of pest-free status. In the case of non-compliance, the consignment can be treated to kill any pests, or it may be returned or destroyed. The application of standards and guidelines, such as the ISPMs, is open to interpretation amongst trading partners and this may lead to disputes that may be resolved through direct negotiation, appeals within trade jurisdictions or through the WTO (MacLeod et al. 2010).

## Regional phytosanitary agreements, regulations and organisations

In addition to the overarching international phytosanitary agreements and cooperating organisations, there are 10 regional plant protection organisations (RPPO's) that were set up under the IPPC (Table 35.1). The RPPOs input the IPPC, for example, through ISPMs development and the drafting of standards. The RPPOs facilitate cooperation on phytosanitary issues on a regional basis and several provide reporting services for emerging pests (Anderson et al. 2004).

There is some overlap in terms of territory covered and membership to these organisations. For example, Australia and New Zealand are parties to both the APPPC and the PPPO, and the USA is party to both the NAPPO and the PPPO. Regional bodies consist of individual states with often differing capabilities and nominated degrees of ALOP. However, member states often cooperate to facilitate arrangements that strengthen the overall protection of the regions from movement of plant pests and hence serve the interest of individual states. Some examples of the type of diversity in membership and approach amongst the RPPO's are outlined in Table 35.1.

*Africa* The Inter-African Phytosanitary Council (IAPSC) is the pre-eminent African Plant Protection Organization with headquarters in Cameroon. IAPSC has 53 members under the umbrella of the African Union. It coordinates plant protection in Africa with 4 sections responsible for Phytopathology, Entomology, Documentation, Information and Communication, and Administration and Finance. IAPSC carries out its work through the 8 African Regional Economic Communities (RECs). The IAPSC works to manage pest impacts on African crop production, non-compliance with ISPM's trade regulation and equivalents, and improve phytosanitary data collection and analysis. There is a variation in quarantine standards, resources and structures amongst member states of the IAPSC and the council seeks to promote regional cooperation and sharing of information to improve the overall standard of quarantine measures throughout Africa. Some countries have operational quarantine stations but others do not. Key collaboration of the IAPSC is in working to manage banana and cassava in pest diagnostics and control technique methods, germplasm and planting material exchange, along with and harmonization of African countries' phytosanitary systems (Anon 2013c).

**Table 35.1** Regional Intergovernmental Plant Protection Organisations

| Agreement | Region covered | Number of signatories |
|---|---|---|
| Asia and Pacific Plant Protection Commission (APPPC) | Asia and Pacific | 23 |
| Comunidad Andina (CAN) | South America | 4 |
| Comité Regional de Sanidad Vegetal Para el Cono Sur (COSAVE) | South America II | 6 |
| Caribbean Plant Protection Commission (CPPC) | Caribbean | 27 |
| European and Mediterranean Plant Protection Organization (EPPO) | Europe and Mediterranean | 50 |
| Inter-African Phytosanitary Council (IAPSC) | Africa | 53 |
| North American Plant Protection Organization (NAPPO) | North America | 3 |
| Near East Plant Protection Organization (NEPPO) | Middle East | 12 |
| Organismo Internacional Regional de Sanidad Agropecuaria (OIRSA) | Central America | 9 |
| Pacific Plant Protection Organisation (PPPO) | Pacific | 26 |

Biosecurity strategies in Africa include the use of "open quarantine" stations for the safer introduction of new plant genotypes with desirable traits. Rather than rely on costly infrastructure, such as quarantine greenhouses, this approach facilitates the introduction of new material and manages the risks through cultural and procedural strategies of the open quarantine facility. Such facilities were used in East and Central Africa for introductions of cassava material resistant to African cassava mosaic virus in the 1990s (Mohamed 2003). The success of this approach has at times been limited by inadequate diagnostic capability (MacLeod et al. 2010) and so the IAPSC continues to work to improve these systems. In order to meet international phytosanitary standards, some African countries pool their resources to develop pest databases, biosecurity tools and infrastructure. For example through the Centre of Phytosanitary Excellence (CoPE) based in Nairobi, Kenya (Anon 2013c).

*Europe* The European and Mediterranean Plant Protection Organization (EPPO) is the regional plant protection organisation for Europe, with ca. half of the 50 member states being members of the European Union (EU). EPPO is responsible for cooperation and harmonisation of phytosanitary matters in Europe and the Mediterranean area. The organisation seeks to develop an international strategy against the introduction and spread of pests that damage cultivated and wild plants, encourage harmonization of phytosanitary regulations and action, promote the use of modern, safe, and effective pest control methods and provide a documentation service on plant protection. Two working parties carry out a large part of EPPO's technical activities, which are the working party on phytosanitary regulations and the working party on plant protection products. The activities of these working parties concern limiting the spread of quarantine pests between and within countries, respectively. Other activities include the development and maintenance of EPPO Standards, the publication of a scientific journal on regulatory plant protection and informing member states through a monthly newsletter on developments of phytosanitary concern. EPPO also maintains lists of pests recommended for regulation as quarantine pests (A1 and A2 lists), which comprise ca. 310 taxa.

The EU comprises 27 member states that share a common plant health policy (the Plant Health Directive, Council Directive 29/2000/EC; Anon 2000). European Commission Directives require that individual states reflect them in their domestic legislation; for example in Britain the previous Plant Health Directive (77/93/EEC) was legislated in the Plant Health (Great Britain) order 1993 (MacLeod et al. 2010). Although under regular inspection by the European Food and Veterinary Organisation, it is possible that differences in inspection methods exist. Swiss plant health legislation is similar to this piece of EU legislation and Switzerland is from the phytosanitary perspective part of the European Union. One of the central premises of the EU Plant Health Directive is that any consignment of plants imported from non-EU member states must be inspected at the first point of entry into the EU, but that no further phytosanitary inspections take place for movement within the EU. Instead, a Plant Passport is issued for intra-EU trade in certain plants and plant products that certifies that consignments are free of the regulated pests that are listed in the Annexes of the Plant Health Directive. The EU has a "black list" approach to biosecurity, with ca. 210 taxa regulated, primarily on species and genus level; unregulated plant species are allowed into the EU, unless these organisms are found. The EU has post-entry quarantine for a restricted list of plant species; the majority of species are allowed into the Union if the outcome of the import inspections were satisfactory.

Some authors have identified shortcomings with the current European phytosanitary system (MacLeod et al. 2010). These shortcomings include that plants not regulated by the Plant Health Directive are allowed entry into the EU with minimal regulation. Therefore, new pathways are not subject to sufficient analysis and only after pest incursions occur and new pests establish is the risk identified, by which time it may be too late to implement phytosanitary measures at the border. On the other hand, stringent controls are applied to some pathways that are compliant and pose very low risk, for example apples from New Zealand and USA (MacLeod et al. 2010).

A recent evaluation of the Common Plant Health Regime conducted for the European Commission (Anon 2010) concluded that although the Common Plant Health Regime had been broadly successful in safeguarding plant health in the EU, there were a number of improvements that could be made to the system. These were put to the European Commission as recommendations to be incorporated into proposed new plant health legislation to be developed and implemented over the next decade. The two issues broadly identified were that plant health risks have increased due to globalisation and climate change. Fifteen recommendations were made and these included a need to focus on prevention and better risk targeting, broaden the scope to invasive plants with wider/environmental impacts, take complementary measures on imports, in particular for emerging risks, e.g. on new trade in plants for planting, strengthen measures for plants for planting via official post-entry inspections and proceed to import bans where necessary, introduce mandatory general epidemio-surveillance, and the stepping up of emergency action and control/eradication measures. Interestingly, many of the themes of the European review are reflected in a government commissioned review of the Australian quarantine system (Beale 2008).

*Asia and Pacific* The 24 member states of the Asia and Pacific Plant Protection Commission (APPPC) vary considerably in size and level of economic development. The association includes most countries of the region. While IPPC is primarily concerned with transnational phytosanitary measures, APPPC addresses a greater range of plant protection functions within the region. In comparison with the EU member states, which share common legislation, there is a great deal of variation in the number and nature of biosecurity measures implemented by member states of the APPPC.

The economic, social and political diversity of the region is reflected in the variation in terms of the implementation of various ISPMs throughout the APPPC. Another contributing factor is the ALOP nominated by individual states in the region. For example, Australia has a very low level of acceptance of produce exported under ISPM 29. This is because 'areas of low pest prevalence' are not deemed sufficient unless accompanied by other supporting measures which act in combination to reduce the risk to an acceptably low level (Anon 2011). New Zealand has a similar low ALOP and, consequently, these countries require more stringent phytosanitary measures to be applied to live plants than other APPPC member states, or indeed than most, if not all other countries in the World. Such measures include obligatory insecticide and acaricide treatments prior to export to these countries and a post-entry quarantine after passing phytosanitary inspections at the border for all live plants, with exception of tissue cultures.

In general terms Australia, China, Japan (not a signatory), New Zealand and the Republic of Korea have implemented the most ISPMs` of the region. Of 16 survey respondents, 9 countries reported a 'large' degree of implementation, 3 a 'partial' implementation and 4 a 'low' implementation rate (Anon 2011a). Factors reported to contribute strongly to ISPM implementation were relevancy of the ISPM, sufficient resources, and supporting policies and manuals. Conversely, the factors that contributed strongly to a low degree of implementation were insufficient capacity building, and lack of qualified staff and resources. Export certification (ISPM No. 7) and phytosanitary certification (ISPM No. 12) had the most factors that contributed to a high degree of implementation. ISPM No. 4 (pest-free areas), No. 9 (pest eradication) and No. 11 (quarantine pests) listed the most factors that limited their implementation.

ISPMs that most directly affect market access or are most easily implemented at the border to protect internal markets are more commonly implemented. For example, access to overseas markets is almost universally predicated on an export certification system, requiring all consignments of live plants to be accompanied by a phytosanitary certificate that is issued by the NPPO of the exporting country to certify that the consignment is free of regulated pests and that any additional phytosanitary requirements of the importing country have been met. Surveillance systems can be expensive to maintain and the cost is not easy to pass on directly to the relevant stakeholder, whereas charges may be more easily applied and accepted by an exporter for certification, or an importer for inspection fees at the border. Good surveillance is required to detect new incursions and also for the generation of robust pest free production status. However, this is a relatively resource-intensive activity and requires highly trained staff. The fact that many APPPC countries reported a lack of skilled staff and insufficient laboratory capacities may have an influence in the inconsistent uptake of surveillance and pest eradication. For the period 2007–2010 new exotic species were reported in 7 countries; most new pests were reported in New Zealand (75 in 2007/08) (Anon 2011a). The reporting of pests may reflect the effort that has been expended on surveillance for new pests rather than actual invasion rates.

## Biosecurity Failures and Success: Lessons Learnt

The single greatest cause of emerging plant disease epidemics (at 56 %) is the introduction of new pathogens in to a region (Anderson et al. 2004). New pathogen introductions are a more significant cause of emerging plant disease epidemics than changes in farming practices, weather and pathogen evolution. Much can be learnt from examples of disease epidemics resulting from new pest introductions.

Recently, Potter et al. (2011) modelled the response to the Dutch Elm Disease epidemic, caused by *Ophiostoma ulmi* and that occurred in the UK during the 1970s, with the purpose to learn and to define better methods for prevention management and control. The aim was to be able to take those lessons and apply them to the current outbreak of sudden oak death (*Phytophthora ramorum*) in the UK and elsewhere. The simulation was based on historical records in the UK and included biological and policy factors in a spatial model. Although the short-term (4 year) simulation showed a reduction in total disease levels with strong intervention compared with the actual response in the period 1964–1968 the longer term simulations indicated that management scenarios and a no intervention policy eventually converge to a similar final outcome. The conclusion from their simulation is that the biology of Dutch Elm Disease, associated vectors and the available means of control and management at that time could not have prevented the eventual devastating impact of the disease once it was established in the UK. The over-riding message is that for fungal diseases with a similar biology the most effective means of control is to prevent the incursion through strong biosecurity measures.

One of the most recent examples of the world-wide movement of a pest through the international trade in live plants is that of sudden oak death (*Phytophthora ramorum*) (Fig. 35.4). This disease emerged in the 1990s as a devastating forest pathogen of oak and tanoak in the US (Liebhold et al. 2012) and rhododendron, viburnum, beech, and other trees and ornamentals in Europe (Brasier 2008). The disease is thought to have originated in Asia and to have been moved throughout Europe and North America via the live plant trade (Brasier 2008; Ivors et al. 2006; Mascheretti et al. 2008). This is a previously unknown pathogen that has alerted some regulators to some of the flaws in current regulatory systems. For one, protocols are usually predicated on lists of named pests. Often this information is well known for economically important plants. However, for newly traded and lesser known ornamentals there is less available knowledge of pests and what is known

**Fig. 35.4** Symptoms of *Phytophthora ramorum* infection causing sap bleeding (a) and cankers (b) on oak (*Quercus* sp.); and cankers (c) and petiole lesions (d) on tanoak (*Notholithocarpus densiflorus*). Photos used with permission from Daniel Huberli

may be based on inaccurate taxonomy, due to the small amount of research effort expended. The biology and potential host range of these pathogens is often unknown and it is not tested until they encounter new hosts in their transplanted habitat. *Phytophthora ramorum*, probably introduced on rhododendron, now has a known host range of over 100 native and non-native trees and shrubs in the USA and over 30 hosts in the UK and Europe (Brasier 2008). There may also be a lag in the time taken for the pathogen to establish and show the extent of its impact by which time cost-effective control and eradication is really too late, as described by Potter et al. (2011). Many other 'newly escaped' organisms were unknown to science before they escaped including Dutch elm disease, *Phytophthora* disease of alder, and box blight. Other lessons from the *P. ramorum* epidemic outlined by Brasier (2008) are problems with inspection and the implementation of regulations and a lack of penalties for breaches. Partly in response to the sudden oak epidemic and the issues that it illustrates, both the US and the EU have initiated reviews of their plant health legislation, particularly with respect to plants for planting (Anon 2008; Anon 2010).

The modality of the interception of *Fusarium circinatum*, the pitch canker-causing fungus, on Douglas fir (*Pseudotsuga menziesii*) scion imported from the United States to New Zealand confirms how the success story in biosecurity are the outcome of a mix of variables. A proper inspection protocol, the existence of a clear regulation, a wide knowledge (supported by proper training) of staff operating in the sector, created a short circuit capable to intercept an alien species (Ormsby 2004).

## International Variation in the Management of Pests and Pathways

Generally, countries regulate species based on the perceived risk. The most stringent action is prohibition of import of certain species or species from certain origins, if the risks are deemed too great or cannot be effectively managed. There are, however, large differences in the perception of risk and the approaches to plant biosecurity taken by countries around the world and some of these countries are in the process of reviewing their legislation, including the USA, Australia, and the European Union. Against this background, it may be informative to look at some of the phytosanitary measures available and the international differences in their implementation to identify opportunities for improvement.

Plant biosecurity efforts can be managed at three key points in the plant trade network: pre-border, border and post-border. The most effective biosecurity efforts involve phytosanitary measures prior to export to minimise the arrival of IAS in the country (see the preceding section). However, phytosanitary inspections at the point of entry and plant quarantine between those inspections and final release of the goods into the country are important components of the phytosanitary strategy of many countries. The primary aim of these inspections often is to check for compliance with the required phytosanitary measures in the exporting country and the inspections also target the presence of regulated pests that may not have been noticed prior to export. Because of the large volume of trade it is often impossible to inspect all imported live plants and so strategies to identify high-risk commodities and consignments are of great importance. As part of the post-border biosecurity efforts, some countries have included obligations to report new species (for example New Zealand) or to eradicate newly arrived IAS' (e.g. the European Union) in their phytosanitary legislation. Interactions between NPPOs and the public, producers, traders and consumers are a key aspect of many successful biosecurity strategies.

In Australia, preparation for emerging plant disease epidemics is managed through production of diagnostic manuals for economically important plant diseases and these inform post entry quarantine assessment of plant consignments. Contingency plans are prepared through industry bodies and responsible government agencies and these provide guidelines to stakeholders on diagnostics, surveillance, survey strategies, epidemiology and pest risk analysis as well as legally binding plant pest response deeds that set out cost sharing arrangements amongst industry stakeholders and state and federal government.

In countries with a black list approach, i.e. where pests are regulated on an individual basis following Pest Risk Assessment, the phytosanitary measures required for specific plant genera or origins are often based around ensuring the absence of these named pests. However, regulating for specific pests carries the hazard of not controlling for unknown pests, and that inspection and diagnostics at the border may be too slow to detect and identify a pest in a timely manner. A systems approach that attempts to control many pests that may be associated with a live plant pathway is generally accepted as the most effective. In a systems approach the risks are managed through combinations of phytosanitary measures that are put in place at all or components of the production and trade in live plants. The implementation of multiple measures has the advantages that it is likely to impact on several pests and that the level of security/protection is less affected by the failure of a single phytosanitary measure.

The effectiveness of the system is monitored through a quality assurance system that may be independently audited. Examples of certification schemes that may be used for biosecurity export services include the Netherland's Naktuinbouw which is responsible for a range of horticultural plant quality certification services (Anon 2013d) and the certification scheme for plant producers in Ohio to reduce the prevalence of *Phytophthora* (Parke and Grünwald 2012). Although the latter example is focussed on plants for the domestic market in the USA, similar schemes could contribute to reducing the international movement of pests. At present there are, unfortunately, few numerical examples that illustrate the effectiveness of individual components of systems approaches, or combinations thereof, in plant production.

Where the risk cannot be adequately managed for the required ALOP solely through off-shore production processes and certification, additional measures may be required. An example is illustrated of managing pathways to prevent the introduction of pine pitch canker (PPC) (*Fusarium circinatum*) to Australia (Fig. 35.5). Although targeted to a known pathogen of high concern, the systems in place also address a range of other pests that may be on the various pine related pathways.

The EU has a comparatively open system for live plant imports, where most plant genera can be imported and special phytosanitary requirements are defined based on the list of regulated species in the annexes of the plant health legislation (Anon 2000). Most countries outside the EU stipulate that an import permit must be obtained from the NPPO prior to shipment of the consignment in the exporting country. The NPPO generally assesses the risks associated with the proposed product prior to issuing the licence. Recently, the United States has created a new "grey list" of genera for which the risks are not well known, called Not Approved Pending Pest Risk Assessment (Anon 2011). The import of plants of those genera, from the specified origins, is prohibited until a pest risk analysis is carried out that determines whether the species can be imported or not. In Australia, an import risk assessment (IRA) is carried out by Department of Agriculture Fisheries and Forestry (DAFF) for the introduction of a proposed product. This involves an analysis of all known pests on the pathway and an assessment of their potential to establish and likely impact is made. Mitigation measures are identified that include pre-border inspection to ascertain freedom from pest and disease, which must be declared on the Phytosanitary Certificate that accompanies the consignment. For example, nursery stock ('Plants for Planting') is required to come from a PPC free country and requires a phytosanitary certificate stipulating freedom from pest and

**Fig. 35.5** Illustration of how the measures along the biosecurity continuum are applied pre-border, border and post-border to protect Australian horticulture and forestry industries from pine pitch canker (*Fusarium circinatum*) introduction (Maxwell and Brelsford 2012)

disease symptoms and freedom from soil and other contaminating material. There is mandatory requirement for a 100% inspection of the consignment at the border followed by pest treatment when required to manage arthropod pests. Following treatment the plants must be grown in an accredited 'high risk' post entry quarantine greenhouse with measures in place to minimise the risk of pest or pathogen egress. The plants require mandatory inspections over the 2 year period by an accredited plant pathologist. The facilities are audited for compliance by DAFF. Similar measures are required for other high risk nursery stock and in some cases, such as for grape vines (*Vitis* spp.), mandatory testing against known pathogens is required.

Most medium risk nursery stock undergoes a similar pre-border and border process including extensive visual inspection. However, the post entry quarantine period is reduced to 3 months along with inspections by DAFF staff; and there is no requirement for mandatory testing against known pathogens. This is in contrast to most other countries around the world where inspection of medium risk ornamentals is usually followed by release without a period of post entry quarantine assessment (Liebhold et al. 2012; Brasier 2008). In many countries only a percentage of consignments are inspected at the border; and of those consignments only a small percentage of plants are inspected (Liebhold et al. 2012; Brasier 2008).

A broad-spectrum measure aimed at preventing the introduction of a range of soil pests that is part of many countries' legislation is the prohibition of soil imports, or the import of plants with soil attached. However, there are often exceptions to such rules in order to ensure that plants stay alive during transport, including allowing specific plant types to be imported in sterilised growth media, or enough soil may be allowed to keep the plant alive. Various countries stipulate pesticide treatment of consignments of live plants or plant products against arthropods. These include pre-export treatments, such as dipping in insecticide and acaricide baths required for all plants to be imported into New Zealand (Anon 2012a). Although these pesticide treatments are effective, not all countries requires treatment of all imported live plants.

Phytosanitary inspections at the point of entry are a common component of phytosanitary practices, but there are large differences in the prescriptions for the inspections. In Australia, most medium-risk nursery stock undergoes a similar pre-border and border process including 100% inspection. The legislation in New Zealand stipulates inspection of all plants in consignments with up to 600 plants, and inspection of 600 units in larger shipments. In many countries, however, only a percentage of consignments may be inspected at the border (Liebhold et al. 2012; Brasier 2008). Moreover, whereas the inspections in some countries focus on regulated pests, inspectors in other countries look for any organism on the plants. No inspection can ascertain freedom of pests, in particular when considering cryptic life stages, but the sampling intensity affects the maximum infestation level that can be ascertained: the more plants inspected without infestations found, the lower the overall infestation level of the consignment (Venette et al. 2002). Unfortunately only very few countries, including the USA and New Zealand, keep records of the outcome of inspections of sufficient detail to allow an estimate of the level of pest infestation to be made. Moreover, it is often unknown what fraction of pests goes unnoticed during phytosanitary inspections at the point of entry, although this may be large (Liebhold et al. 2012). Such information would be valuable for improving inspection policy.

Many countries have some degree of quarantine requirement for live plant imports. Some countries stipulate that high or medium risk plants must be kept in a quarantine facility for a specified time after they have been brought into the country ("post-entry quarantine"). Whether post-entry quarantine is required depends on the perceived risk associated with the plant species and its origin. In Australia and New Zealand, all live plants must go into post-entry quarantine. In New Zealand, the standard minimum period is 3 months, but this can be longer depending on the plant species (Anon 2012a). During this period, the plants are regularly inspected by inspectors from the Ministry for Primary Industries and if any pests are found the consignment is either treated or destroyed. Post-entry quarantine has been found to be particularly effective for the detection of pathogens, since plants may not be symptomatic at the time of import and can develop these during their time in post-entry quarantine. This strict procedure is in contrast to most other countries around the world where inspection of medium risk ornamentals is usually followed by release without a period of post entry quarantine assessment (Liebhold et al. 2012; Brasier 2008).

## Conclusion

Historically, measures to manage the risk of pest introductions associated with horticultural trade have been made in response to identified threats and improvements in the understanding of those pest threats. The current biosecurity framework in many countries is based in legislation that was first promulgated in the early 1900s and since amended. This legislation often evolved to control major pests of high value agricultural crops after they became known. The growth in international trade has undergone exponential shifts firstly through the industrial revolution and then through the current era of globalisation. As the trade in horticultural products has increased and market structure evolved, so too has the number of pests introduced to new areas grown. However, although measures have been put in place to manage the risk of pest introductions, those measures have not kept pace with the rate of increase in horticultural exports. Although technologically more sophisticated today, the main principles governing international phytosanitary requirements have not altered significantly in 100 years. The international framework for biosecurity measures is based on the identification of known pests and managing those pests. Increasingly, previously unknown pests are emerging from the trade in ornamental plants and these are invading before the regulatory system can adapt to control them. The implementation of ISPM's is variable across the world depending on resources and nominated ALOP. Many countries are recognising that their biosecurity legislation is in need of modernising and that more effective systems need to be developed to cope with the volume and structural changes in plant trade. This has been seen in the recommended review of the EU's legislation (Anon 2012), the US regulatory reviews (Anon 2008) and Australia's recent legislative review by Beale (2008).

**Acknowledgments** "AMV has received funding from the European Union Seventh Framework Programme FP7 2007–2013 (KBBE 2009–3) under grant agreement 245268 ISEFOR and RE was financially supported through a grant from the Swiss Secretariat for Science, Education and Research to join the EU COST Action PERMIT."

## References

Alpert P, Bone E, Holzapfel C (2000) Invasiveness, invisibility and the role of environmental stress in the spread of non-native plants. Perspect Plant Ecol Evol Syst 3:52–66

Anderson PK, Cunningham AA, Patel NG et al (2004) Emerging infectious diseases of plants: pathogen pollution, climate change and agrotechnological drivers. Trends Ecol Evol 19:535–544

Anon (1999) The horticultural sector in the Ile-de-France region. Lien Horticole 36(45):7–8

Anon (2000) Official Journal of the European Communities COUNCIL DIRECTIVE 2000/29/EC of 8 May 2000 on protective measures against the introduction into the Community of organisms harmful to plants or plant products and against their spread within the community

Anon (2002) FAO. Guide to the international plant protection convention. FAO, Rome, pp 20

Anon (2004) International Association of Horticultural Producers (AIPH). International Association of Horticultural Producers. International Statistics Flowers and Plants 2004, vol 52. Institut für Gartenbauokonomie der Universität, Hannover. http://www.aiph.org/site/index_en.cfm. Accessed 15 April 2013

Anon (2008) APHIS. USDA register. http://www.aphis.usda.gov/import_export/plants/plant_imports/downloads/q37_whitepaper.pdf. Accessed 15 April 2013

Anon (2010) Food Chain Evaluation Consortium. Evaluation of the community plant health regime: final report prepared for the European Commission Directorate General for Health and Consumers. European Union registar. http://ec.europa.eu/food

Anon (2011) FAO. FAO register. http://www.apppc.org/index.php?id=1110810&L=0. Accessed 15 Feb 2013

Anon (2011a) FAO. RAP PUBLICATION 2011/11 Plant protection profiles from Asia-Pacific countries (2009–2010), 3rd edn. RAP publication 2011/11 Asia and Pacific Plant Protection Commission and Food and Agriculture Organization of the United Nations regional office for Asia and the Pacific, Bangkok, p 572

Anon (2012) Wardian case. http://en.wikipedia.org/wiki/Wardian_case. Accessed 27 April 2013

Anon (2012a) UNCTAD. Handbook of statistics. United Nations Conference on Trade and Development, New York

Anon (2013) DAFF ICON, DAFF register. http://www.daff.gov.au/aqis/import/plants/ICON. Accessed 15 Feb 2013

Anon (2013a) MPI biosecurity New Zealand, MPI register. www.biosecurity.govt.nz/regs/imports/plants/nursery. Accessed 15 March 2013

Anon (2013b) IPPC register. https://www.ippc.int/IPP/En/default.jsp. Accessed 21 March 2013

Anon (2013c) IAPSC register. http://r4dreview.org/2011/04/iapsc-protecting-africas-plant-health. Accessed 15 March 2013

Anon (2013d) Naktuinbouw. naktuinbouwregister. http://www.naktuinbouw.nl/en. Accessed 13 March 2013

Beale (2008) Review of Australian Quarantine, DAFF register. http://www.daff.gov.au/about/annualreport/annual-report-2008-09/annual-report-2008-09/special-report-review-australian-quarantine-biosecurity-bealereview

Brasier CM (2000) The rise of hybrid fungi. Nature 405:134–135
Brasier CM (2008) The biosecurity threat to the UK and global environment from international trade in plants. Plant Pathol 57(5):792–808
DeBary A (1861) Die gegenwartig herrschende Kartoffelkrankheit, ihre Ursache und ihre Verhutung. Arthur Felix, Leipzig, Germany
Dehnen-Schmutz K, Holdenrieder O, Jeger MJ, Pautasso M (2010a) Structural change in the international horticultural industry: some implications for plant health. Sci Hortic 125(1):1–15
Dehnen-Schmutz K, MacLeod A, Reed P, Mills PR (2010b) The role of regulatory mechanisms for control of plant diseases and food security—case studies from potato production in Britain. Food Sec 2:233–245
Desprez-Loustau M-L (2009) The alien fungi of Europe. In: DAISIE (ed) Handbook of alien species in Europe. Springer, Berlin, pp 15–28
Dixon GR (1987) The practicalities and economics of micropropagation for the amenity plant trade. Micropropagation in horticulture: practice and commercial problems. In: Proceedings of the Institute of Horticulture Symposium, University of Nottingham, School of Agriculture, 24–26 March 1986, pp 183–196
Dixon GR (2000) Changing fortunes. Horticulturalist 9:7–9
Ebbels DL (1979) A historical review of certification schemes for vegetatively-propagated crops in England and Wales. ADAS Q Rev 32:21–58
Ercsey-Ravasz M, Toroczkai Z, Lakner Z, Baranyi J (2012) Complexity of the International Agro-Food Trade Network and its impact on food safety. PLoS ONE 7(5):e37810
FCEC (2010) (Food Chain Evaluation Consortium) Evaluation of the Community Plant Health Regime: Final Report Prepared for the European Commission Directorate General for Health and Consumers. European Union register. http://ec.europa.eu/food
Findlay R, O'Rourke KH (2007) Power and plenty: trade, war, and the world economy in the second millennium. Princeton University, Princeton, USA
Guo Q, Rejmanek M, Wen J (2012) Geographical, socioeconomic, and ecological determinants of exotic plant naturalization in the United States: insights and updates from improved data. NeoBiota 12:41–55
Goodwin SB, Cohen BA, Fry WE (1994) Panglobal distribution of a single clonal lineage of the Irish potato famine fungus. Proc Natl Acad Sci USA 91:11591–11595
Goodwin BJ, McAllister AJ, and L. Fahrig (1999) Predicting invasiveness of plant species based on biological information. Conservation Biology 13:422–426
Goodwin SB, Sujkowski LS, Fry WE (1995) Rapid evolution of pathogenicity within clonal lineages of the potato late blight disease fungus. Phytopathology 85:669–676
Guzmán I, Arcas N, Guelfi R et al (2009) Technical efficiency in the fresh fruit and vegetable sector: a comparison study of Italian and Spanish firms. Fruits 64:243–252
Hufnagel L, Brockmann D, Geisel T (2004) Forecast and control of epidemics in a globalized world. PNAS 101(42):15124–15129
Hulme PE (2009) Trade, transport and trouble: managing invasive species pathways in an era of globalization. J Appl Ecol 46(1):10–18
Ivors KL, Garbelotto M, Vries IDE et al (2006) Microsatellite markers identify three lineages of Phytophthora ramorum in US nurseries, yet single lineages in US forest and European nursery populations. Mol Ecol 15:1493–1505
Jeger MJ, Pautasso M, Holdenrieder et al (2007) Modelling disease spread and control in networks: implications for plant sciences. New Phytol 174:179–197
Kenis M, Rabitsch W, Auger-Rozenberg M-A, Roques A (2007) How can alien species inventories and interception data help us prevent insect invasions? B Entomol Res 97:89–502
Liebhold AM, Brockerhoff EG, Garrett LJ et al (2012) Live plant imports: the major pathway for forest insect and pathogen invasions of the US. Front Ecol Environ 10(3):135–143
Liu J, Liang SC, Liu FH et al (2005) Invasive alien plant species in China: regional distribution patterns. Divers Distrib 11:341–347
Lunati F (2007) L'ombra dell'Olanda sul futuro della floricultura italiana. Informatore Fitopatologico 57(7–8):3–6

Mack R, Simberloff D, Lonsdale W, Evans H, Clout M, Bazzaz F (2000) Biotic invasions: causes, epidemiology, global consequences, and control. Ecol Appl 10:689–710
MacLeod A, Pautasso M, Jeger MJ et al (2010) Evolution of the international regulation of plant pests and challenges for future plant health. Food Security 2(1):49–70
Mascheretti S, Croucher PJP, Vettraino A et al (2008) Reconstruction of the sudden oak death epidemic in California through microsatellite analysis of the pathogen *Phytophthora ramorum*. Mol Ecol 17:2755–2768
Maxwell A, Brelsford H (2012) A pitch for biosecurity: minimising the risk of Pine Pitch Canker (*Fusarium circinatum*) introduction into Australia. Proceedings of the Australian Soil Disease Conference, Fremantle Western Australia, Sept 2012
Mitchell CE, Power AG (2003) Release of invasive plants from fungal and viral pathogens. Nature 421:625–627
Mohamed RA (2003) Role of open quarantine in regional germplasm exchange. In: Legg JP, Hillocks RJ (eds) Cassava brown streak virus disease: past, present and future. Proceedings of an International Workshop, Mombasa, Kenya, 27–30 October 2002. Natural Resources International Ltd., Aylesford, p 100, pp 63–65
Neven D, Reardon T (2004) The rise of Kenyan supermarkets and the evolution of their horticulture product procurement systems. Develop Policy Rev 22(6):669–699
Niederhauser JS (1991) The Potato Association of America and international cooperation 1916–1991. Am Potato J 68:237–239
Nowicki M, Foolad MR, Nowakowska M, Kozik EU (2012) Potato and tomato late blight caused by Phytophthora infestans: an overview of pathology and resistance breeding. Plant Dis 96(1):4–17
O'Riordain F (1999) Directory of European plant tissue culture laboratories, 1996–97. COST Action 822. Commission of the European Communities, Brussels
Ormsby M (2004) Pitch Canker in Quarantine—a biosecurity success story. Biosecurity 51:10
Park KS, McLaughlin EW (2000) The US wholesale produce industry: structure, operations and competition. Acta Hort 524:197–204
Parke J, Grünwald N (2012) A systems approach for management of pests and pathogens of nursery crops. Plant Dis 96:1236–1244
Paterson A (2000) The plant hunters: two hundred years of adventure and discovery around the world. Stud Hist Gard Des L 20(3):258–259
Perrings C, Dehnen-Schmutz K, Touza J et al (2005) How to manage biological invasions under globalization. Trends Ecol Evol 20:212–215
Pimentel D, Zuniga R, Morrison D (2005) Update on the environmental and economic costs associated with alien-invasive species in the United States. Ecol Econ 52:273–288
Potter C, Harwood T, Knight J et al (2011) Learning from history, predicting the future: the UK Dutch elm disease outbreak in relation to contemporary tree disease threats. Philos Trans R Soc London Ser B 366(1573):1966–1974
Prévost IB (1807) Memoir on the immediate cause of bunt or smut of wheat, and of several other diseases of plants, and on preventives of bunt. Paris, pp 80
Pyšek P, Lambdon PW, Arianoutsou M et al (2009) Alien vascular plants of Europe. In: DAISIE (ed). Handbook of alien species in Europe. Springer, Berlin, pp 43–61
Roques A (2007) Old and new pathways for invasion of exotic forest insects in Europe. In: Evans H, Oszako T (ed) Alien invasive species and international trade. Warsaw, pp 80–88
Roques A, Rabitsch W, Rasplus J-Y et al (2009) Alien terrestrial invertebrates of Europe. In: DAISIE (ed) Handbook of alien species in Europe. Springer, Berlin, pp 63–79
Sakai AK, Allendorf FW, Holt JS et al (2001) The population biology of invasive species. Ann Rev Ecol Syst 32:305–332
Santini A, Ghelardini L, De Pace C. et al. (2013) Biogeographical patterns and determinants of invasion by forest pathogens in Europe. New Phytologist 197: 238–250.
Sax DF (2001) Latitudinal gradients and geographic ranges of exotic species: implications for biogeography. J Biogeog 28:139–150
Schumann GL (1991) Plant diseases: their biology and social impact. APS, St. Paul, p 397

Smith R, Baker RHA, Malumphy CP et al (2007) Recent nonnative invertebrate plant pest establishments in Great Britain: origins, pathways, and trends. Agric For Entomol 9:307–326

Usinger R (1964) The role of Linnaeus in advancement of entomology. Annu Rev Entomol 9:1–17

Venette R, Moon R, Hutchison W (2002) Strategies and statistics of sampling for rare individuals. Annu Rev Entomol 47:143–174

Virtue J, Bennett S, Randall R (2004) Plant introductions in Australia: how can we resolve 'weedy' conflicts of interest? In: Sindel BM, Johnson SB (eds) Procedings of Fourteenth Australian Weeds Conference. Weed Society, NSW, p 718

Vitousek PM, D'Antonio CM, Loope LL, Rejmanek M, Westbrooks R (1997) Introduced species: a significant component of human-caused global change. New Zeal J Ecol 21:1–16

Waage JK, Woodhall JW, Bishop SJ et al (2008) Patterns of plant pest introductions in Europe and Africa. Agric Syst 99(1):1–5

Westphal MI, Browne M, MacKinnon K et al (2008) The link between international trade and the global distribution of invasive alien species. Biol Invasions 10:391–398

Wilcove DS, Rothstein D, Dubow J, Phillips A, Losos E (1998) Quantifying threats to imperiled species in the United States. Bioscience 48:607–615

# Chapter 36
# Horticulture and Art

**Jules Janick**

**Abstract** One of the unique characters of horticulture as an agricultural discipline is that it has an esthetic component. There are two approaches to the esthetics of horticulture: (1) art in horticulture, the direct use of plants alone and in groups as pleasing visual objects, and (2) horticulture in art, the use of horticultural objects as a basic component of artistic expression. Art in horticulture revolves around plants as beautiful objects, individually and en masse. This concept has generated distinct disciplines such as flower arranging and the floral arts, garden design and development, and landscape design and architecture. Horticulture in art, refers to the depiction of horticultural plants in connection with various manifestations of the visual arts such as sculpture and mosaics, drawings and painting, and embroidery and tapestry. The depiction of plants is one of the great themes in artistic expression as exemplified in their widespread use in the decorative patterns in the design of innumerable objects from floor and ceiling patterns, silverware, pottery and ceramics, coins and banknotes, to heraldry.

**Keywords** Floral arts · Gardens · Landscape architecture · Mosaics · Painting · Sculpture · Tapestry

## Introduction

Horticulture is unique among the agricultural disciplines in that it has an esthetic dimension (Janick 1984). There are two approaches to consider in the relation of horticulture to artistic expression: (1) art in horticulture, the direct use of plants alone or en masse as pleasing visual objects, and (2) horticulture in art, the use of horticulture objects and subjects as a basic component of artistic expression in various art forms such as drawings and painting, sculpture, mosaics, photography, and tapestry (Janick 2007). The relation of horticulture and art has created unique disciplines including the floral arts, garden design, landscape architecture, and still life painting. The depiction of plants is one of the great themes in artistic expression and

J. Janick (✉)
Department of Horticulture and Landscape Architecture, Purdue University,
West Lafayette, IN 47907-2010, USA
e-mail: janick@purdue.edu

G. R. Dixon, D. E. Aldous (eds.), *Horticulture: Plants for People and Places, Volume 3*, 1197
DOI 10.1007/978-94-017-8560-0_36, © Springer Science+Business Media Dordrecht 2014

has spawned the use of plants in decoration of innumerable objects such as floor and ceiling patterns, sculptural columns, silverware, ceramics, banknotes, and heraldry.

## Art in Horticulture

Horticultural plants are often considered beautiful and pleasing objects in themselves based on a combination of shape, texture, color, form, design, symmetry, as well as fragrance and taste (Janick 1984). Plants also may be viewed individually or arranged as components in a larger context and became an essential part of three artistic disciplines: the floral arts, garden design, and landscape architecture. There is no clear distinction between these three components except that of scale. The growing of horticultural plants for esthetic purposes in the home and in the landscape has developed a huge part of horticulture now referred to as the "green industry."

### *The Esthetic Value of Plants*

Our perception of beauty is strongly affected by our emotional feelings and by our cultural attitudes towards objects. Thus, things that are feared such as snakes or spiders are thought by some as ugly despite having many attributes we ascribe to beautiful objects. Basically, the things that have been accepted as beautiful for long periods of time, and which are more or less universally admired, have a basic simplicity and harmony of form and function. Thus, our concept of beauty is made up of two parts; sensory stimulation and a cultural component.

Most plants have an inherent capacity to visually stimulate. The most obvious feature is their coloring, not only the brilliant hues of flowers, fruits and leaves, but the muted tones of stems and bark. Green of course is the most common environmental color and our positive response is probably more than coincidental since it also psychologically is the most restful. Structure and shape (form) of plants shows tremendous variation from turf and creeping ground covers, to shrubs, and trees of various sizes and shapes. Symmetry makes random shapes orderly. All plants show some types of symmetry a common feature of plant growth which in inherently pleasing (Fig. 36.1a). However, the use of plants in asymmetrical patterns or arrangements also produces visual interest. (Fig. 36.1b).

### *Gardens*

With the possible exception of arctic peoples, human cultures have developed in plant-dominated environments. Plants provide food for people and their animals, as well as fiber, shelter, and shade. Our dependence upon plants has influenced and molded our esthetic consideration of them. And no doubt plants have been

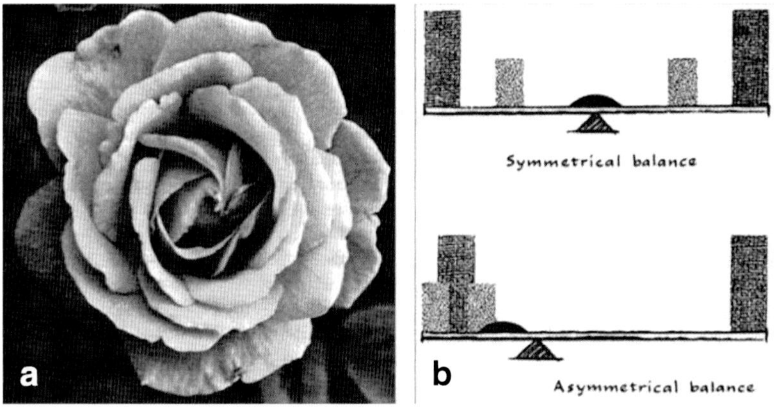


**Fig. 36.1** Symmetry and balance: **a** symmetry in the rose; **b** graphic representation of symmetrical and asymmetrical balance. (Source: Janick 1984)

culturally accepted as beautiful partially because they are useful. In modern cultures only a relatively few people are directly involved with the growing of plants, but we all depend on them. At present, the production and management of ornamentals, known collectively as the green industry, remains one of the important parts of modern horticulture. Horticulture has a place in all our lives.

Civilizations create gardens (Groening 2007). The origin of the garden is rooted in the human desire to be surrounded by plants, both useful food plants such as, trees for shade, fruits, vegetables, and spices for sustenance and pleasure, as well as plants that are esthetically pleasing based on appearance and fragrance. Thus we speak of pleasure gardens and kitchen gardens. In many cases it is often difficult to separate the purely functional from the esthetic. The first gardens in recorded history are found in ancient cultures of Egypt, Mesopotamia, and China but gardens have been greatly influenced in England, Greece, Japan and Persia (Thacker 1979). The two opposing traditions in gardens—formalism and naturalism—originated in Egypt and China, respectively.

**Formalism** The orderly, non-natural arrangement of plants represents an essentially artificial environment using plants as structural material. The formal garden represents the human dominance over nature. Formalism is achieved by orderly placements of plants, emphasis on symmetry, severe plant pruning and training. Formalism was developed in ancient Egypt where the natural vegetation was scarce, and the garden in a sense represents an artificial oasis. The dry climate demanded irrigation which in turn demands orderly arrangement of plantings. The Egyptian garden was enclosed, typified by water and shade, with pools, and orderly arranged plantings was copied everywhere (Fig. 36.2). Although altered by local variations in plants and climate, formalism spread to Persia, Syria, and India and ultimately to the Rome empire. It is still a major force in modern public gardens throughout the world.

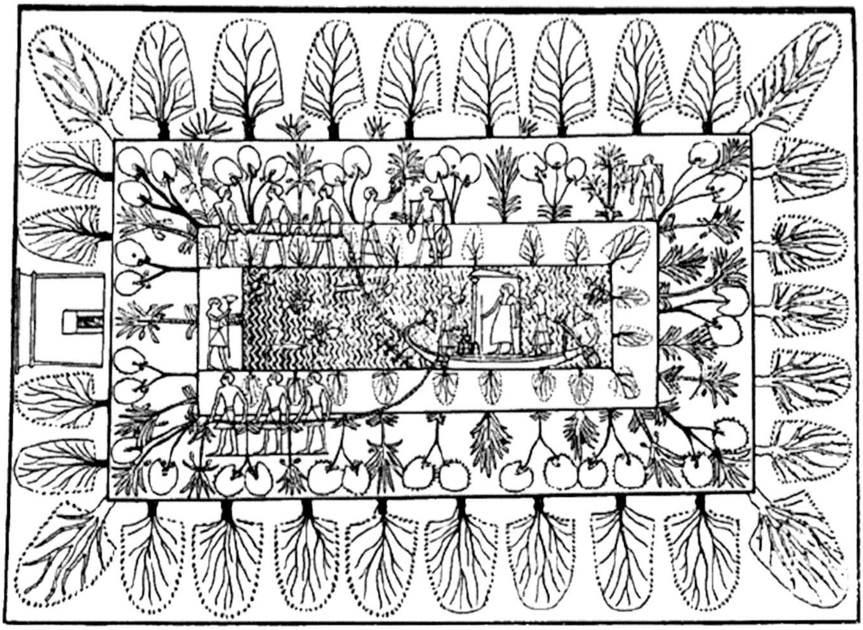

**Fig. 36.2** Formalism in garden design as represented in an ancient garden plan for a wealthy Egyptian estate. (Source: Singer et al. 1954)

**Naturalism** Naturalism is an attempt to live with rather than dominate nature. The concept of naturalism is to emulate the natural world and to achieve the effect of being in a happy accident of nature. Unlike the formal tradition where the plants are pruned to geometric shapes, in naturalism, the free form is emphasized and exaggerated. Although the separation between gardens and landscapes in formalism is severe, in naturalism it is vague and indistinct. The landscape blends into the garden. If formalism is the straight line of geometry, naturalism is the free curve.

The concept of naturalism originated in China, and reached its highest development in Japan where there were beautiful natural landscapes to copy (Fig. 36.3). Naturalism also developed independently in the West, specifically in England where the natural landscape—verdant meadows and rolling hills—were emulated. However, the methods to achieve naturalism are as artificial as those of formalism. It involves severe training and pruning, and is combined with the wide use of many natural materials such as stones and wood. In the Eastern tradition, plants further assumed symbolic significance.

**Combinations** The fusion of Eastern naturalism and Western formalism took place in eighteenth century England where the influence of Asian cultures coincided with a movement away from formalism to take advantage of the English landscape. The marriage was not always successful. Some English gardens became interspersed with Chinese pagodas amid fake antique Gothic ruins. This influence of English gardens survives today in the use of curved walks, artificial wishing wells, and herbaceous borders.

**Fig. 36.3** A naturalistic garden in Kyoto, Japan. (Photograph by Jules Janick)

The contemporary trend in gardens is to develop a meaningful design for living. Freed from the confines of "formalism" or "naturalism" modern gardens strive to reach esthetic expression through the capacity for both abstraction and utility. Plants and people, as in the past, make good companions. We have turned full circle with the concept of the garden and now consider it primarily as a vital need in our society and not merely as an esthetic mix.

## Landscape Architecture

Landscape architecture in its broadest sense is concerned with the relationship between people, plants and the landscape and is involved with all aspects of land use. The profession deals with site development, building arrangement, grading, paving, plantings, gardens, playgrounds, and pools. It is concerned with the individual home and the entire community. Thus it deals with parks and parkways, shopping centers, and urban planning. Landscape architecture is ultimately concerned with the allocation of space and the interaction of people and the environment. If the landscape architect must be first an artist, he or she must also be a horticulturist and a civil engineer. Although landscape architecture was in the past intimately associated with architecture—two opposing sides of the same door—the two have become rather distinct professions. The objectives of the landscape architect have been to functionally and esthetically integrate people, buildings, and site.

During the Renaissance, the grand period of the West's cultural revival, the concept of the garden was transformed from relative insignificance to a magnificent splendor befitting the age. The grounds design became the important concept, while the plant was treated rather impersonally as merely an architectural material. The plant was pruned, clipped, and trained to conform to the design plan. Even architecture became subservient to the landscape plan, the landscape engulfing and dominating the stately palaces or grand residences, especially those of royalty or high ranking dignitaries. The resultant "noble symmetry" included courtyards, terraces, statuary, staircases, cascades, and fountains. The emphasis was on long symmetrical

**Fig. 36.4** Seventeenth century gardens at Versailles designed by André Le Nôtre

**Fig. 36.5** Sidewinder, a landscape form of Patrick Dougherty consisting of red maple and black willow saplings constructed by students and faculty at Purdue University, 2011. (Photo courtesy Ann Hildner)

vistas and promenades. The small enclosed garden remained but only within the walls of the buildings, as a component part of the grand plan. Formalism reached its peak in seventeenth century France in the Age of Louis XIV (1635–1715). The master architectural gardens of André Le Nôtre (1613–1700) still remain unsurpassed examples of this concept of design predominating over nature (Fig. 36.4).

A modern trend in landscape architecture considers the landscape itself as an art form (Sovinsky 1995). This is achieved by various installations and constructions many of which use different plant forms. A splendid example is the installation of a creation consisting of red maple and black willow saplings at Purdue University by the artist Patrick Dougherty (b. 1945) in 2011 entitled Sidewinder (Fig. 36.5). Many of the willow saplings have rooted which has created a living sculptural form that can be entered permitting an intimate interaction between observer, plant, and form.

## Floral Arts

The floral arts include the decorative use of flowers and plants in various arrangements usually but not always on a small scale, from individual flowers in a vase, corsages, container plantings, and large floral floats. Floral design bears about the same

**Fig. 36.6** Bonsai or tray culture is an oriental art form achieved through pruning and controlled nutrition. (Photograph by Jules Janick)

relationship to landscape architecture, as a string quartet to a symphony orchestra. The principles are the same but the scale is reduced. The arranging of flowers and decorative parts of plants has long been used for home decoration. In Japan, flower arrangement (*ikebana*) has a continuing tradition that has been an integral part of cultural life for over thirteen centuries. Unlike the occidental concept, the Oriental tradition emphasizes the element of line over form and color. In the classical concept, line is symbolically partitioned into a representation of heaven (vertical), earth (horizontal), and humanity (diagonal and intermediate). The chief aim is to achieve a beautiful flowing line. To accomplish this, the most ordinary materials may be used. The concept of naturalism is expressed throughout. Symmetry is avoided.

The floral arts are still an important component of Japanese life. There are many different styles and schools: *ikenobo*, classical arrangements, *rikka*, large ornate upright reproduction of the landscape by means of flowers and plants, *nageire*, simple naturalistic arrangements, and *morbiana*, expressive scenic arrangements with greater use of foliage and flowers. Other typical Oriental types of artistic expression involve growing plants. *Bonsai*, the culture of miniature potted trees, dwarfed by pruning and controlled nutrition, is a spectacular example of the horticultural arts. Living trees, some over a 100 years old and yet less than a meter in height are gown in containers arranged to resemble a portion of a miniature landscape (Fig. 36.6). *Bonseki* is the construction of a miniature landscape out of stone, sand and living vegetation.

In both the East and West, flowers are now an important part of cultural life. Flowers and potted plants are readily purchased in the market place, both in special shops and supermarkets, and are in common use as part of normal living. Flowers

**Fig. 36.7** Topiary: **a** hedge sculpture in Portugal, photograph by Jules Janick; **b** aborisculptpural forms made by grafting. (Source: Mudge et al. 2009, Fig. 9.2)

are emphasized in special occasions such as formal dining, decoration in religious holidays, appropriate remembrances (weddings, funerals birthdays, anniversaries, get well gifts) and as gifts for special remembrance (they are prominent at Valentine's Day and Mother's Day). Corsages were once important parts of proms and formal dances. In many parts of the world street floral displays are part of the culture.

The plant itself may become the basis for artistic construction. Certain woody shrubs can be trained and pruned in a great variety of shapes, limited only by the imagination of the horticulturist (Fig. 36.7). Plant sculpture, known a topiary, exploits the plasticity of the growing plant to create various shapes, including animals and architectural facades. In addition, many unnatural architectural forms can be achieved with the aid of grafting.

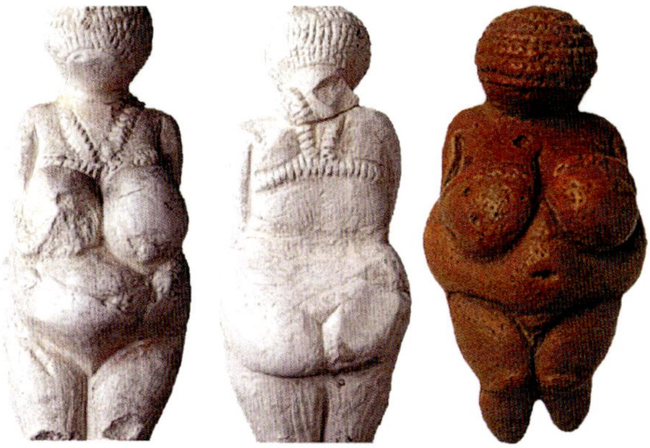

**Fig. 36.8** Woman figures of the Paleolithic period showing evidence of textile technology

## Horticulture in Art

Horticultural plants are a major component of artistic expression. There are numerous sources of plant iconography: cave paintings, ancient mosaics, sculpture, carvings and inlays, frescos, tapestries, illustrated manuscripts, herbals, books, and photography. Furthermore, works of art involving plants from prehistory and antiquity to the present constitute an alternate source of information on plants and crops (Janick 2007; Janick et al. 2011). Plant iconography becomes a valuable resource for investigations involving genetic and taxonomic information, as well as crop history including evolution under domestication, crop dispersal, and lost and new traits. Crop images are one of the unequivocal tools for assessing the historical presence of botanical taxa in a particular region and are an especially valuable resource for determining morphological changes of crops from antiquity to the present. Although a plethora of ancient plant images exists, they are widely scattered among libraries and museums, and are often difficult to locate and to access. Recently, the digitization of information by some of the major world libraries has greatly facilitated the search for ancient illustrations, although they still remain expensive to publish because of copyright issues.

### *Sculpture*

Prehistoric stone sculptures of voluptuous women known as Venuses dating 25,000–30,000 years ago indicate a keen interest of early humans in fertility that still engender an emotional impact. A number of them show evidence of clothing made from local plant sources (Fig. 36.8) that indicate the development of weaving and textile technology.

The ancient Near East cultures, known as Mesopotamian civilization, are largely based on Semitic populations that existed between the Tigris and Euphrates Rivers that expanded to the area known as the Fertile Crescent, which includes parts of present day Israel, Jordan, Lebanon, Syria, Iraq, and Iran. A second Neolithic

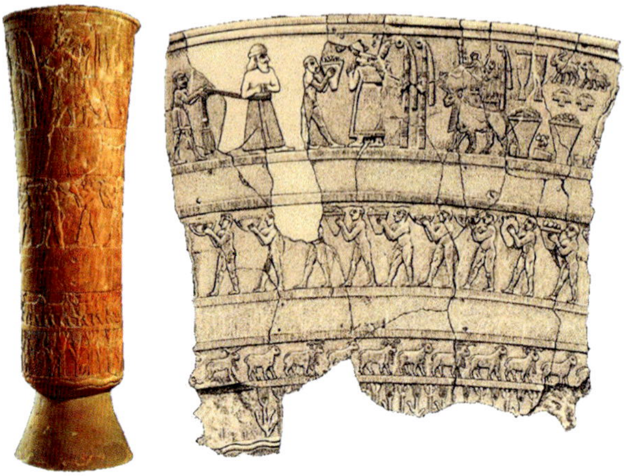

**Fig. 36.9** Uryuk Vase ca. 4th millennium BCE with wedding attendants offering fruit in a wedding ceremony between a priest king and the goddess Innana. (Source: Janick et al. 2011)

**Fig. 36.10** Date palm pollination depicted by Assyrian bas reliefs, 883–859 BCE. The pollinator assumes the form of a godlike figure (genie) and the date palm has been transformed into a symbolic tree. (Source: Paley 1976)

Revolution between 6000 and 3000 BCE (the Bronze Age) involved the change from villages to permanent urban centers and the development of a settled agriculture coinciding with the beginning of fruit culture. This is well documented in the decorations of a vase of the late 4th millennium BCE (Fig. 36.9), found in Uryuk (biblical Erekh), an ancient city on the Euphrates north of present-day Basra, Iraq, that is associated with Sumerian civilizations, where writing was invented. It portrays barley and sesame above a watery matrix, domesticated sheep, and attendants bearing baskets of fruit to a wedding between a priest king and the goddess Inanna (Istar). Evidence of agricultural technology includes the refinement of a plow with a seed drill from a cylinder seal and date palm pollination from a bar relief (Fig. 36.10).

Plants in sculptured form are found in Egyptian, Greek, Roman, pre-Columbian American, Indian, and European Renaissance art. In ancient Egypt, the papyrus and lotus were symbols of the upper and lower Nile region; and the reunification of

**Fig. 36.11** Intertwining lotus symbolizing the reunification of upper and lower Egypt. (Source: Janick 2002)

**Fig. 36.12** Sculpted acanthus leaves ca. 450 BCE from a column in the Delphi Museum. (Photograph by Jules Janick)

**Fig. 36.13** Bar relief of snake melon (*Cucumis melo* var. flexuosus from Merida Spain, fourth century CE. (Photograph by Jules Janick)

**Fig. 36.14** Precolumbian ceramic jars from Peru **a** peanut, **b** potato, **c** squash, **d** cacao pod. (Source: Leonard 1973)

**Fig. 36.15** Fruits of eggplant, pepper, tomato, and cucumber adorning the bronze doors of Pisa Cathedral at the Piazza dell Duomo. (Photo by Jules Janick)

Egypt in the third millennium BCE is shown in illustrations where these two plants are intertwined (Fig. 36.11) and these forms are also found in architectural columns. A greek column at Delphi about 400 BCE representation an acanthus leaf (Fig. 36.12). Roman bas relief of snake melon from Merida, Spain is identified by its leaves and striated fruit (Fig. 36.13). In pre-Columbian American ceramics celebrate the domestication of indigenous crops such as potato, peanut, and cacao (Fig. 36.14). The cathedral bronze doors in Pisa, Italy dated 1601, are rich in sculpted food crops that surround the panels of religious scenes and include eggplant, cucumber, and tomato (Fig. 36.15).

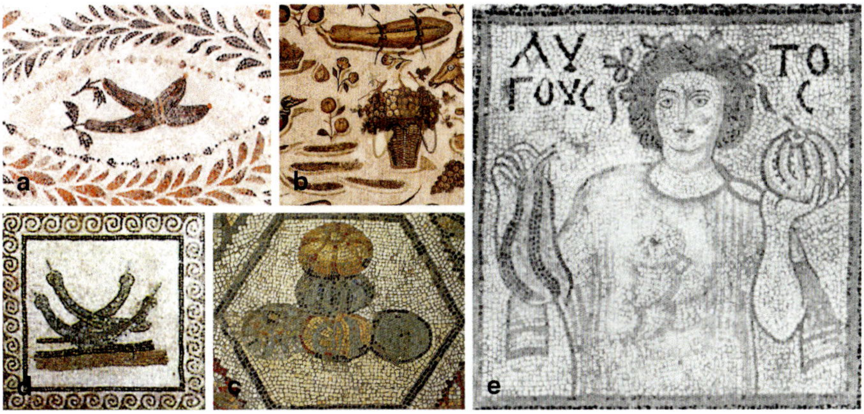


**Fig. 36.16** Cucurbits in Roman mosaics: **a** Snake melon (*Cucumis melo* Flexuosus Group) from Tunisia second century, **b** immature and mature snake melon showing fruit splitting Tunisia third century, **c** round striped melon (*C. melo*) Tunisia fourth century, **d** bottle gourd (*Lagenaria siceraria* showing characteristic swelling on the peduncular end, **e** youth holding bottle gourd in right hand and watermelon (*Citrullus lanatus*) in left hand Tegea Episkopi, Peloponnese. Late fourth to fifth century. (Source: Janick et al. 2007)

**Fig. 36.17** Apple culture in mosaics, Saint-Roman-en-Gal, third century, Vienne, France; **a** detached scion grafting; **b** fruit harvest; **c** juice extraction. (Source: Janick 2007)

## *Mosaics and Inlays*

The assemblage of images from small pieces of colored glass, stone, or gems referred to as mosaics, date to the third millennium BCE. Mosaics were popular in ancient Greece and Rome and survive in Christian and Islamic art up to the present time. Mosaics were prominent as decorations on floors, walls, and ceilings of private residences and public buildings, especially churches, mosques, palaces or mansions and constitute some of the glories of ancient, medieval, and Renaissance art in the West. Mosaic art spread throughout the Roman Empire and is particularly rich in areas that today are in Italy, Tunisia, Libya, Syria, and Turkey. Roman mosa-

**Fig. 36.18** Floral motifs in the Taj Mahal, seventeenth century: **a** stone inlays (pietra dura) of chrysanthemum; **b** bas reliefs (dado) showing iris in the center and in descending order columbine, daffodil, columbine, windflower, tulip, windflower, poppy capsule, delphinium and daffodil. (Source: Janick et al. 2010)

ics included rich scenes of horticultural plants that included cucurbits such as the snake melon, bottle gourd, and watermelon (Fig. 36.16) A third century panel from St. Roman-en gal, in Vienne, France depicts fruit culture scenes and contains the first image of detached scion grafting (Fig. 36.17).

Mughal mosaics and motifs are found among the decorations of the Taj Mahal, constructed in Agra, India, by Shah Jahan (1592–1666) from 1632 to 1658 as a memorial to his wife known as Mumtaz Mahal (1592–1666) (Janick et al. 2010). Islamic decoration restricts graven images of humans but is rich in botanical subjects and includes floral inlays known as *pietra dura* and bas relief sculptures known as *dados*. The plant images are dominated by ornamental geophytes (bulb crops) common to the region (Fig. 36.18).

## *Paintings*

Paintings from antiquity to the present have often used plants and crops as themes for their esthetic and/or symbolic value. Cave paintings are rich in animal forms but crude depictions of plants can be found (Fig. 36.19). The ancient technology of agriculture can be vividly reconstructed from the artistic record, paintings and sculpture in tombs and temples dating onward from 3000 BCE. Agricultural activities were favorite themes of artists, who illustrated lively scenes of daily life that adorn the tombs of the pharaohs and dignitaries. The artistic genius engendered by Egyptian civilization, the superb condition of many burial chambers, and the dry climate have made it possible to reconstruct a detailed history of agricultural technology. Ancient Egypt is shown to be the source of much of the agricultural technology of the Western World. Illustrations of these artifacts can be gleaned from four key references: Keimer (1924), Singer et al. (1954), Darby et al. (1977), and Manniche (1989).

Examples of the presence of plant images from ancient Egypt are shown in a brief sampling of the artistic record. This includes grape harvest and wine mak-

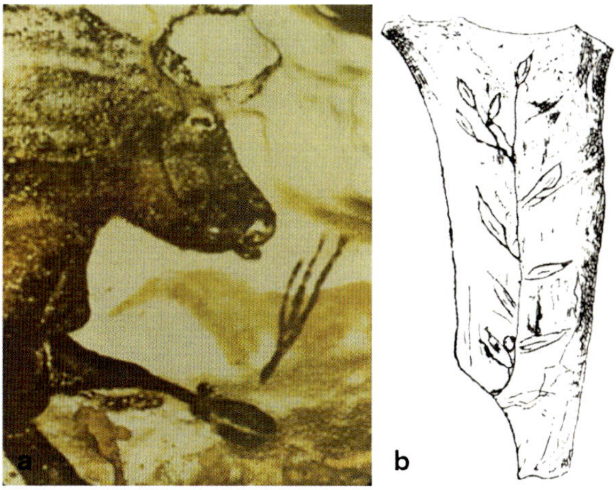

**Fig. 36.19** Paleolithic images of plants, 17,000–30,000 years ago: **a** aurock with a primitive plant image; **b** more sophisticated imaged carved on a reindeer horn. (Source: Tyldesleay and Bahn 1983)

**Fig. 36.20** Egyptian wine making: **a** grapes collected from a round arbor with grapes crushed by stomping before storage in amphorae; **b** Pressing grapes in a bag press; **c** bag press encased in a frame. (Source: Janick 2002a)

**Fig. 36.21** Roman fruit paintings from Pompeii (first century): **a** figs; **b** peach. (Source: Jashemski 1979)

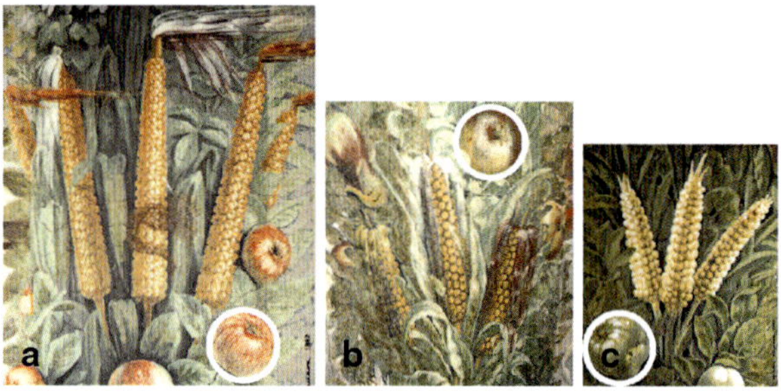


**Fig. 36.22** The first images of maize in Europe in the Loggia of Psyche Villa Farnesina 1515–1518 painted by Giovanni da Udina. The *encircled* apples provide an estimate of size. (Source: Janick and Caneva 2005)

ing (Fig. 36.20) and a collection of cucurbits The absence of images of cucumber (*Cucumis sativus*) supports the conclusion that the many reference to cucumbers in English translations of ancient texts should be understood as being snake melons, *Cucumis melo* L. subsp. *melo* Flexuosus group (Janick et al. 2007a).

Frescoes, paintings on freshly applied plaster on walls and ceilings, are well preserved since the pigments seep into the plaster. The frescoes of Pompeii and Herculaneum in Italy have been preserved as a result of the eruption of Vesuvius in the year 79 CE and are valuable resources for ancient depictions of plants. Examples include images of fig and peach from Pompeii (Fig. 36.21).

Paintings of plants increased during the Italian Renaissance. The Roman residence (now known as Villa Farnesina) of the wealthy Roman financier Agostino Chigi, decorated between 1515 and 1518, is a splendid source of crop images. The ceiling of the Loggia of Cupid and Psyche illustrate scenes from *Metamorphoses* (*The Golden Ass*) by Apuleius, a second century CE Roman author, painted in fresco by Raphael Sanzio (1483–1520) and his assistants, including Giovanni Martini da Udina (1470–1535) who was responsible for the festoons that are a fantastic source of crop images. The thousands of images of 163 species in 49 botanical families include some of the first illustrations of New World plants (Janick and Caneva 2005; Janick and Paris 2006a). Included are the first images of maize showing three distinct ear types (Fig. 36.22).

The early paintings of the baroque artist Michelangelo Merisi (1571–1610), also known as Caravaggio, are particularly rich in the inclusion of fruits and vegetables (Janick 2004a). Furthermore, the photorealistic style makes it possible to distinguish diseases and examples of insect predation (Fig. 36.23). This genre of Baroque paintings known as *natura morta* (still life) emphasizing fruits vegetables,

**Fig. 36.23** Fruit basket by Michelangelo Merisi, known as Caravaggio, showing evidence of disease and insect injury including fig anthracnose, quince scab, codling moth on apple, oriental fruit moth damage on peach, leaf roller damage on pear, grape mummies, and grasshopper injury. (Source: Janick 2004)

**Fig. 36.24** Two fruit renaissance fruit markets: **a** Produce seller (1567) by Pieter Aertsen; **b** Fruit seller, (1570) by Vincenzo Campi. (Source: Janick et al. 2011)

and flowers is a rich source of information (Zeven and Brandenburg 1986). Baroque painters found scenes of everyday life intriguing subjects to paint, and fruit and vegetable markets increasingly became a common subject. Two example of fruit market paintings are shown in Fig. 36.24. The Flemish painter Pieter Aertsen (1508–1575) entitled the *Produce Seller* (1567) is rich is *Brassica* crops including head cabbage (7 green and 1 red) as well as cauliflower and various cucurbits

**Fig. 36.25** Sixteenth and Seventeenth century horticultural paintings: **a** Giuseppe Arcimboldo (1528–1593), **b** Giovanna Garzoni, (1600–1670), and **c** Bartolomeo Bimbi (1648–1723). (Sources: Ferino-Pagden 2007; Meloni et al. 2000; Consiglio Nationale delle Ricerche 1982, respectively)

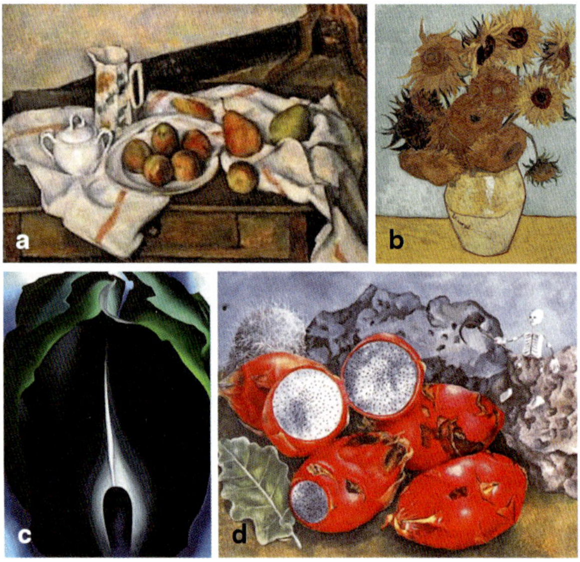


**Fig. 36.26** Nineteenth and twentieth century horticultural paintings: **a** apples and pears by Paul Cezanne, **b** sunflowers by Vincent Van Gogh; **d** jack-in-the pulpit by Georgia O'Keeffe; **c** pitayas by Freda Kahlo

including bottle gourd, melon, pumpkin, and cucumber and also includes Belgium waffles! *The Fruit Seller* by Vincenzo Campi (1580) displays a plethora of fruits and vegetables in an Italian market—included in the upper right, a box of pears and young squash, *Cucurbita pepo* subsp. *pepo* Cocozelle group, with flowers attached, still a common commodity in Mediterranean countries (Janick and Paris 2005; Paris and Janick 2005). Other noteworthy painters of horticultural crop images include Guiseppi Arcimboldo (1521–1593), Giovanna Garzoni (1600–1679) and Bartolomeo Bimbi (1648–1723) (Fig. 36.25). Note that Bimbi includes a key to

**Fig. 36.27** Fruit of loquat (*Eriobotrya japonica*), and a mountain bird by an anonymous Chinese artist (1127–1279)

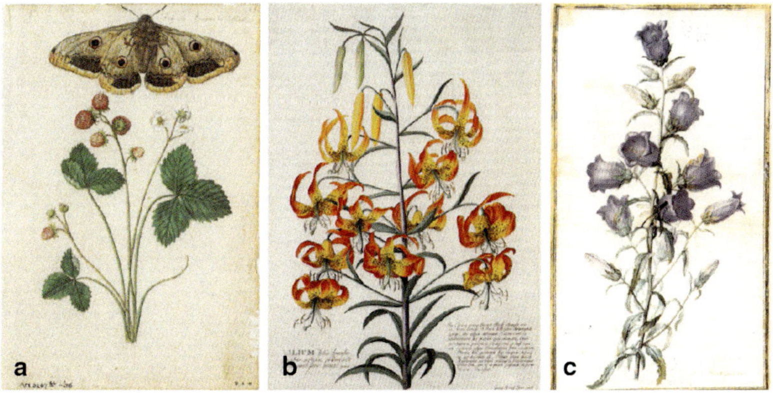


**Fig. 36.28** Botanical illustrations: **a** wide strawberry by Jacques Le Moyne (1533–1588); **b** lily by George Dionysus Ehret (1710–1770); and **c** Canterbury bells by Pierre-Joseph Redoute (1759–1840)

the cultivar names. The inclusion of fruits and vegetables continues to be a popular subject in the nineteenth century and twentieth century evidenced by the horticultural paintings of Paul Cezanne (1839–1906), Vincent Van Gogh (1853–1890), Georgia O'Keeffe (1887–1986), and Freda Kahlo (1907–1954) (Fig. 36.26).

The illustrations of botanical and horticultural plants become a specialized art form in its own right. Plants and flowers are also found in early Oriental art (Fig. 36.27). Masters of the genre include Jacques Le Moyne de Morgues (1533–1588), Georg Dionysus Ehret (1708–1770), and Pierre Joseph Redouté (1759–1840) (Fig. 36.28). Plant forms also became a favorite subject of photography (Fig. 36.29). Finally

**Fig. 36.29** Photograph of cabbage ('Larry's perfection') by Charles Jones (1866–1959). (Source: Sexton and Johnson 1998)

**Fig. 36.30** Embroidery of capsicum pepper, Peru, 400–500. (Source: Andrews 1995)

plants themselves have become art material. In Korea dried flowers have been used to create special pictures including plants, landscapes, and animal forms.

## Embroidery and Tapestries

The elaboration of textiles into decorated patterns is an ancient technique in various cultures and horticultural plants are a common motif. In Peru, during the early Nazca period (400–600 CE) an embroidery shows a man holding two capsicum peppers with two fruits on a cord around his neck (Fig. 36.30). Tapestry, a form of woven textile, became an extremely popular art form in the Middle Ages and

**Fig. 36.31** The unicorn in captivity, one of the most famous medieval tapestries is resplendent with plants. (Source: Metropolitan Museum of Art, New York)

**Fig. 36.32** Repeated floral motifs in a seventeenth century Persian carpet. (Source: Michell 2007, p. 26)

Renaissance and was carried out extensively in the seventeenth century in Flanders, one of the regions and communities of Belgium. The Hunt of the Unicorns, a seven piece tapestry from 1495 to 1505 is especially rich in horticultural imagery (Fig. 36.31). Garden themes are common motifs in carpet weaving, an essential part of Persian art and culture (Fig. 36.32)

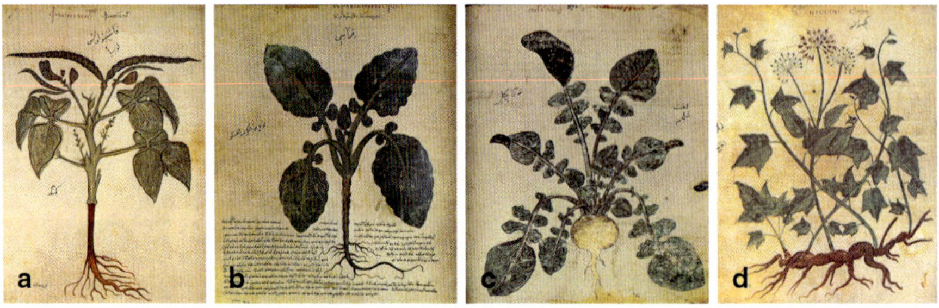

**Fig. 36.33** Horticultural illustrations of the *Juliana Anicia Codex* 512; **a** cowpea, **b** nonheading cabbage, **c** turnip, **d** English ivy. (Source: Janick and Hummer 2012)

## *Illustrated Manuscripts*

The *Juliana Anicia Codex (JAC)* or *Codex Vindobonenis*, 512 CE, a magnificent, illustrated manuscript from late antiquity found in Constantinople is based on the famous non-illustrated herbal Περί ὕλης ιατρικής (Latinized as *De Materia Medica, On Medical Matters*) originally written about 65 CE by the Roman army physician Pedanios Dioscorides (20–70) born in Anazarbus, Cilicia in what is now southeastern Turkey. The *JAC*, made as a presentation volume for the daughter of the Roman emperor Anicius Olybrius, contains descriptions, medical uses, and illustrations of 383 plants listed alphabetically and can now be accessed through a two-volume facsimile edition, *Der Wiener Dioskurides* (1998, 1999). Four examples of horticultural images from the *JAC* are included in Fig. 36.33.

A late medieval example of crop illustrations can be found in a series of lavish versions of an eleventh century manuscripts known as the *Tacuinum Sanitatis* (*Tables of Health*) which probably were prepared as royal gifts in Europe. There are six major works (one is divided) in libraries in Liège, Vienna, Rome, Paris, and Rouen, which were commissioned by northern Italian nobility during the last decade of the fourteenth century and the course of the fifteenth century (Paris et al. 2009; Daunay et al. 2009; Janick et al. 2009). The text is based on an eleventh-century Arabic manuscript, *Taqwim al-Sihha bi al-Ashab al-Sitta* (*Rectifying Health by Six Causes*), written as a guide for healthy living by the Christian Arab physician known as Ibn Butlan (d. 1063). Vivid agricultural imagery includes scenes of the harvest of vegetables, fruits, flowers, grains, and culinary and medicinal herbs, accompanied by a brief summary of the health aspects of each subject. Each of the manuscripts are drawn by different artists but are obviously related. The *Vienna codex Ser. N. 2644* contains the most accurate depictions, which include 9 cereals, 26 vegetables, 33 fruits, 3 flowers, and 21 culinary and medicinal herbs. The illustrations show crops at the optimal state of maturity and, moreover, are a rich source of information on life in the feudal society, with nobles engaged in play and romance while laborers worked on the estate. Illustrations of eggplant and watermelon from the *Tacunum Sanitatis* are presented in Fig. 36.34.

**Fig. 36.34** Eggplants and watermelon in the *Tacuinum Sanitatis*, *Vienna 2644*, an illustrated version of a Latin of an eleventh century Arabic manuscript by Ibn Butlan (d. 1603). (Source: Paris et al. 2009)

Many illustrated medieval and renaissance books are filled with horticultural illustrations. A French Royal prayer book known as *Les Grandes Heures d'Anne de Bretagne* (*Manuscript Latin 9474*) contains prayers with illustrated margins, and full page monthly calendars and paintings of religious themes (Paris et al. 2006). This stunningly illustrated manuscript was prepared for the personal use of Anne de Bretagne (1477–1514), twice Queen of France as wife of Charles VIII (1470–1498) and Louis XII (1462–1515), by the famous artist Jean Bourdichon (1457–1521), probably painted between 1503 and 1508, about a decade from the return of explorer Christopher Columbus (1451–1506) of Spain. There are miniature paintings of plants and small animals, mostly insects, on each page that can be searched on www.hort.purdue.edu/newcrop/bilimoff/default.html. Well over 300 plant species are included in this prayerbook. This work contains the first European illustration of a non-esculent gourd of *Cucurbita pepo* subsp. *texana* (Fig. 36.35). The seed source for this gourd cannot be determined but could have been obtained from various sources. Seeds from the voyages of Columbus were transmitted in 1494 by Peter Martyr D'Angheria, Tutor to the Spanish royal household, to Cardinal Asconio Sforza secretary of state to the Vatican and seed could have reached France this way. Other possibilities include the voyages of Europeans including Amerigo Vespucci (1454–1512) who entered the Gulf of Mexico as early as 1498, or from various Bretons or Norman who reached the Americas by 1503, returning with parrots and Brasilwood.

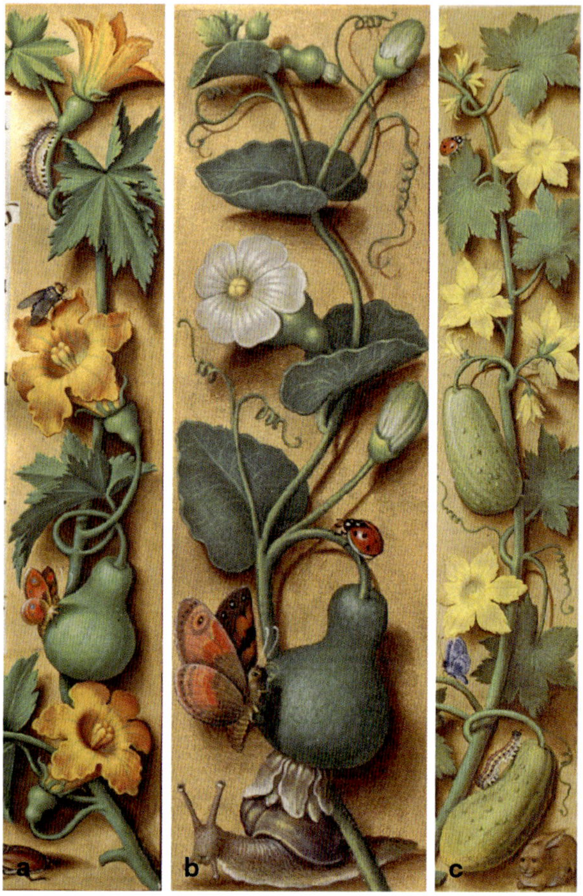


**Fig. 36.35** Cucurbits of *Les Grandes Herues d'Anne de Bretagne* (1505–1538) by Juan Bourdichon: **a** gourd (*Cucurbita pepo*), **b** bottle gourd (*Lagenaria siceraria)*, and **c** cucumber (*Cucumis sativa*). (Source: Paris et al. 2006)

## *Printed Herbal Woodblocks*

Herbals, botanical works emphasizing the medical uses of plants, are one of the most important sources of plant iconography (Eisendrath 1961). A splendid introduction to the field can be found in Agnes Arber's 1938 book on herbals. Facsimile editions exist for a number of printed herbals, including the 1542 herbal of Leonhard Fuchs (Meyer et al. 1999), the 1597 *Herball* of John Gerard(e) (John Norton, London) and the 1633 edition amended by Johnson (Dover Publ.). The illustrations of Renaissance herbals are derived mostly from woodcuts and sometimes from original painted drawings. However many herbals copy parts of text and woodcuts from previous herbals or are based on an exchange of woodblocks by printers, thus they often contain errors in identification. Woodcuts of sweet potato (*Ipomoea batatas*) and potato (*Solanum tuberosum*), the first printed illustration of potato in Europe, from the famous English *Herball* of John Gerard(e) (1597) are presented in Fig. 36.36. Gerard is responsible for the confusion between potato, *Solanum tuberosum* (Indian

**Fig. 36.36** Woodcuts from the 1597 herball of John Gerard(e): **a** sweet potato (*Ipomoea batatas*); **b** potato (*Solanum tuberosum*)

**Fig. 36.37** A bouquet of peony, poppy, foxglove pansy and French marigold by Jean-Louis Prévost (1760–1810). Florilegias and botanical illustrations were used as a source of motifs for designers of china and textiles. (Source: Saunders 1995)

name *papas*) and sweet potato, *Ipomea batatas* (Indian name *batatas*), because he labeled his printed illustration of potato, the first one to be published in Europe, *Battata Virginiana sive Virginianorum & Pappus, Potatoes of Virginia*; Virginia being the area where the tubers he grew in his garden came from.

A study of the iconography of the Solanaceae (Daunay et al. 2008) shows the richness of information found in herbals. In the seventeenth and eighteenth centuries, botanical art became a sensation and many Royal collections of plant images called florilegias were made for their sheer beauty and for conveying the knowledge of exotic plants brought back by travelers around the world (Fig. 36.37). They became the source of floral art for commercial uses such as fabrics and wallpaper.

## Conclusions

It is clear that horticulture and art are intimately entwined. Horticultural plants, flowers, and fruits are considered beautiful objects in themselves and it natural that they become motifs for artistic expression to adorn the body, the table, the home, and as principle objects in the canvas of the landscape. Horticultural plants have also become subjects in traditional art forms such as paintings and sculptures but also tapestries, weavings, mosaics, and photographs. The study of the images of plants (plant iconography) involves both art and history. Furthermore, plant iconography is an outstanding resource for research on crop evolution and genetic diversity. This is especially true in prehistory where images are older than writing. Plant iconography provides a valuable resource for investigations involving genetic and taxonomic information, as well as crop history.

## References

Andrews J (1995) Peppers: the domesticated capsicums. Univ. Texas Press, Austin

Arber A (1938) Herbals: their origin and evolution. A chapter in the history of botany 1470–1670, 2nd edn. Cambridge University Press, Cambridge

Consiglio Nazionale delle Ricerche (1982) Agrumi, frutta, e uve nella Firenze di Bartolomeo Bimbi Pittore Mediceo. Italia

Darby WJ, Ghalioungui P, Grivetti L (1977) Food: the gift of Osiris, vol 2. Academic Press, London

Daunay M-C, Laterrot H, Janick J (2008) Iconography and history of Solanaceae: antiquity to the 17th century. Hortic Rev 34:1–111 (+ 31 plates)

Daunay M-C, Janick J, Paris HS (2009) Tacuinum Sanitatis: horticulture and health in the late middle ages. Chron Hortic 49(3):22–29

Der Wiener Dioskurides (1998, 1999) 2 Vol. Akademische Druck-u.Verlagsanstalt, Graz

Eisendrath ER (1961) Portraits of plants. A limited study of the 'icones'. Annals of the Missouri Botanical Garden 48:291–327

Ferino-Pagden S (2007) Arcimboldo 1526–1593. Skkira Editore, Milan

Gerarde J (1597) The herball or generall historie of plantes. John Norton, London

Groening GD (2007) Gardens as elements in an urbanizing world. Acta Hortic 759:109–123

Janick J (1984) Esthetics of horticulture. In: Horticultural science, 4th edn. W.H. Freeman and Co, New York, pp 681–707
Janick J (2002) Ancient Egyptian agriculture and the origins of horticulture. Acta Hortic 582:23–39
Janick J (2004) Caravaggio's fruit: a mirror on baroque horticulture. Chron Hortic 44(4):9–15
Janick J (2007) Art as a source of information on horticultural technology. Acta Hortic 759:69–88
Janick J, Caneva G (2005) The first images of maize in Europe. Maydica 50:71–80
Janick and Hummer (2012) The 1500th anniversary (512–2012) of the Juliana Anicia Codex: An illustrated Dioscordean recension. Chron Hortic 52(3):9–15
Janick J, Paris HS (2005) Baby squash in the Italian market, 1580. Cucurbit Netw News 12(1):4
Janick J, Paris HS (2006) The cucurbit images (1515–1518) of the Villa Farnesina, Rome. Ann Bot 97:165–176
Janick J, Paris HS, Parrish D (2007) The cucurbits of Mediterranean antiquity: identification of taxa from ancient images and descriptions. Ann Bot 100:1441–1457
Janick J, Daunay M-C, Paris HS (2009) Horticulture and health: ancient medieval views. In: Nath P, Gaddagimath PB (eds) Horticulture and livelihood security. Scientific Publishers, Jodhpur, pp 39–52
Janick J, Kamenetsky R, Puttaswamy SH (2010) Horticulture of the Taj Mahal: gardens of the imagination. Chron Hortic 50(3):30–33
Janick J, Daunay M-C, Paris HS (2011) Plant iconography—a source of information for archaeogenetics. In: Byulai G (ed) Plant archaeogenetics. Nova, New York, pp 143–159
Jashemski WF (1979) The gardens of Pompei, Herculaneum and the villas destroyed by Vesuvius. Caratzas Brothers Publications, New Rochelle
Keimer L (1924) Die Gartenpflanzen im Alten Ägypten. Hoffmann & Campe, Hamburg
Leonard JN (1973) First farmers. Time Life Books, New York
Manniche L (1989) An ancient Egyptian herbal. Univ. Teas Press, Austin
Meloni, Trkulja S, Fumagalli E (2000) Still lives: Giovanna Garzoni. Bibliotheque de l'Image, Paris
Meyer FG, Trueblood EE, Heller JL (1999) The great herbal of Leonhart Fuchs, vol 2. Stanford Univ. Press, Stanford
Michell G (2007) The majesty of Mughal decorations. Thames & Hudson, New York
Mudge K, Janick J, Scofield S, Goldschmidt EE (2009) A history of grafting. Hortic Rev 35:437–493
Paley SM (1976) King of the world: Ashur-nasir-pal II of Assyria 883–859 BC. The Brooklyn Museum, New York
Paris HS, Janick J (2005) Early evidence for the culinary use of squash flowers in Italy. Chron Hortic 45(2):20–22
Paris HS, Janick J, Daunay M-C (2006) First images of Cucurbita in Europe. Proc. Cucurbitaceae 2006, Universal Press, Raleigh, pp 363–371
Paris HS, Daunay M-C, Janick J (2009) The Cucurbitaceae and Solanaceae illustrated in medieval manuscripts known as the Tacuinum Sanitatis. Ann Bot 103:1187
Saunders G (1995) Picturing plants; an analytical history of botanical illustration. University of California Press, Berkeley
Sexton S, Johnson RF (1998) Plant kingdoms: the photographs of Charles Jones. Smithmark Publishers, New York
Singer C, Holmyard EJ, Hall AR (1954) A history of technology, vol 1. Fall of ancient empires. Oxford University Press, London
Sovinsky R (1995) Paradise regained: "Avant gardeners" have spearheaded a 90s kind of art form. Arts Indiana Arts Indiana 17(5):21–23
Thacker C (1979) The history of gardens. University of California Press, Berkeley
Tyldesley JA, Bahn PG (1983) Use of plants in the European paleolithic: a review of the evidence. Quarternary Sci Rev 2:53–83
Zeven AC, Brandenburg WA (1986) Use of paintings from the 16th to the 19th centuries to study the history of domesticated plants. Econ Bot 40:397–408

# Chapter 37
# Scholarship and Literature in Horticulture

**Ian J. Warrington and Jules Janick**

**Abstract** Horticulture and its related sciences have produced a rich diversity of literature ranging from highly specialised scientific journals and scholarly books to detailed manuals for producers, from technical and popular books on gardening and cooking to encyclopaedias on highly specialised topics, and from newspaper and magazine articles to entries on the World Wide Web. These publications span a period of over 200 years. Included are food crops such as fruit, nuts, vegetables, and condiments; ornamentals and landscaping plants including trees, shrubs, and cut flower and bedding crops; turf grasses; medicinal plants and the use of plants for human wellbeing and therapy. This chapter presents selected examples of how horticulture has contributed to scholarship and literature. It also includes examples of how horticulture has been recorded in classical literature and become an integral part of many every-day sayings.

**Keywords** Scholarship · Literature · Art · Teaching · Science · Technology · Poetry · Aphorisms

## Introduction

Scholarship associated with horticulture is multifaceted. The literature relating to the science and technology of horticulture is diverse and prolific, and goes back to antiquity. Horticultural information forms a vital part of human knowledge. It ranges from highly specialised scientific journals and treatises to detailed manuals for producers, from technical and popular books on gardening and cooking to

I. J. Warrington (✉)
Massey University, Palmerston North, New Zealand
e-mail: ianjw@xtra.co.nz

J. Janick
Department of Horticulture and Landscape Architecture, Purdue University, West Lafayette, IN 47907-2010, USA
e-mail: janick@purdue.edu

G. R. Dixon, D. E. Aldous (eds.), *Horticulture: Plants for People and Places, Volume 3*,
DOI 10.1007/978-94-017-8560-0_37, © Springer Science+Business Media Dordrecht 2014

encyclopaedias on highly specialised topics, and from newspaper and magazine articles to entries on the World Wide Web. All fill a niche in covering those topics that define horticulture. Included are food crops such as fruit, nuts, vegetables, and condiments; ornamentals and landscaping plants including trees, shrubs and cut flower and bedding crops; turf grasses; medicinal plants and the use of plants for human wellbeing and therapy. Incorporated in this literature are all of the elements of science and the details of techniques and technologies that are associated with horticulture: plant breeding and cultivar selection, seed technology, pruning and training, harvesting, quality assurance, disease and pest control, irrigation, fertilisation, and postharvest management. Horticultural science contributes directly and very significantly to primary knowledge about all phases of plant growth and development. The role of the horticulturist is to apply that knowledge for the betterment of humankind through achieving sustainable and efficient production of food and ornamental crops, and to enhance human wellbeing in a myriad of different ways.

Horticulture is part of the literary tradition. Many of the beloved plants of horticulture—particularly tree and herbaceous fruits, and various ornamentals and especially flowers—are embedded in the writings of all cultures, including various holy books such as the Hebrew bible, New Testament, Qur'an, and sacred texts of Hinduism and Buddhism. There are frequent allusions to horticultural plants by many famous writers in the West from Shakespeare to James Joyce and in the East from scholars in countries such as China and Japan. Horticulture references in prose and poetry are a distinct part of our humanitarian heritage (Palter 2002). The object of this chapter is to provide a very brief overview of the enormous subject of scholarship and literature in horticulture. The related topic of art and horticulture is covered in a separate chapter of this volume.

## Scientific Journals

The outputs of what can be deemed to be "modern" horticultural science have been published for more than 200 years in many different and diverse scientific journals, books, industry magazines and the popular press. Many of the current scientific journals that are used to publish results from horticultural research have their origins in the horticultural science societies that were formed in many countries in the nineteenth century following the modernisation of agriculture. A further phase occurred in the twentieth century with the emergence of scientific research as a critical and respected activity within modern society. These journals are typically hosted by scientific societies who organise the editorial policies, peer review processes and publishing, distribution, and marketing. The majority of these publications cover most aspects of horticultural science including all horticultural crops (fruit, vegetables, cut flowers, ornamental and landscape plants, viticulture, turfgrasses, mushrooms, medicinal plants, and others) and all of the technologies associated with horticultural research (Table 37.1).

**Table 37.1** Disciplines associated with horticultural science

| |
|---|
| Plant sciences: botany, ecology, genetics, plant physiology, plant molecular biology, plant breeding, soil science, taxonomy |
| Production technologies: irrigation, growing media, hydroponics, plant nutrition, pruning and training |
| Computing and engineering: greenhouse design, grading and handling, mechanization, precision agriculture, storage |
| Plant protection: entomology, plant pathology, weed science |
| Enterprise management: enterprise structure and operations, sales and marketing, supply chain components and practices |
| Behavioral sciences: communication and information transfer, sensory evaluation, market research |

Horticultural science societies have retained the control and management of scientific journals in spite of the activities of major publishing houses in this business. These societies regard such publishing as providing a critical service to their members, a means of keeping publishing costs down through the voluntary efforts of members in editorial, reviewing, and publishing activities, and a means of showing independence in the control of the dissemination of scientific findings.

Notwithstanding this independent role of the horticultural science societies, these same societies have had to make a number of changes, particularly over the past two decades, to remain relevant and competitive. This has included relatively simple changes such as the inclusion of coloured photographs, the adoption of new formats, and an increase in page number and size. The most significant change, however, has been the shift to include on-line availability of each issue of the journal, often before the physical version is printed and distributed. Many societies ensure that archived issues are available to download for a nominal fee or free of charge to members (see www.Pubhort.org and www.ashs.org). A sample of the journals that specialise in horticultural science is shown in Table 37.2.

As horticultural science is not a single discipline but is comprised of a great number of related and inter-dependent disciplines, it is not surprising that the outputs of horticultural science are not only published in journals that specialise in horticultural science but in many other scientific journals that focus on related disciplines. A small sample of such journals is summarised in Table 37.3.These journals too are either managed by specialised professional societies or, more typically, by large, multi-national publishing companies.

A major development of the past two decades has been the emergence of impact and related factors for scientific journals and of citation ratings for individual articles. These measures have been developed by organisations such as the ISI Web of Knowledge, provided by the private company Thompson Reuters. The merits or otherwise of these metrics (Janick 2008) have been extensively explored across many disciplines but the measures are now well established and are used to assess the relative rankings of journals, the importance of specific scientific papers, and even the weighting that should be given to an individual scientist's publications when submitted in support of applications for academic promotion. The basis of the metrics underpinning measures such as impact factors rely heavily on publica-

**Table 37.2** Selected international journals specialising in horticultural science

| Journal title | Country of origin | Year established | Publisher |
|---|---|---|---|
| Acta Horticulturae | Belgium | 1963 | International Society for Horticultural Science |
| Acta Horticulturae Sinica | China | 1962 | Chinese Society for Horticultural Science |
| American Journal of Enology and Viticulture | USA | 1950 | American Society for Enology and Viticulture |
| American Journal of Potato Research | USA | 1923 | Springer |
| European Journal of Horticultural Science | Germany | 1929: Formerly Gartenbau-wissenschaft | Verlag Eugen Ulmer Stuttgart (German Society for Horticultural Science) |
| Fruits | France | 1945: Formerly Fruits d'Outre Mer | EDP Sciences |
| HortScience | USA | 1968 | American Society for Horticultural Science |
| HortTechnology | USA | 1991 | American Society for Horticultural Science |
| Horticulture Reviews | USA | 1979 | Wiley |
| Indian Journal of Horticulture | India | 1942 | Horticulture Society of India |
| International Journal of Fruit Science | USA | 2006: Formerly Small Fruits Review (2000–2005) and Journal of Small Fruit and Viticulture (1992–2000) | Taylor and Francis |
| Journal of the American Pomological Society | USA | 1946: Formerly Fruit Varieties and Horticultural Digest (1946–1972) and Fruit Varieties Journal (1973–1999) | American Pomological Society |
| Journal of the American Society for Horticultural Science | USA | 1903: Formerly Proceedings of the American Society for Horticultural Science (1903–1968) | American Society for Horticultural Science |
| Journal of Horticultural Science and Biotechnology | United Kingdom | 1919 | JHSB Trust |
| Journal of the Japanese Society for Horticultural Science | Japan | 1925 | Japanese Society for Horticultural Science |
| Korean Journal of Horticultural Science and Technology | Republic of Korea | 1998 | Korean Society for Horticultural Science |

**Table 37.2** (continued)

| Journal title | Country of origin | Year established | Publisher |
|---|---|---|---|
| New Zealand Journal of Crop and Horticultural Science | New Zealand | 1900: Formerly known as the New Zealand Journal of Experimental Agriculture (1900–1988) | The Royal Society of New Zealand (Taylor and Francis) |
| Postharvest Biology and Technology | The Netherlands | 1991 | Elsevier |
| Potato Research (Journal of the European Association for Potato Research) | Europe | 1958 | Springer |
| Rivista di Frutticoltura e di ortofloricultura | Italy | 1937 | Edagricole |
| Scientia Horticulturae | The Netherlands | 1973 | Elsevier |

**Table 37.3** Selected international journals with significant content relating to horticultural science

| Journal title | Country of origin | Year established | Publisher |
|---|---|---|---|
| Annals of the Entomological Society of America | USA | 1908 | Entomological Society of America |
| Annals of Botany | UK | 1887 | Oxford University Press |
| Canadian Journal of Plant Science | Canada | 1920 | Agricultural Institute of Canada |
| Communications in Soil Science and Plant Analysis | USA | 1970 | Taylor and Francis |
| Critical Reviews in Plant Sciences | USA | 1983–84 | Taylor and Francis |
| Economic Botany | Germany | 1947 | SpringerLink |
| Journal of Economic Entomology | USA | 1908 | Entomological Society of America |
| Journal of Natural Products | USA | | American Society Pharmacognosy; American Chemical Society |
| Journal of Plant Nutrition | USA | 1979 | Taylor and Francis |
| Journal of the Science of Food and Agriculture | USA | 1950 | Wiley-Blackwell |
| Physiologia Plantarum | USA | 1948 | Wiley-Blackwell |
| Phytopathology | USA | 1911 | American Phytopathology Society |
| Plant Physiology | USA | 1926 | American Society of Plant Biologists |

tion volume (number of papers per journal issue, especially those that are recently published) and readership (citation) numbers. In many if not all areas of horticulture, these measures have to be interpreted very carefully when consideration is given to the very small numbers of scientists who work internationally on topics such as blueberry production, or the postharvest storage of melons, or the breeding of tomatoes. On that basis, and given that there is a large number of journals covering the diversity that is horticulture, the rankings for horticultural science journals and for the papers that they publish are always going to be low. Furthermore the numbers of journals and journal articles that are published annually in horticulture are comparatively small within the total sciences in general or even within the plant sciences. For example, the Web of Science lists only 31 journals involved with horticultural science whereas there are 84 involved with the plant sciences (see http://admin-apps.webofknowledge.com). Within the horticultural science journals listed by ISI, 3156 papers were published in 2011. However, nine of 31 journals published fewer than 50 papers per annum; the highest number (421) appeared in *Scientia Horticulturae*.

## Textbooks

Teaching of horticultural science degree programs in universities throughout the world has brought with it the need to have specialised texts on horticulture, on horticultural science, and on many of the specific related sciences and technologies that are included in such training programmes. A sample of such textbooks (published in English) is as follows:

### *General Horticulture and Horticultural Science*

*Horticulture*. R Gordon Halfacre and John A Barden (1979).
*Horticultural Science*. 4th edn. Jules Janick (1986).

### *Fruit*

*Temperate-Zone Pomology. Physiology and Culture*, 3rd edn. Melvin N Westwood (1993).
*Modern Fruit Science: Orchard and Small Fruit Culture*. Norman F Childers et al., G Steven Sibbett and Justin R Morris (1995).
*Introduction to Fruit Crops (Crop Science)*. Mark Rieger (2006).

### *Viticulture*

*Biology of the Grapevine*. G Mullins, Alain Bonquet and Lorry E Williams (1992).
*Wine Science: Principles and Application*. Ronald S Jackson (2008).

### *Vegetables*

*Vegetable Crops*. Dennis R Decoteau (2000).

### *Postharvest Biology*

*Postharvest Biology*. Stanley J Keys and Robert E Paull (2004).
*Postharvest Biology and Technology of Fruits, Vegetables and Flowers*. Gopinadhan Paliyath Dennis P Murr, Avtar K Handa and Susan Lurie (2008).
*Postharvest Technology of Horticultural Crops*. Adel Kader (2002).

## Floriculture

*Introduction to Floriculture*, 2nd edn. Roy A Larson (1992).
*Floriculture: Principles and Species*, 2nd ed. John M Dole and Harold F Wilkins (2004).

## Greenhouses Greenhouse Operation and Management

*Greenhouse Operation and Management*, 7th edn. Paul V Nelson (2011).
*The Commercial Greenhouse*, 3rd edn. James Boodley and Steven E Newman (2008).

## Plant Propagation

*Hartmann & Kester's Plant Propagation: Principles and Practices*, 8th edn. Hudson T Hartmann, Dale E. Kester, Fred T Davies and Robert Geneve (2010).

## Turfgrass Management

*Turfgrass Management*, 9th edn. A.J. Turgeon (2011).
*Turfgrass Management*, 4th. edn. R. Emmons (2007).

## Parks Management

*Designs for Parks and Recreation Spaces*. TD Walker (1987).
*Arboriculture: Integrated Management of Landscape Trees, Shrubs and Vines*, 4th edn. RW Harris, JR Clark and NP Matheny (2003).

## Landscape Design

*The Essential Garden Design Workbook*. R Alexander (2009).
*Garden Design Workbook*. J Brookes (2001).
*Introduction to Landscape Design*. JL Motloch (2000).
*Landscape Design: A Cultural and Architectural History*. Elizabeth Barlow Rogers (2001).
*RHS Encyclopedia of Garden Design*, Chris Young (2009).

Many of the texts that are outlined above are used in different countries around the world to teach the specialised elements of horticulture and horticultural science, including those where English is not the first language. However, in addition to the above, a number of texts on elements of horticulture are published in a range of languages including, for example, Italian (see Baldini 1996; Fabbri 2001; Sansavini et al. 2012; Marenghi 2005; Vezzosi 1998), German (Link 2002, 2011; Bettin 2011; Wonneberger et al. 2004; Friedrich and Fischer 2000; Keppel 1998;

Crüger et al. 2002; Wohanka 2006), Korean (Lee and Lee 2011; Lee et al. 2007) and Japanese (Mizutani 2002; Abe et al. 1979; Suzuki et al. 1993). Recently, Indian publishers have been releasing a large number of texts relating to horticulture (e.g., Chundawat and Sen 2002).

## Books, Monographs, Encyclopedias

Amongst the earliest recorded publications relating to horticulture are those by Theophrastus (372–288 BCE.), a Greek philosopher who has been termed the "father of botany" (Mitchell 2011). Theophrastus published a number of texts of which very few survive, but those that do include: *History of Plants* (Grk: *Ιστορία των φυτών*, Ltn: *Historia de Plantes*) and *Causes of Plants* (Grk: *Τα αίτια των φυτών*, Ltn: *De causis plantarum*) which were the greatest and amongst the earliest treatises of their kind in the ancient world. These provide an encyclopedic knowledge and analysis of plants (Gundersen 1918). Significantly and remarkably, Theophrastus wrote about growth, propagation, and the development of plants (for example, from seeds, grafting, and budding); environmental effects on fruits, trees, and other plants; meteorology and geology, and their relation to plant growth; the consequences of plant spacing to growth; he noted the movements of flowers and leaves at certain times of the year or day (now termed tropisms); the proper techniques for cultivation of some plants (including soil choice, trimming, watering, fertilizer choice/method, and weeding). He also noted that plants which grow too close together will both deplete the pool of available nutrients, and that "artificial and unnatural" forces impact on plants (such as through decay and disease), and finally on the odour and taste of different plants (now the subjects of sensory science) (Sengbusch 2003).

Earlier, Homer in The Odyssey (900 BCE) described fig, apple, pear and grape cultivation in the orchards of Alcinous (Roach 1985).

The Romans were also strongly engaged in horticulture where the most notable publication is *De Materia Medica* (77 CE) by Dioscorides who identified many plants and described their medicinal value.

Horticultural crops were also featured strongly in the writings from Eastern cultures. For example, the culture of pears is recorded as far back as 2,500–3,000 years ago and peaches and plums are also mentioned from these times. Notable is description of the use of rootstocks and the selection of productive scions (Shen 1980).

Early plant scientists in the modern era who had a close involvement with horticultural crops included Carl Linnaeus (1707–1778; the new science of plant nomenclature), Charles Darwin (1809–1882; descriptions of geotropism and phototropism) and Gregor Mendel (1822–1884; establishing the foundations of modern genetics and breeding, using the garden pea as the main subject of study).

A number of the classic papers in horticultural science have been collated and represented by Janick (1989).

One of the most famous treatises in horticulture is *The Cyclopedia of American Horticulture* by Liberty Hyde Bailey and Wilhelm Miller (1900), later reorganized and expanded as *The Standard Cyclopedia of Horticulture* (1914). This huge work of 3,639 pages still remains a most useful resource for horticulture.

## Fruits and Vegetables

Hundreds of books have been published on various horticultural crops and the different management practices that are associated with them. Many of these refer to the science that underpins current practices. Treatises on various aspects of horticultural crop management, including the description of tree training and pruning, grafting, and cultivar selection, for a crop such as apple, can be traced back to the early to mid seventeenth century (*see* Juniper and Mabberley 2006). These have texts with wonderful titles such as:

> *A Treatise on Fruit Trees Shewing their Manner of Grafting, Pruning,and Ordering, of Cyder and Perry, of Vineyards in England* by J. Beale (1653).
>
> *A Treatise of Fruit-Trees Shewing the Manner of grafting, Setting, Pruning, and Ordering of them in All Respects: According to Diverse New and Easy Rules of experience; Gathered in the Space of Twenty Years* by R. Austen (1657).
>
> *Systema Agriculturae, the Mystery of Husbandry Discovered by J. Worlidge (1669); and The Art of pruning Fruit-Trees…with an Explanation of Some Words Which Gardiners Make Use of in Speaking of Trees. And a Tract of the Use of the Fruits of Trees, for Preserving Us in Health, or for Curing Us When We Are Sick. Translated from the French Original, Set forth in the Last Year by a Physician of Rochelle* by N. Venette (1685).

A typical characteristic of early fruit growing was the very large number of different cultivars that were grown on individual orchards. This provided a range of choices for the consumer throughout the season, diversity within the orchard to better cope with losses due to extreme weather and pest events, and the opportunity to store some better suited cultivars beyond the end of the production season. For fruit crops such as apple, this applied as much to fresh as it did to cider cultivars. One of the consequences of these practices was the publication of impressive texts that defined, in great detail, the characteristics of individual cultivars—the so-called "pomonas". Good examples of such publications include:

> *Pomona Londinensis Containing Coloured Engravings of the Most Esteemed Fruits Cultivated in British Gardens* by W. Hooker (1818); and
> *The Herefordshire Pomona Containing Original Figures and Descriptions of the Most Esteemed Kinds of Apples and Pears* by R. Hogg. and H.G. Bull (eds) (1876–1885).

Some of the most famous pomonas in the twentieth century were organized by the New York State Experiment Station in a series known as the "*Fruits of New York*" including grapes (Hedrick 1908), plums (Hedrick 1911), cherries (Hedrick 1915), peaches (Hedrick 1917), pears (Hedrick 1921) and small fruits (Hedrick 1925). There are also pomonas in many other countries that are too numerous to list here.

A similar range of early references for vegetable production was also published and goes back to the seventeenth century, including:

> *Kalendarium Hortense* by John Evelyn (1664);
> *Directions for the Gardiner* by John Evelyn (1686) (see Campbell-Culver 2009); and
> *History of cultivated vegetables: comprising their botanical, medicinal, edible, and chemical qualities; natural history; and relation to art, science, and commerce* by H Phillips (1821).

Modern books often form part of a series that is devoted to different fruit and vegetable crops—such as CABI Publishing's "Crop Production Science in Horticulture" series with issues specialising in blueberries (Retamales and Hancock 2012), raspberries (Funt and Hall 2012), grapes (Creasy and Creasy 2009), olives (Therios 2008), bananas and plantains (Robinson and Galán Saúco 2010), peach (Layne and Bassi 2008), peppers (Bosland and Votava 2012), onions and other edible alliums (Brewster 2008), lettuce, endive and chicory (Ryder 1999), brassicas and related crucifers (Dixon 2007), tropical fruits (Paull and Duarte 2012), and citrus (Albrigo et al. 2012).

In another CABI series covering "botany, production and uses", volumes are available, for example, on apples (Ferree and Warrington 2003), peppers (Russo 2012), avocados (Schaffer et al. 2013), mango (Litz 2009), and peach (Layne and Bassi 2008).

Other CABI texts cover topics such as *Principles of Tropical Fruit Production* (Midmore 2012), *Principles of Fruit and Nut Production* (Andrews 2013), and *Vegetable Production and Practices* (Welbaum 2013). They have also published a major *Encyclopaedia of Fruit & Nut Crops* (Janick and Paull 2006).

CRC Press offers texts on subjects such as propagation (Beyl and Trigiano 2008), tissue culture (Trigiano and Gray 2010), and organic farming (Barker 2010), a handbook on plant nutrition (Barker and Pilbeam 2006) and a dictionary of plant breeding (Schlegel 2009).

The publisher Elsevier (Academic Press) has a large catalogue of texts relating to fruits and vegetables within which there is an extensive series on diseases and pests including those on fruit crops (Alford 2007), lettuce (Blancard et al. 2006), mushrooms (Gaze and Fletcher 2007), peas and beans (Biddle and Cattlin 2007), tomatoes (Blancard 2012) and vegetables (Koike et al. 2006).

## *Amenity Horticulture*

The fascination with plants for use as ornamentals, for cut flowers, and for landscaping in the western world stretches back to antiquity and the Renaissance. The design of ornamental gardens occurred in early civilisations, even preceding the Greeks and Romans, and included the use of plants for both food production and for purely leisure and aesthetic purposes. The Hanging Gardens of Babylon are likely the most recognised in this respect. The Egyptians, Persians, Greeks and Romans, followed by Byzantium and Moorish cultures all contributed elements to the design of managed gardens. In addition, Chinese and Japanese influences were very important in eastern cultures. In the thirteenth through to the sixteenth centuries, the development of formal gardens was very significant in France, Italy and Spain through to the development of the Italian Renaissance garden and the very formal style of the Gardens of Versailles. These developments were succeeded by the English and French landscape gardens in the eighteenth century following which a number of other influences emerged in the nineteenth and twentieth centuries (see Rogers 2001).

In the eighteenth and nineteenth centuries, as some sections of society became more affluent, as voyages of global exploration dramatically increased, and as societies went through a particular fascination with the natural world, opportunities for new developments in landscape horticulture were markedly enhanced. It was an era when the activities of apothecaries, who were heavily reliant on plants and plant extracts for their profession, and the early horticulturalists converged. In addition, new wealth enabled landscaping on a massive scale for both private and public gardens. The voyages of discovery allowed the collection and display of plants which came to be admired and even celebrated in ways never before possible.

The origins of many current publications can be traced back to these times. *The Curtis Botanical Magazine* (now the *Kew Magazine*), with its wonderful coloured hand drawings of ornamental plants, was established in 1777. This history is described in considerable detail by Desmond (1987) including descriptions of the activities of the plant collectors of that time.

Throughout modern history, large treatises have been published to outline the botanical descriptions and horticultural uses of ornamental plants, flowers, trees and shrubs. Earlier versions of such publications included *An encyclopaedia of gardening; comprising the theory and practice of horticulture, floriculture, arboriculture, and landscape-gardening, including all the latest improvements; a general history of gardening in all countries* by J.C. Loudon (1828).

More recent volumes include the Royal Horticultural Society's series that includes *Plants and Flowers* (Brickell 2010), *Perennials* (Rice 2006), *Gardening* (Brickell 2007), *and the A–Z of Garden Plants* (Brickell 2003); or the Reader's Digest *Gardener's Encyclopaedia of Plants and Flowers—the Definitive Reference Work for Australia & New Zealand* (Macoboy et al. 2010); or The American Horticultural Society's *New Encyclopaedia of Gardening Techniques* (American Horticultural Society 2009).

Separate encyclopaedias are devoted to trees and shrubs—such as *Dirr's Encyclopedia of Trees and Shrubs* (Dirr 2011); the *Timber Press Encyclopedia of Flowering Shrubs* (Gardiner 2012); *The Hillier Gardener's Guide to Trees and Shrubs* (Kelly 2004); and *Techniques du Jardinier L'encyclopédie* (Bureaux 2011).

A characteristic of horticulture publications is that books can be identified that cover almost any specific group of ornamental plants ranging, for example, from roses: *The Ultimate Rose Book* (Macoboy 2007), the American Rose Society *Encyclopedia of Roses* (Quest-Ritson and Quest-Ritson 2010); to perennials (*Encyclopedia of Perennials* (Rice 2006), *Rodale's Illustrated Encyclopedia of Perennials* (Phillips and Burrell 1999), *Armitage's Garden Perennials* (Armitage 2011)); orchids (*Botanica's Orchids: over 1200 species* (Botanica 2002), *The*

*Illustrated Encyclopedia of Orchids* (Pridgeon 1992)); and bulbs (*Bulbs* (Bryan 2002), and *The Complete Practical Handbook of Garden Bulbs: How to create a spectacular flowering garden throughout the year in lawns, beds, borders, boxes, containers and hanging baskets* (Brown 2009)).

Notwithstanding these very large books on various aspects of ornamental plants, there are equally a considerable number of very small but authoritative texts in this field, such as The Timber Press Pocket Guide Series that includes volumes on topics such as conifers (Bitner 2010), hostas (Grenfell and Shadrack 2007), bamboos (Meredith 2009), palms (Riffle 2008) and shade perennials (Schmid 2004). Similarly, the Expert Books' series includes separate books on lawns, flowering shrubs, trees and shrubs, greenhouses, house plants, fruit, vegetables and herbs (e.g., Hessayon 1991 and 1997).

Landscape architecture is a highly specialised area of horticulture that has its own extensive literature which records developments over recent centuries, links those developments with changes in societies and across different cultures, and relates the management of green spaces to changes in building architecture and art.

Harvard University Press has published an interesting series on landscape, history and cultures that form the Dumbarton Oaks Colloquium Series in the History of Landscape Architecture. Titles include the following:

*Botanical Progress, Horticultural Innovations, and Cultural Changes* (Conan and Kress 2007);
*Sacred Gardens and Landscapes—Ritual and Agency* (Conan 2007); and
*Perspectives on Garden Histories* (Conan 1999).

Other titles from the same publisher, but not in the series, include:

*Gardens, City Life, and Culture: A World Tour* (Conan and Wangheng 2008); and
*Gardens and Cultural Change—A Pan American Perspective* (Conan and Quilter 2008).

A significant characteristic of the last 60 to 80 years has been the urbanisation of societies. This has brought with it the development of urban landscaping within major cities where special attention has been given to landscaping freeways and city streets through to the development of roof-top gardens that are used primarily for "green spaces" but also for the production of food. Publications specific to these applications of horticulture include:

*Rooftop Gardens: the Terraces, Conservatories, and Balconies of New York* (Calicchio and Amon 2011); and
*Skygardens: Rooftops, Balconies and Terraces* (Nielsen 2004).

## Horticulture and Health

The present-day emphasis on horticulture and health has an ancient tradition based on medicinal uses of plants and diet (Daunay et al. 2009; Janick and Hummer 2010). Plant cures have long been a basic component of medicine and there are treatises found in ancient Sumer, Egypt, Greece, China, India, and Mesopotamia. The Greek herbal of Pedanious Dioscoredes of Anazarba written in the year 65 lists health-giving properties of 500 plants and an illustrated version from 512 known as the *Juliana Anicia Codex* survives and became the basis of the illustrated herbal tradition in the Renaissance (Arber 1965; Janick and Hummer 2012). Botany, horticulture, and medicine were essentially in step during the eighteenth century when each turned scientific and from this juncture botanical works would essentially ignore medicinal uses while medicinal works were devoid of plant lore. However, the medicinal use of plants continues as an alternate form of medicine and remains popular to the present day despite the questionable efficacy of many popular herbs and the reliance of a number of herbal recommendations on superstition and astrology. The *Journal of Natural Products* (original *Lloydia*), the official journal of the American Society of Pharacognosy (now co-published by the American Chemical Society), contains many articles on the medicinal chemistry of horticultural plant products, many of pharmaceutical interest.

The importance of horticulture to the overall mental health and wellbeing of humankind has received increasing attention over the past two decades as positive aspects of horticultural therapy have received more attention and as the negative aspects of high density housing, the absence of "green spaces" in modern cities has increased, and as the relationships between human behaviour and the natural environment have become better understood. Authors such as Kaplan and Kaplan (1982), Kellert and Wilson (1995) and Lewis (1996) have explored and defined these issues.

## Extension Publications

### *Twentieth Century*

An iconic feature of government and state-funded extension services throughout the twentieth century was the large number of well prepared and well presented extension bulletins and advisory booklets that were produced for producers in various horticultural industry sectors. These were a particular feature of national departments of agriculture in countries such as the USA, the United Kingdom, New Zealand, Australia and Canada. Their preparation and distribution was also a key element within the activities of agricultural and horticultural faculties in US Land Grant Universities, especially in states such as California, New York, Michigan, Ohio, Massachusetts, North Carolina, Georgia, Florida, Texas, and others. These publications, which were usually distributed free of charge, covered many different production-related subjects. They were developed over a number of years and revised frequently by specialists who had often spent decades working in their respective fields. Consequently, these extension publications were the most current, up-to-date and validated sources of reliable information that were available to local producers. They were also prepared independently of any commercial influences.

Throughout the 1990s and 2000s a number of these publications became unavailable in printed form but were rapidly transferred to on-line web-based versions. A number of these remain but the range now available is markedly reduced.

The disbanding of government support for agricultural advisory services in the later part of the twentieth century and the early part of the twenty-first century in many countries has seen an immediate loss of such services and with that a cessation of the publication of such helpful and at times critical information for producers. In some instances, this loss of independent state-funded information has been replaced with proprietary sources such as those from fertilizer and seed companies, as well as from private consultants (but in that case only to fee-paying clients).

Some of the professional societies that are associated with horticultural science do continue to publish technical material which is of direct relevance to industry. For example, the American Society for Horticultural Science publishes a number of works on different crops such as watermelons (Maynard 2001) as well as those on subjects such as weed management (McGiffen 1997) and organic composting (Tyler 1996). The Entomological Society of America publishes volumes on subjects that include, for example, a handbook on turfgrass insect pests (Brandenburg and Villani 1995) while the American Phytopathology Society publishes texts on plant diseases such as those on chrysanthemum (Horst and Nelson 1997), rhododendron (Coyier and Roane 1986) and herbaceous perennials (Gleason et al. 2009).

## *Industry Publications*

Many industry groups around the world publish magazines on a regular basis that disseminate information to producers about the application of recent research discoveries, technical details about new products and new cultivars, market intelligence, and industry politics. Plant nursery catalogues are often a very good source of information about the origins and timings of new plant introductions into a country or region as well as informing about the introduction of new cultivars of a particular crop. Examples of such publications are shown in Table 37.4. Some of these have been produced for many decades including the *Gardener's Chronicle*—now *Horticulture Week* - which began publication in 1841 and the *American Fruit Grower* which began in 1880.

## Gardening

Gardening provides an important leisure activity for many people around the world. The results of these endeavours include a beautification of the landscape for both personal and public enjoyment, the production of fruit and vegetables for personal

**Table 37.4** Selected industry magazines devoted to fruit, vegetables, nursery production and gardening

| Publication title | Country of origin | Publishing entity | Year established | Publication frequency |
|---|---|---|---|---|
| **Fruit** | | | | |
| Good Fruit Grower | USA | Washington State Fruit Commission | 1946 | 17 times a year |
| American/Western Fruit Grower | USA | Meister Media Worldwide | 1880 | Monthly |
| Fruit Growers News (FGN) (formerly Great Lakes Fruit Growers News) | USA | Great American Media Services | 1974 | Monthly |
| The Orchardist of New Zealand | New Zealand | Horticulture New Zealand | 1926 | Monthly (11 issues) |
| The Fruit Grower | United Kingdom | ACT Publishing | 1986 | Monthly |
| European Fruit Magazine | Poland | Plantpress Ltd | 2009 | Monthly (published in german, dutch and english) |
| The Fruit Grower | United Kingdom | ACT Publishing | 1986 | Monthly |
| **Vegetables** | | | | |
| **NZ Grower** | New Zealand | Horticulture New Zealand | 1945 | Monthly (11 issues) |
| American Vegetable Grower | USA | Meister Media Worldwide | 1908 | Monthly |
| Vegetable Grower News (VGN) | USA | Great American Media Services | 1966 | Monthly |
| The Vegetable Farmer | United Kingdom | ACT Publishing | 1989 | Monthly |
| **Nursery production and gardening** | | | | |
| Australian Horticulture | Australia | Rural Press Ltd, Australia | 1903 | Monthly |
| Commercial Horticulture | New Zealand | The Reference Publishing Company | 1967 | Bi-monthly |
| American Nursery Magazine | USA | American Nurseryman Publishing co. | 1904 | Monthly |
| Canadian Garden Centre & Nursery Magazine | Canada | Annex Business Media | 2006 | Monthly |
| The Garden | United Kingdom | The Royal Horticultural Society | 1964 | Monthly |
| The Plantsman | United Kingdom | The Royal Horticultural Society | 1994 | Quarterly |
| Horticulture Week | United Kingdom | Haymarket Publications | 1841 as Gardeners' Chronicle and absorbed the Grower in 2006 | Weekly |

consumption, physical exercise and recreation for the participants, and in the case of community gardens, social interaction amongst those involved. It can range in scale from many hectares where the managed spaces surround private homes or where the garden is an integral part of the larger landscape, to small green spaces on rooftops of modern apartments. Gardening as a leisure activity and as an art form is practised in many different countries throughout the world and is as strong in many eastern countries as it is in western ones.

Those involved in gardening receive their information and inspiration from a number of sources including membership of garden clubs, television programs, magazines and newspapers, and of course from books. A number of such books are covered in the sections above that refer to fruit crops, vegetables, and ornamental plants. Examples of other significant texts include:

*Complete Idiots Guide to Small-Space Gardening* (McLaughlin 2012);
*The Complete Gardener* (Don 2009);
*The Blooming Great Gardening Book—A Guide for All Seasons* (Whysall 2000);
*The New Encyclopedia of Gardening Techniques* (American Horticultural Society 2009); and
*RHS How to Grow Practically Everything* (Royal Horticultural Society 2010).

Particular styles of gardening can be recognised as being influenced by specific countries such as Japan. As a consequence, books have been published that describe in detail the history and nature of those gardens (e.g. Keane (2007); Ohashi (2000)) while others cover the ways in which those styles and principles can be applied in a western context (e.g. Kawaguchi 2008). Similarly, French gardens have had a major influence on the design of gardens elsewhere in the world (e.g. Babelon and Chamblas-Ploton (2001) and Smithen (2002)).

## Food Guides and Cookbooks

An unusual but significant segment of literature that relates to horticulture is that of food guides and cookbooks that focus on the use of fruits and vegetables in human diets. Cookbooks have an ancient tradition (Darby 2003) and are a very useful source of information on horticultural crops. With the growing awareness of the critical importance of horticultural crops for providing essential minerals, vitamins and other active compounds in diets and for enhancing health, this segment of literature has grown markedly in recent years. This awareness has been enhanced within many communities through "5 plus a day" nutrition programmes and through the promotion of specific crops with high concentrations of known or claimed health-enhancing compounds such as antioxidants and bioflavonoids.

Some texts provide general information about such properties across a range of crops (Heaton 1997; Watson and Preedy 2009) while others are very specific—for example, for crops and products such as olives and olive oil (Preedy and Watson 2010) or oriental vegetables (Larkcom 2008). Other texts provide information on cooking in general with fruit and vegetables or for maintaining a vegetarian diet (Bittman 2007; Madison 2007).

## Horticulture in Literature

References to horticultural crops and practices in literature are so widespread that only a brief sampling can be provided here. In the following, we use examples of horticultural allusions from Sumer, Homer's *Ulysses*, Laws of Hammurabi, various "bibles" including the Hebrew bible, the New Testament, and the Qu'ran, Arab poetry, different authors such as Shakespeare, Jane Austin, and James Joyce as well as references which are the source of popular and well-used sayings that allude to horticulture.

### *Ancient Sumer*

The Disputation between the Hoe and the Plow, dated from ca. 2500 BCE is perhaps the first poetic statement contrasting the state of the ordinary folk with the exalted and wealthy based on agricultural metaphors. The common man represented as the hoe argues his status against the rich and mighty (represented as the plow). The following translation is found in Hallo (2002). Only three of about 25 "stanzas" are presented.

> Hoe picked a quarrel with the Plow. Hoe and Plow—this is their dispute.
> Hoe cried you to Plow
>
> O Plow, you draw furrows—what is your furrowing to me?
> You make clods—what is your clod making to me?
> You cannot dam up water when it escapes.
> You cannot heap up earth in the basket.
> You cannot press clay or make bricks.
> You cannot lay foundations or build a house.
> You cannot strengthen an old wall's base.
> You cannot put a roof on a man's house
> O Plow, you cannot straighten a street.
> O Plow, you draw furrows—what is your furrowing to me?
> You make clods—what is your clod-make to me?
>
> The Plow cries out to the Hoe "I am Plow, I was fashioned by the great owers,
> assembled by noblest hands!
> I am the might registrar of God Enlil!
> I am the faithful farmer of Mankind!
> At the celebrations of my harvest-festival in the field,
> Even the King slaughters cattle for me, adding sheep!
> Drums and tympans sound! The king himself takes hold
> of my handle-bars;
> My oxen he harnesses to the yoke:
> Great noblemen walk at my side;

The nations gaze at me in admiration,
Land watches me in Joy!

## *Ancient Greece*

Homer in the Odyssey, nineth century BCE, refers to the Garden of Alcinöus owned by the King of the Phaeacians, a legendary country, which is rich in descriptions of fruit trees.

> And without the courtyard by the door is a great garden, of four plough-gates, and a hedge runs round on either side. And there grow tall trees blossoming, pear-trees and pomegranates, and apple-trees with bright fruits, and sweet figs, and olives in their bloom. .... Pear upon pear waxes old, and apple on apple, yea, and cluster ripens upon cluster of the grape, and fig upon fig. ......These were the splendid gifts of the gods in the palace of Alcinöus.
> The Odyssey Book VII, Hedrick 1921.

## *Mesopotamia*

The Laws of Hammurabi (1750 BCE) predate the mosaic ten commandments. Laws 64 and 65 relate to pollination of date palm are clearly quite sophisticated legally and might be considered the beginning of agricultural economics.

> 64. If a man give his orchard to a gardener to pollinate (the date palms), as long as the gardener is in possession of the orchard, he shall give to the owner of the orchard two thirds of the yield of the orchard, and he himself shall take one third.
>
> 65. If the gardener does not pollinate the (date palms in the) orchard and thus diminishes the yield, the gardener (shall measure and deliver) a yield of the orchard to (the owner of the orchard in accordance with) his neighbor's yield.

## Biblical References

The Hebrew bible is rich in allusions to viticultural practices and wine making.

> Now will I sing to my wellbeloved a song of my beloved touching his vineyard. My well beloved had a vineyard in a very fruitful hill. And he fenced it, and gathered out the stones thereof, and planted it with the choicest vine, and built a tower in the midst of it, and also made a winepress therein: and he looked that it should bring forth grapes, and it brought forth wild grapes. And now…judge…betwixt me and my vineyard. What could have been done more to my vineyard, that I have not done in it? Wherefore, when I looked that it should bring forth grapes, brought it forth wild grapes?
> Isaiah 5:1–7 & 10

Protection of grapes from birds and thieves is a common feature of the early cultivation of wine, and the construction of walls and towers is associated with vineyards in ancient Israel. Various techniques were developed for over-wintering, including covering sprawling vines with soil, techniques that still exist in Afghanistan. Grapes were preserved by sun drying to produce raisins, or by transforming grape juice to wine. The culture of grapes and the technology of wine making are common themes in biblical writings, and become infused in Jewish and Christian religious practices and social encounter. Wine was associated with blessings and joy, although drunkenness was frowned upon. Grapes and raisins are highly prized in the Qu'ran and although wine is prohibited in Islam "*rivers of wine*" are promised in Paradise.

Olive, along with grape, is the most mentioned fruit in the Hebrew bible and their importance permeated the western world. The olive tree became a symbol of beauty, freshness, fertility, wealth, fame, and peace. Its importance is reflected in the widespread use of oil for religious purposes such as consecration ceremonies (anointing) in Judaism and Christianity; the word messiah (Christ) literally means "the anointed one." Although grafting is not referred to in the Hebrew bible, grafting of olive is mentioned in the New Testament:

> And if some of the branches be broken off, and thou, being a wild olive tree, wert graffed in among them, and with them partakest of the root and fatness of the olive tree…For if thou were cut out of the olive tree which is wild by nature, and were graffed contrary to nature in a good olive tree: how much more shall these, which be the natural branches, be graffed into their own olive tree?
> Romans 11:17 & 24

Fig is another iconic Mediterranean fruit that some believe was the original tree of knowledge in the Garden of Eden.

> A certain man had a fig tree planted in his vineyard; and he came and sought fruit thereon, and found none. Then said he unto the dresser [cultivator] of his vineyard, Behold, these three years I come seeking fruit on this fig tree, and find none; cut it down; why cumbereth it the ground? And he answering said unto him, Lord, let it alone this year also, till I shall dig about, and dung it: And if it bear fruit, well: and if not, then after that thou shalt cut it down.
> Luke 13:6–9

Finally, horticultural metaphors are associated with love making in the *Song of Songs*:

> An oh, may your breasts be like clusters
> Of grapes on a vine, the scent
> Of your breath like apricots,
> Your mouth good wine–

## *Arab Poetry*

Medievel Arab poetry often uses horticultural allusions. In 1123, the eggplant (*Solanum melongena*) inspired the poet Ibn Sara of Santarem (now Portugal) to write the following:

> Spheroid
> Fruit, pleasing

To Taste, fattened
By water gushing in all
The gardens, glossy cupped
In its calyx, ah heart
Of a lamb in
A vulture's claws
translated by C. Middleton and L. Garzon Falcon (1997).

Compare this to a modern poem on the same subject by an anonymous author.

Who am I?
My skin is black and glossy,
It can be white as snow.
Sometimes I'm plump and saucy.
My roots go down below.

I reach out for the burning sun,
But grovel in the dirt.
My daggers will pierce anyone,
I draw blood and hurt.

My flesh is bitter, spicy,
But kiss me just the same.
Caress me, be not icy,
I dare you speak my name.

## *Chinese Poetry*

Chinese poets referred to many different horticultural crops and ornamental plants in early writings. Two examples follow.

During the Tang dynasty (618–906 CE), the lychee was celebrated and treated as a delightful exotic fruit in poetry and art and enjoyed great prestige. The lychee was so greatly favored by Emperor Xuan-zong's concubine, Yang gui-fei, that he had couriers on speedy horses from Szechwan province deliver fruit to the capital of Chang-an.

From Changan the palace embroidered the scene,
On the mountain top palace gates opened one by one,
One horse rider kicking up red dust, the concubine laughs,
No one knew it was the lychee express arriving!

The green trees of Xinfeng covered with dust,
As the emissaries to Yuyang returned with favorable news.
The sounds of Rainbow Feather Shawl embraced the peaks,
And graceful dancing feet trampled the nation.
Passing the Hua Qing Palace by Du Mu (803–852)

Another important Tang Dynasty poet was Meng Haoran who wrote strongly of pastoral life and leisure.

Preparing me chicken and rice, old friend,
You entertain me at your farm.
We watch the green trees that circle your village

And the pale blue of outlying mountains.
We open your window over garden and field,
To talk mulberry and hemp with our cups in our hands.
...Wait till the Mountain Holiday—
I am coming again in chrysanthemum time.
Stopping at a Friend's Farmhouse by Meng Haoran (689 or 691–740)

## *Shakespeare*

William Shakespeare (1533–1603) is considered the greatest writer in English if not in any tongue. His plays and poems are a rich source of horticultural information in the Elizabethan period (Ellacomber 1884). Of all nature's images, the greatest number is devoted to horticulture (Spurgeon 1935). The bard displays an intimate knowledge of plant growth, propagation, grafting, pruning, manuring, weeding, ripeness, and decay. Almost 200 plants are referenced. The following two garden scenes are rich in horticultural imagery.

In Richard II, the mismanagement of England is reflected in a conversation between two gardeners.

Go, bind thou up yon dangling apricocks,
Which, like unruly children, make their sire
Stoop with oppression of their prodigal weight;
Give some supportance to the bending twigs.
Go thou, and like an executioner,
Cut of the heads of too fast growing sprays,
That look too lofty in our commonwealth:
All must be even in our government.
You thus employed, I will go root away
The noisome weeds, with without profit suck
The soil's fertility from wholesome flowers.
Richard II. III.iv

In a rural scene in Bohemia from *The Winter's Tale*, Perdita, unknown to her the daughter of King Leontes, King of Sicilia, has being abandoned and is being brought up by shepherds due to the supposed infidelity of his wife by his friend, Polixenes. In this scene, the falsely accused Polixenes, King of Bohemia, checks out Perdita for his son who has fallen in love with her. In a famous repartee concerning streaked gillyvors (variegated carnation) which Perdita assumes to be due to either unnatural breeding or grafting, a philosophical discussion ensues on the nature of what is natural and what is unnatural. Perdita, as many today, will have none of it. This controversy still resonates in horticulture.

Perdita
Sir, the year growing ancient,
Nor yet on summer's death, nor on the birth
Of trembling winter, the fairest
Flowers o' the season
Are our carnations and streak'd gillyvors
Which some call nature's bastards: of that kind
Our rustic garden's barren; and I care not

To get slips of them.

Polixenes
Wherefore, gentle maiden,
Do you neglect them?

Perdita
For I have heard it said
There is an art which in their piedness shares
With great creating nature.

Polixenes
Say there be;
Yet nature is made better by no mean
But nature makes that mean: so, over that art,
Which you say adds to nature, is an art
That nature makes. You see, sweet maid, we marryA gentler scion to the wildest stock,
And make conceive a bark of baser kind
By bud of nobler race; this is an art
Which does mend nature, change it rather, but
The art itself is nature.

Perdita
So it is.

Polixenes
Then make your garden rich in gillyvors,
And do not call them bastards.

Perdita
I'll not put
The dibble in earth to set one slip of them;
No more than were I painted, I would wish
This youth should say, 'twere well, and only therefore
Desire to breed by me.
The Winters Tale IV.iv

Other notable extracts from Shakespearian works include:

That which we call a rose
By any other name would smell as sweet
Romeo and Juliet II.ii

When I have plucked the Rose,
I cannot give it vital growth again.
It needs must wither. I'll smell it on the tree
Othello V. ii

My salad days
When I was green in judgement
Anthony and Cleopatra I.v

Mine eyes smell Onions, I shall weep anon
All's Well that Ends Well V.iii

This is the state of man: today he puts forth
The tender leaves of hopes, to-morrow blossoms,
And bears his blushing honors thick upon him:
And third day comes a frost, a killing frost

> And, when he thinks, good easy man, full surely
> His greatness is a-ripening, nips his root,
> And then he falls, as I do
> Henry VIII, III,ii.

> Rough winds do shake the darling buds of May.
> Sonnets XVIII.

## *Jane Austen*

This quintessential and still beloved nineteenth century British author of manners, is known for her table talk and horticulture and gardens are frequently referenced. In *Mansfield Park*, she alludes to a well known apricot named 'Moor Park' and provides information on the cost of the tree:

> Sir it is a Moor Park, we bought it as a Moor Park, and it cost us—that is, it was a present from Sir Thomas, but I saws the bill—and I know it costs seven shillings, and we charged as a Moor Park.

> You were imposed on, ma'am replied Dr. Grant. "these potatoes have as much the flavour of a Moor Park, as the fruit from that tree. It is an insipid fruit at the best, but a good apricot is eatable, which none from my garden are."

> "The truth is, ma'am" said Mrs. Grant, pretending to whisper across the table to Mrs. Norris, "that Dr Grant hardly knows the nature of taste of our apricot is, he is scarcely ever indulged with one, for it is so valuable a fruit, with a little assistance, and ours is such a remarkable large fair sort that what with early tarts and preserves, my cook contrives to get them all."

## *James Joyce*

Finally we include a reference to melons in *Ulysses,* Joyce's masterpiece that describes a single day, June 16, 1904, in Dublin. This choice verbally wild snippet can be considered a sampling of the rich use of horticultural imagery in sensual and erotic literature:

> He kissed the plump mellow yellow smellow melons of her rump, on each plum melonous hemisphere, in their mellow yellow furrow, with obscure prolonged provocation melonsmellonous osculation.

## *Aphorisms and Proverbs*

Many references are made to horticulture in common speech, proverbs, and aphorisms. These include references to fruits, vegetables, and flowers as well as to various horticultural practices such as pollination, grafting, and weeding. These sayings and proverbs have many different and varied sources. The meaning and origins of the proverbs can be found in texts such as Pickering (2001).

Selected examples include the following:

Life is just a bowl of cherries
-by Ray Henderson, song with lyrics by Lew Brown (1931)

Rose is a rose is a rose.
-Gertrude Stein: Sacred Emily.

Cauliflower is nothing but a cabbage with a college education.
-Mark Twain: Pudd'nhead Wilson (1894)

An apple a day keeps the doctor away (English)

If apples bloom in May, you may eat them night and day (English)

April showers bring forth May flowers (English)

Apples, pears and nuts spoil the voice (Italian)

The apple never falls far from the tree (German)

Like tree, like fruit (English)

A good tree brings forth good fruit (English)

The best wine comes out of an old vessel (English)

Cabbage twice cooked is dead (Greek)

There is a devil in every berry of the grape (English)

If you would enjoy the fruit, pluck not the flower (English)

Great trees keep down little ones (English)

The higher the tree, the sweeter the plum (English)

Old friends and old wine are best (English)

Soon ripe, soon rotten (Roman)

Walnuts and pears you plant for your heirs (Greek)

The rotten apple injures its neighbours (English)

One generation plants the trees; another gets the shade (Chinese)

No matter how tall the tree is, its leaves will always fall to the ground (Chinese)

Flowers leave a part of their fragrance in the hands that bestow (Chinese)

A flower cannot blossom without sunshine nor a garden without love (Chinese)

A beautiful flower is incomplete without its leaves (Chinese)

Peach and chestnut bear fruit three years after germination while persimmon takes eight years to bear first fruit after germination. (A long time is necessary to accomplish something valuable). (Japanese)

It is stupid to prune ornamental cherry trees while it is stupid not to prune Japanese apricot trees. (Japanese)

## The Future

The scholarship and literature of horticulture, although scattered, had long been preserved in part in specialized libraries. These include, but are not limited to, such collections at the Royal Horticultural Society in London (Lindley Library), St George's Chapel Windsor Archives, The British Library, Natural History Museum Library, The Garden History Museum, Royal Botanic Garden Kew, Royal Botanic Garden Edinburgh, the National Agricultural Library of the United States in Beltsville Maryland, The Dumbarton Oaks collection in Washington DC, The Arnold Arboretum Horticulture Library at Harvard, Jamaica Plain, Massachussets, The library of the Missouri Botanical Garden in Saint Louis, The Lloyd Library in Cincinnati, Ohio, and the German Horticultural Library in Berlin. Many of these libraries have very special collections, for example the National Library in Beltsville contains a huge record of nursery catalogues used in the United States. In the past it was often difficult and expensive to access the horticultural literature but the recent digitization of all scientific and horticultural literature is transforming and easing this situation.

The publication of scientific journals and books has undergone major changes in the past decade with the advent of web-based search options and the personal computer. Most if not all scientific journals are now available on-line either prior to being printed or certainly soon afterwards. As a consequence, research findings are not subject to the same delays due to printing and distribution requirements. Many subscribers now choose to elect for electronic on-line delivery (usually at a lower cost) rather than delivery of a printed version of the same material. The same has occurred with the evolution of e-books.

The direct use of the world-wide web as a repository of technical information is expanding rapidly. Such material typically includes text and graphics, but increasing includes video material and sophisticated imagery. For example, a current search on "apple grafting" will result in identifying close to 1 million entries on the web ranging from encyclopaedic entries such as Wikipedia (http://en.wikipedia.org/wiki/Grafting) which includes pertinent published references, YouTube videos of grafting practices (http://www.youtube.com/watch?v=LTqG8-OhElY), and over 400 photographic and diagrammatic images of grafts, grafting tools, and different grafting methods. However, printed material will remain important as a means of summarising, validating, and interpreting the enormous literature of horticulture in an authoritative and informed way.

The literature of horticulture is rich and diverse. It has impacted our lives for centuries and is likely to do so for centuries to come.

**Acknowledgements** Dr Ryutaro Tao and Dr Zhu Jinyu are thanked for identifying the literature sources cited from Japan and China, respectively.

## References

Abe S, Okada S, Kawata J, Higuchi S, Machida H, Tanaka H, Iida I (1979) Floriculture. Asakura Publishing Co., Tokyo

Albrigo LG, Timmer LW, Rogers M (2012) Citrus, 2nd edn. CABI Publishing, Wallingford

Alexander R (2009) The essential garden design workbook. Timber Press, Portland

Alford D (2007) Pests of fruit crops. Academic Press, Burlington

American Horticultural Society (2009) New encyclopedia of gardening techniques. Mitchell Beazley, USA

Andrews PK (2013) Principles of fruit and nut production. CABI Publishing, Wallingford
Arber A 1965. Herbals: their origin and evolution. A chapter in the history of botany. 1470–1670. 3rd edn. Cambridge Science Classics, Cambridge UK (1st edn. in 1912)
Armitage AM (2011) Armitage's garden perennials. 2nd edn. Timber Press Inc, Portland
Austen R (1657) A Treatise of fruit-trees shewing the manner of grafting, setting, pruning, and ordering of them in all respects: according to diverse new and easy rules of experience; gathered in the space of twenty years. Oxford, UK (1st edn. 1653)
Babelon J, Chamblas-Ploton M (2001) The French garden. Vendome Press, New York
Bailey LH (1900) Cyclopedia of American horticulture, vol 1–4. Macmillan, NY
Bailey LH (1914) The standard cyclopedia of horticulture. Macmillan, NY
Baldini E (1996) Arboricoltura general. Clueb, Bologna
Barker AV (2010) Science and technology of organic farming. CRC, USA
Barker AV, Pilbeam DJ (eds) (2006) Handbook of plant nutrition. CRC, USA
Beale J (1653) A treatise on fruit trees shewing their manner of grafting, pruning, and ordering, of cyder and perry, of vineyards in England. Oxford, UK
Bettin A (ed) (2011) Kulturtechniken im zierpflanzenbau. Ulmer Verlag, Stuttgart
Beyl CA, Trigiano RN (eds) (2008) Plant propagation concepts and laboratory exercises. CRC, USA
Biddle A, Cattlin N (2007) Pests, diseases and disorders of peas and beans. Academic, USA
Bitner RL (2010) Timber Press pocket guide to conifers. Timber Press, Portland
Bittman M (2007) How to cook everything vegetarian: simple meatless recipes for great food. Wiley, USA
Blancard D (2012) Tomato diseases. Academic, The Netherlands
Blancard D, Lot H, Maisonneuve B (2006) A color atlas of diseases of lettuce and related salad crops. Academic, The Netherlands
Boodley J, Newman SE (2008) The commercial greenhouse. 3rd edn. Delmar Cengage Learning, USA
Bosland P, Votava E (2012) Peppers—vegetable and spice capsicums, 2nd edn. CABI Publishing, Wallingford
Botanica (2002) Botanica's orchids: over 1200 species. Laurel Glen Publishing, California
Brandenburg R, Villani M (eds) (1995) Handbook of turfgrass insect pests. Entomological Society of America, Maryland, USA
Brewster JL (2008) Onions and other vegetable alliums. CABI Publishing, Wallingford
Brickell C (2003) The Royal Horticultural Society A—Z encyclopedia of garden plants, 3rd edn. Dorling Kindersley, UK
Brickell C (ed) (2007) The Royal Horticulture Society encyclopedia of gardening, 3rd edn. Dorling Kindersley, UK
Brickell C (ed) (2010) The Royal Horticultural Society encyclopedia of plants and flowers. Dorling Kindersley, UK
Brookes J (2001) Garden design workbook. Dorling Kindersley, Melbourne
Brown K (2009) The complete practical handbook of garden bulbs: How to create a spectacular flowering garden throughout the year in lawns, beds, borders, boxes, containers and hanging baskets. Southwater, UK
Bryan JE (2002) Bulbs. Timber Press Inc, Portland
Bureaux C (2011) Techniques du jardinier l'encyclopédie. De Vecchi, France
Calicchio DL, Amon R (2011) Rooftop gardens: the terraces, conservatories, and balconies of New York. Rizzoli, New York
Campbell-Culver M (ed) (2009) Directions for the gardiner and other horticultural advice: John Evelyn. Oxford Univ. Press, Oxford, UK
Chundawat BS, Sen NL (2002) Principles of fruit culture. Agrotech Publishing Academy, India
Childers NF, Sibbett GS, Morris JR (1995) Modern fruit science: orchard and small fruit culture. Dr Norman F Childers Publ., USA
Conan M (1999) Perspectives on garden histories. Dumbarton Oaks Research Library and Collection, USA

Conan M (2007) Sacred gardens and landscapes–ritual and agency. Dumbarton Oaks Research Library and Collection, USA
Conan M, Kress WJ (2007) Botanical progress, horticultural innovations, and cultural changes. Dumbarton Oaks Research Library and Collection, USA
Conan M, Quilter J (2008) Gardens and cultural change: a pan-American perspective. Dumbarton Oaks Research Library and Collection, USA
Conan M, Wangheng C (2008) Gardens, city life, and culture: a world tour. Dumbarton Oaks Research Library and Collection, USA
Coyier DL, Roane MK (1986) Compendium of rhododendron and azalea diseases. APS, USA
Creasy GL, Creasy LL (2009) Grapes. CABI Publishing, Wallingford
Crüger G, Backhaus GF, Hommes M, Smolka S, Vetten H-J (eds) (2002) Pflanzenschutz im gemüsebau. Ulmer Verlag, Stuttgart
Darby A (2003) Food in the ancient world from A to Z. Routledge, London
Daunay M-C, Janick J, Pair HS (2009) Tacuinum sanitatis: horticulture and health in the late Middle Ages. Chronica Horticulturae 49(3):22–29
Decoteau DR (2000) Vegetable crops. Prentice Hall, USA
Desmond R (1987) A celebration of flowers; two hundred years of Curtis's Botanical Magazine. Collingridge Books, England
Dirr MA (2011) Dirr's encyclopedia of trees and shrubs. Timber Press, Portland
Dixon GR (2007) Vegetable brassicas and related crucifers. CABI Publishing, Wallingford
Dole JM, Wilkins HF (2004) Floriculture: principles and species, 2nd edn. Prentice Hall, USA
Don M (2009) The complete gardener. Dorling Kindersley, UK
Ellacomber HN (1884) The plant-lore & garden-craft of Shakespeare. W. Satchell and Co London. Reprinted 1973. AMS Press, New York
Emmons R (2007). Turgrass science and management. Delmar Cengage Learning, USA
Evelyn J (1664) Kalendarium hortense or the gard'ners almanac: directing what he is to do monthly throughout the year, and what fruits and flowers are in prime. Printed for G Huddleston, London
Fabbri A (2001) Produzioni vegetali. Edagricole, Bologna
Ferree DC, Warrington IJ (eds) (2003) Apples—botany, production and uses. CABI, Wallingford
Friedrich G, Fischer M (eds) (2000) Physiologische grundlagen des obstbaus. Ulmer Verlag, Stuttgart
Funt RC, Hall HK (eds) (2012) Raspberries. CABI, Wallingford
Gardiner J (2012) The Timber Press encyclopedia of flowering shrubs. Timber Press, Portland
Gaze R, Fletcher J (2007) Mushroom pest and disease control. Academic, The Netherlands
Gleason ML, Daughtrey ML, Chase AR, Moorman GW, Mueller DS (2009) Diseases of herbaceous perennials. APS, USA
Grenfell D, Shadrack M (2007) Timber Press pocket guide to hostas. Timber Press, Portland
Gundersen A (1918) A sketch of plant classification from Theophrastus to the present. Torea 18(11):214
Halfacre RG, Barden JA (1979) Horticulture. McGraw-Hill, USA
Hallo WH (2002) The context of scripture: monumental inscription from the biblical world. Vol II. Brill, Leiden
Harris RW, Clark JR, Matheny NP (2003) Arboriculture: integrated management of landscape trees, shrubs and vines, 4th edn. Prentice Hall, USA
Hartmann HT, Kester DE, Davies FT, Geneve R (2010) Hartmann & Kester's plant propagation: principles and practices, 8th edn. Prentice Hall, USA
Heaton DD (1997) A produce reference guide to fruits and vegetables from around the world: nature's harvest. CRC, USA
Hedrick UP, assisted by Booth NO, Taylor OM, Wellington R, Dorsey MJ (1908) The grapes of New York. Report of the New York Agricultural Experiment Station. II. JB Lyon, Albany
Hedrick UP, assisted by Wellington R, Taylor OM, Alderman WH, Dorsey MJ (1911) The plums of New York. Report of the New York Agricultural Experiment Station. II. JB Lyon, Albany
Hedrick UP, assisted by Howe GH, Taylor OM, Tubergen CB, Wellington R (1915) The cherries of New York. Report of the New York Agricultural Experiment Station. II. JB Lyon, Albany

Hedrick UP, assisted by Howe GH, Taylor OM, Tubergen CB (1917) The peaches of New York. Report of the New York Agricultural Experiment Station. II. JB Lyon, Albany
Hedrick UP, assisted by Howe GH, Taylor OM, Francis EH, Tukey HB (1921) The pears of New York. Report of the New York Agricultural Experiment Station. II. JB Lyon, Albany
Hedrick UP, assisted by Howe GH, Taylor OM, Berger A, Slate GL, Einset O (1925) The small fruits of New York. Report of the New York Agricultural Experiment Station. II. JB Lyon, Albany
Hessayon DG (1991) The fruit expert. Expert Books, UK
Hessayon DG (1997) The vegetable and herb expert. Expert Books, UK
Hogg R, Bull HG (eds) (1876–1885) The Herefordshire pomona containing original figures and descriptions of the most esteemed kinds of apples and pears. Jakeman & Carver, Hereford, UK
Hooker W (1818) Pomona Londinensis containing coloured engravings of the most esteemed fruits cultivated in British gardens. Vol 1., Published by the author and printed by James Moyes, Hatton Garden, London
Horst RK, Nelson PE (1997) Compendium of chrysanthemum diseases. APS, USA
Jackson RS (2008) Wine science: principles and applications, 3rd edn. Academic, USA
Janick J (1986) Horticultural science. W H Freeman, New York
Janick J (ed) (1989) Classic papers in horticultural science. Prentice Hall, USA
Janick J (2008) The tyranny of the impact factor. Chronica Horticulturae 4(2):3–4
Janick J, Hummer K (2010) Healing, health and horticulture: introduction to the workshop. HortScience 45(11):1584–1586
Janick J, Hummer K (2012) The 1500th Anniversary (512–2012) of the Juliana Anicia Codex: an illustrated Dioscoridean recension. Chronica Horticulturae 52(3):9–15
Janick J, Paull RE (eds) (2006) The encyclopedia of fruit and nuts. Cambridge University, Cambridge
Juniper BE, Mabberley DJ (2006) The apple. Timber Press, Portland
Kader A (2002) Postharvest technology of horticultural crops, 3rd edn. University of California Agriculture and Natural Resources, USA
Kaplan S, Kaplan R (1982) Humanscape: environments for people. Ulrichs Books, USA
Kawaguchi Y (2008) Serene gardens: creating japanese design and detail in the western garden. New Holland Publishers, UK
Keane MP (2007) Japanese garden design. Tuttle Publishing, USA
Kellert SR, Wilson EO (eds) (1995) The biophilia hypothesis. Island, Washington
Kelly J (ed) (2004) Hiller gardener's guide to trees and shrubs. David & Charles, UK
Keppel H (ed) (1998) Obstbau: Anbau und Verarbeitung. Verlag, Germany
Keys SJ, Paull RE (2004) Postharvest biology. Exon, USA
Koike S, Gladders P, Paulus A (2006) Vegetable diseases. Academic, The Netherlands
Larkcom J (2008). Oriental vegetables. The complete guide for the gardening cook. Kadansha USA, New York
Larson RA (1992) Introduction to floriculture, 2nd edn. Academic, The Netherlands
Layne DR, Bassi D (2008) The peach—botany, production and uses. CABI Publishing, UK
Lee J-M, Choi G-W, Janick J (eds) (2007) Horticulture in Korea. Korean Society for Horticultural Science, Suwon
Lee SW, Lee JM (2011) Horticulture and life, 2nd edn. Donghwa, Republic of Korea
Lewis CA (1996) Green nature/human nature: the meaning of plants in our lives. University of Illinois, USA
Link H (ed) (2002) Lucas' Anleitung zum Obstbau. Verlag, Stuttgart
Link H (ed) (2011) Ertragssteigerung im obstbau. Verlag, Stuttgart
Litz RE (2009) The mango—botany, production and uses. CABI Publishing, UK
Loudon JC (1828) An encyclopaedia of gardening; comprising the theory and practice of horticulture, floriculture, arboriculture, and landscape-gardening, including all the latest improvements; a general history of gardening in all countries. Printed for Longman, Ross, Orme, Brown, and Green, London
Macoboy S (2007) The ultimate rose book. Harry N Abrams, USA

Macoboy S, Rodd T, Spurway M (2010) Gardeners' encyclopedia of plants and flowers. Readers' Digest (Australia) Pty Ltd, Australia
Madison D (2007) Vegetarian cooking for everyone. Clarkson Potter, USA
Marenghi M (2005) Manuale di viticoltura. Edagricole, Bologna
Maynard DN (ed) (2001) Watermelons: characteristics, production and marketing. American Society for Horticultural Science, USA
McGiffen ME (1997) Weed management in horticultural crops. American Society for Horticultural Science, USA
McLaughlin C (2012) Complete idiots guide to small-space gardening. Penguin Group, USA
Middleton C, Falcon LG (translators) (1997) Eggplant. p. 462. In: Wasahburn IK, Major JSA (eds) World poetry. An anthology of verse from antiquity to our times. W.W. Norton, New York
Meredith TJ (2009) Timber Press pocket guide to bamboos. Timber Press Inc, USA
Midmore DJ (2012) Principles of tropical horticulture. CABI Publishing, UK
Mitchell F (2011) Theophrastus (371–287 BC). In: Origins of Botany. The University of Dublin, Ireland
Mizutani F (ed) (2002) Pomology. Asakura Publishing Co, Japan
Motloch JL (2000) Introduction to landscape design. John Wiley and Sons, New York
Mullins MG, Bouquet A, Williams LE (1992) Biology of the grapevine. Cambridge University, UK
Nelson PV (2011) Greenhouse operation and management, 7th edn. Prentice Hall, USA
Nielsen S (2004) Skygardens: rooftops, balconies and terraces. Schiffer Publishing, USA
Ohashi W (2000) Japanese gardens of the modern era. Japan Publications Trading, Japan
Paliyath G, Murr DP, Handa AK, Lurie S (2008) Postharvest biology and technology of fruits, vegetables and flowers. Wiley, New York
Palter R (2002) The Duchess of Malfi's apricots, and other literary fruits. University of South Carolina Press, Columbia
Paull RE, Duarte O (2012) Tropical fruits, vol 2, 2nd edn. CABI Publishing, UK
Phillips H (1821) History of cultivated vegetables: comprising their botanical, medicinal, edible, and chemical qualities; natural history; and relation to art, science, and commerce (two volumes). H Colburn and Co., London
Phillips E, Burrell CC (1999) Rodale's illustrated encyclopedia of perennials. Rodale Press, USA
Pickering D (2001) Cassell's dictionary of proverbs. Cassell, London
Preedy VR, Watson RR (eds) (2010) Olives and olive oil in health and disease prevention. Academic, The Netherlands
Pridgeon AM (1992) The illustrated encyclopedia of orchids. Timber Press Inc., Portland
Quest-Ritson C, Quest-Ritson B (2010) The American rose society encyclopedia of roses. DK
Retamales JB, Hancock JF (2012) Blueberries. CABI Publishing, Wallingford
Rice G (ed) (2006) Royal horticulture society encyclopedia of perennials. Dorling Kindersley, London
Rieger M (2006) Introduction to fruit crops (Crop Science). CRC, New York
Riffle RL (2008) Timber press pocket guide to palms. Timber Press, Portland
Roach FA (1985) Cultivated fruits of Britain, their origin and history. Blackwell, New York
Robinson JC, Galán SV (2010) Bananas and plantains. CABI Publishing, Wallingford
Rogers EB (2001) Landscape design: a cultural and architectural history. Harry N. Abrams, New York
Royal HS (2010) RHS how to grow practically everything. Dorling Kindersley, UK
Russo VM (ed) (2012) Peppers—botany, production and uses. CABI Publishing, Wallingford
Ryder EJ (1999) Lettuce, endive and chicory. CABI Publishing, Madison
Sansavini S, Costa G, Gucci R, Inglese P, Ramina A, Xiloyannis C (2012) Aboricoltura generale. Patron Editore, Bologna
Schaffer B, Wolstenholme BN, Whiley AW (2013) The avocado: botony, production and uses. CABI Publishing, Wallingford
Schlegel RHJ (2009) Dictionary of plant breeding, 2nd edn. CRC, New York
Schmid WG (2004) Timber Press pocket guide to shade perennials. Timber Press, Portland
Sengbusch P (2003) Botany: the history of a science. Universität Hamburg, Germany. http://www.biologie.uni-hamburg.de/b-online/e01/01.htm. Accessed 5 Aug 2012

# Chapter 38
# A Short History of Scholarship in Horticulture and Pomology

**Silviero Sansavini**

**Abstract** Literature about horticulture (fruits, flowers and vegetables) has long been produced by georgic writers and experts in agriculture. Generally it comprises testimony of experience and the desire to disseminate knowledge about plants and how to grow them. This has been true since ancient times and is particularly true of the Mediterranean area. Often, authors copied from their predecessors and the contents of a book were handed down over the centuries. This was the case until the mid-eighteenth century, when agriculture became the subject of scientific investigation and began to benefit from a knowledge of biology.

This chapter summarizes the writings of the Greeks (Theophrastus), Latin writers (Columella and Virgil), and then, moves forward about a thousand years, to the work of authors in the Middle Ages, often from Arabia and Italian sources (Ibn Al-Awwan and de' Crescenzi), followed by the Renaissance (Aldrovandi) and then modern French (De la Quintinye) and Dutch (De Vries) and then to the more recent French and Italian sources (du Breuil and Gallesio).

In the nineteenth century, after the revolutions in the understanding of the workings of the universe (Copernicus, Galileo, Newton) and the fall of the old schools of philosophy and medicine, Europe began to produce numerous works on pomology and horticulture. The first studies were carried out on applied biology and physiology, together with investigations into reproduction and the fixing of characteristics by seed propagation and grafting. Frequently, the works described varieties (pomology), soil management, fertilization, irrigation and the influence of the environment on production.

The great leap in the quality and quantity of horticultural products came with the discovery of the synthesis of nitrogenized fertilizers and a proper understanding of photosynthesis, involving the acquisition by plants of carbon and the synthesis of organic matter. Knowledge was also acquired concerning the roles of water and the soil in plant growth and fruition. In these areas, horticulture benefited from the research of the Europeans (particularly in Germany) and Americans. The author has

S. Sansavini (✉)
Dipartimento di Scienze Agrarie, University of Bologna,
Viale Fanin, 46-40127, Bologna, Italy
e-mail: silviero.sansavini@unibo.it

G. R. Dixon, D. E. Aldous (eds.), *Horticulture: Plants for People and Places, Volume 3*, 1255
DOI 10.1007/978-94-017-8560-0_38, 

chosen about thirty authors, as representative of their age and for their significant contributions. The choice is subjective, of course, and some branches of knowledge may have been under-represented, but overall the hope is to achieve a fair balance, including some wrong turns and mistakes.

The green revolution changed agriculture—including fruit growing—in the second half of the twentieth century, introducing genetic research in pomology and agronomy. In addition to breeders, professionals included pomologists and geneticists, working along the lines established by Darwin and Mendel, and making positive contributions from the first decades of the twentieth century onwards. Agronomists, physiologists and biochemists have also contributed, enabling the application of new technologies in orchard management. In modern times, bio-technologists, molecular biologists, mechanical and plant engineers enabled hitherto unthinkable results to be achieved at the end of the twentieth century, in line with the new philosophy of eco-sustainable growth, preventing an excessive dependence on polluting chemicals and the use of non-renewable energy sources, which might have compromised expectations and the future of the next generations. Our investigations end at the beginning of the twentieth century. What follows would deserve a chapter of its own (Sansavini, L'agricoltura verso il terzo millennio attraverso i grandi mutamenti del XX secolo, pp. 307–382, 2002 and Sansavini, European horticultural challenges in a global economy: role of technological innovations, Chronica Hort. 53(4): pp. 6–14, 2013).

**Keywords** History · Agriculture · Horticultural science · Pomologists · Plants

Foreword: Towards a history of the literature of pomology (Antonio Saltini)

*A history of the literature of agronomy is based on written texts which do not necessarily have to be specialist works. Poetry, narrative literature and journalism, for example, can all provide us with valuable insights into the relationship between human societies and agricultural resources. There is one proviso however. Primary importance must be assigned to the nature and the source of the written text itself. Let me explain why. Societies have always had an artisan-farmer culture, a system of traditions and long-standing practices comparable to a real science but one which was passed on by word of mouth and not in writing. Anthropologists who have tried to reconstruct these worlds see things from a different viewpoint and have not consequently succeeded in producing descriptions which meet our purposes. We are obliged to fall back on the specialist writings of the time.*

*Up until the Renaissance, the cultivation of fruit trees and vegetables was described in the general writings on agronomy. Later, it was to branch off and become a discipline in its own right. From this time onwards, the major, classical works of the great agronomists had chapters dedicated specifically to pomology and the subject began to assume the status of a separate science.*

*We find the first real expression of thought on the cultivation of fruit trees in the Odyssey (8 to 9th bc) which describes the many cultivars already grown and propagated, by grafting, to maintain their individual characteristics. All that remains of*

*Greek science are the meanderings about plant propagation written by Theophrastus (Third century bc) later to be repeated by Virgil (First century bc). The ideas of Theophrastus on reproduction were countered by those of Lucretius (First century bc). He propounded ideas about reproduction which were based on extraordinary insights into genetics which included observations on the constancy and variability in the progeny of every species. Columella (First century ad) provided an admirable summary of all the Greek and Latin literature on agronomy. His chapters on the cultivation of the vine and the olive were masterful. His writings on fruit trees a little less so. Pliny took up the same themes but with less illuminating insights. Later, Palladius (Second or third century ad) wrote some colourful pages on horticulture during the decline of the empire.*

*In the Middle Ages we can pass over the peregrinations of Pietro de Crescenzi whose main work was published between 1304 and 1309. More striking in the 1350s were the perceptive writings of Ibn al-Awwām, an Arab author who followed in the footsteps of the great Arabian biologists. During the Renaissance, pomology was the subject of particularly detailed, broad ranging treatment in the works of the two great agronomists of the time—Agostino Gallo (1st edition 1564, definitive edition 1569) in Italy and Olivier De Serres (1st edition 1600) in France. Gallo illustrated the cultivation techniques used in the Italian brolo or vegetable garden. He covered most aspects of planting, pruning, grafting, working the soil, manuring and the ripening calendar of the main species. De Serres described the techniques for growing fruit in the gardens of a French villa and provided related advice. De Serres also illustrated the latest methods, providing, for example, precise instructions for growing fruit on flat frames against a wall. Gallo was the first to propose growing citrus fruit in greenhouses, an idea to be copied by his French counterpart.*

*In the 1600s various English authors wrote about fruit growing. Evelyn, for example, writing in Pomona (printed in 1664) pioneered the cultivation of apples and pears for making cider. Undoubtedly the most important work of the time (published posthumously in 1690) remains that by Jean Baptiste de La Quintinie, superintendent of the new orchards at Versailles. Here, with the assistance of the many gardeners and labourers available, he conducted experiments with the espalier technique of De Serres. He perfected the espalier method primarily on the basis of a deep understanding of the biology of plant growth and regrowth gained through close observation. The espalier method he described was hugely labour intensive. His mastery of everything to do with fruit trees was such that he can be considered the founder of modern day, scientific fruit growing. His principles and methods were to be employed for a century to come without any significant changes.*

*In the 1700s significant writings about fruit growing, vine cultivation and horticulture were penned by the Hungarian Ludwig Mitterpacher. The Habsburg government ordered the original Latin edition (1777) to be translated into Italian. Notes were added to the Italian edition (1784) about the types of cultivation typical to the Kingdom of Lombardy-Venetia, written by agronomists from Lombardy. Noteworthy are the pages on the cultivation of the olive tree around the lakes of Insubria.*

*The 1700s also saw the publication of the great pomological atlases. Every nation in Europe printed these monumental illustrated works cataloguing and classifying the fruit varieties grown in their countries. I have made a list of these and although I have not examined their contents closely I do not think they contain any biological ideas which could be described as original. Italy was the last European country to print its own Pomona Italiana (1817–1839), authored by Giorgio Gallesio. Prior to the Pomona Italiana, Gallesio had written a work which proposed a completely new approach to the biology of fruit growing and propagation (Traité du citrus 1811). Even though many of Gallesio's ideas proved to be unfounded they were held in great consideration by Darwin when writing his Variation of Animals and Plants under Domestication (1868), another work to be considered in the history of the literature of pomology.*

*In the 1800s, after Gallesio, fruit growing science in Italy went through a long period of lethargy contrasted only by the works of the Roda brothers, royal gardeners at the Castle of Racconigi. Their highly original guide went through numerous reprints, some of them unauthorised, between 1845 and 1880. Volumes have been written about the work of the two brothers as park directors but nothing about their work concerning the renewal of fruit growing. Nothing, that is, until a recent thesis supervised by Mignani and Saltini for the 150th Anniversary of Italian Unity. This author may not have delved deep enough into the subject matter but did succeed in highlighting the stimulating innovations proposed in the manuals and guides proposed by these two Piedmontese gardeners.*

*The dawn of the 1900s saw the arrival of works by the distinguished pomologist Hugo de Vries. Unfortunately I have not studied his texts. The Italian Girolomo Molon claimed the merit for having understood, from international publications, that a radically new form of fruit farming was taking shape in California. He undertook the long journey to America to study the farmers, their methods and the potential. The diary of his travels, published in 1918, is more the work of a diligent journalist faithfully recording what he saw, rather than the work of professional scientist. Nevertheless, it should be said that in mapping out what was to become the future of fruit farming he made an invaluable contribution, in general terms rather than in terms of pure agronomy, to the history of horticultural literature.*

## 1st Part: The Latin/ Greek World

### *Lucius Junius Moderatus Columella (AD 4–ca. AD 70)*

The author would like to start this brief history of scholarship in horticulture with a Latin author, Lucius Junius Moderatus Columella. At the start of his career, filled with an all consuming passion for the pruning and management of fruit trees, the author often liked to quote one of Columella's maxims from *De Re Rustica* (Book V, Chapter IX): *Nam veteris proverbii meminisse conveni, eum qui aret olivetum, rogare fructum; qui stertore, exorare; qui caedat, cogere.* ("Remember the old proverb which says: he who ploughs the olive grove, asks for fruit; he who manures it, begs for fruit; he who prunes it, forces it to yield fruit").

Columella, Spanish by birth, lived in Rome in the first century AD and wrote with a style which was both highly didactic and eclectic in the subject matter cov-

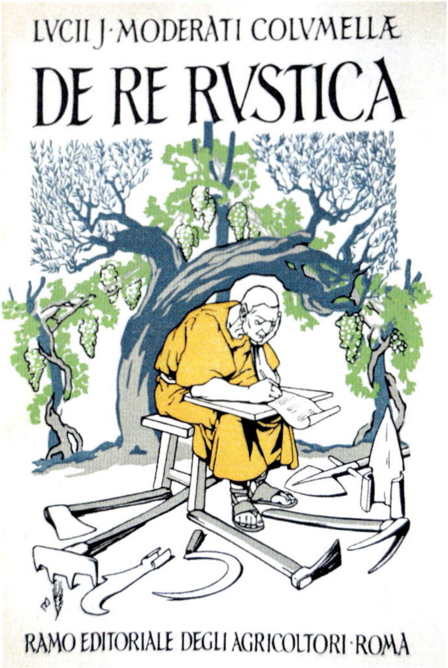


**Fig. 38.1** Frontispiece of L.M. Columella's "*De re rustica*", vol. 1, REDA, Rome 1948

ered. In his works he succeeded in transcribing and documenting his experiences as a farmer, a horticulturalist and an animal breeder. In *De re rustica* (Fig. 38.1) (Calzecchi-Onesti 1948), as an enthusiastic farmer he covered the universe of fruit species and the principal types of fruit trees including the fig, almond, pomegranate and walnut. He reserves a special place for the olive tree, with a guide to its cultivation, its cultivars and the three types of grafting to be used: "cleft" performed after the winter pollarding; "insertion of the scion under the bark" in spring; and the "single eye" or "budding" graft in summer. Today, these continue to be the three most widely used grafting techniques. Columella was also aware that the olive tree seldom bears well two years in succession and consequently maintained that, in order to compensate for this biological phenomenon which endangered the profitability of a farm in the poor years, it was necessary to force half of the rows with different cultivation methods so that every year the olive farmer could be assured of at least half of the harvest.

Columella's knowledge and wisdom centres primarily on the vine. In Books III and IV (Calzecchi-Onesti 1948) he reviews propagation, an argument related to mass selection (a practice already adopted by the best vine dressers), planting, growing and pruning.

Noteworthy, in his instructions about how to perform these operations, is his frequent insistence on writing "…unlike the ancients, I counsel …" where the ancients were the leading personalities such as Virgil and Cato who lived one or two centuries before him. It is also worth noting how Columella was always trying to convince his reader of the economic advantages of the operation he was proposing. He disputed, for example, the practice of mixing varieties when planting the vineyard. This, he maintained, required more work, increased harvesting costs and also worsened the quality of the wine, given the different ripening times of the grapes. According to Columella, vineyards were much more profitable than other forms of cultivation. Each *iugero* of vineyard (2,514.82 $m^2$—28,000 $ft^2$) yielded 600 *urnae* each holding 12.90 L of wine (Book III, Chapter III). For him the vineyard had to be tended well so that it achieved a higher yield and so that, as he says, "... still profit prevails over pleasure". Thus, if you look after your vineyard or "Wheresoever the god has turned his goodly head", the gods will be kinder to you. (*Sed haec quamvis plurimum delectent, utilitas tamen vincit voluptatem* (...) *et quod de sacro numine poeta dicit: "Et quocunque deus circum caput egit honestum"*—But though for all these give the greatest delight, still profit prevails over pleasure. (For the head of the household comes down the more willingly to feast his eyes upon his wealth in proportion to its splendour;) and, as the poet says of the sacred deity "Wheresoever the god has turned his goodly head", truly, wherever the person and eyes of the master are frequent visitors, there the fruit abounds in greater measure (Book III, Chapter XXI).

The author would also like to quote some of Columella's advice which continues to be valid today. For example, to propagate the vine he suggests using "cuttings" from the stock in preference to vine shoots. In this case the cutting he advises is mallet-shaped, the shape being the result of cutting part of the old wood along with the stock so that it is kept joined to the cutting; he also recommends that the cutting itself is reduced to the minimum.

When it comes to growing he writes that "pruning must be done with plants made from both 'leaf-bud cuttings' and from 'shoots'." "The vine dresser must leave more or less stock according to the nature of each plant thereby to 'encourage' it to produce with longer 'heads' rather than to 'slow it' (the vine) with a rather 'short' pruning."

Also very interesting are his descriptions of vine plants, especially those he liked best, the specialist varieties, where at least 16,000 to 20,000 cuttings covered the cultivated area. At the same time he did not underestimate the attractiveness and utility of the hedgerow type of vine cultivation, the classic tree-lined row of vines dividing one field from another. These have survived to the present day and are still one of the attractive characteristics of the Emilian countryside. Here, the vines are "wedded" to a "living support" usually elm or ash trees (Book V, Chapter VI) whose fronds were also used as animal feed. In the Po valley area of Italy, in the last decades of the 1900s, field maple (*Acer campestre*) was mainly used for these *piantata* (or *alberata*), or lines of trees supporting the vines.

Curiously, Columella the farmer was also concerned that farm estates at the end of the season should dry apples, pears (and plums) in the open air because "they serve in no small measure for the sustenance of slaves" (*Eorum si est multitudo, non minimam partem cibariorum per hiemen rusticis vindicant* ...) during the winter months (Book XII, Chapter XIV).

**Fig. 38.2** Statue of Pomona, goddess of fruits (Galleria degli Uffizi, Florence)

## *Virgil (Publius Vergilius Maro) (70 BC—19 BC)*

The poet Virgil was from a farming family and lived in the first century BC. As one of Rome's greatest poets he had a major influence on Latin culture. It is enough to quote the Aeneid, an epic poem dedicated to philosophical meditation and in particular to Epicureanism. He found the *hortus epicureo*, the Garden of Epicurus, a private retreat and an ideal place to rediscover the tranquillity lost in life's tempests (and in the bustle of Rome). The Epicurean garden was the inspiration for two of his best known works, the "Bucolics", a pastoral work based on the Greek model, and the "Georgics", an elegy to life in the fields. Virgil was a great and impassioned lyricist of the natural landscape and the works of the countryside. Pomona was the goddess of the fruits (Fig. 38.2).

In the Georgics (Virgilio Marone 1908), Virgil wanted to introduce the idea of Mother Earth, (today Terra Madre is the name of a global movement of small farmers defending their traditions and combating the advance of multinationals in agriculture), made fruitful by human effort, according to a scientific conception of Nature which had already inspired Lucretius. It was Virgil who, in the work of the farmer, found a great "epos", an epic narrative combining human ethics and the sense of anguish linked to the hardships imposed by Nature which all too often endangered crops and harvests. At the same time, it is Nature which induces humanity to perform grandiose works as a sort of reaction to the sufferings it imposes. Thus he foretold the destiny of Aeneas, the founder of Rome.

Book II of the Georgics is dedicated to the cultivation of trees, vines and the olive. The *incipit* announces: "*Hactenus arvorum cultus et sidera caeli/ nun te, Bacche, canam, nec non silvestria tecum/ virgulta et prolem tarde crescentis olivae*". (So much for the cultivation of fields, and the stars in the sky/ Now I'll sing you, Bacchus, not forgetting the saplings/ of woodlands and the children of slow-growing olives).

Virgil immediately reports that the olive produces much later than the vine. He recognises the importance of genetic variability in species and how this requires different approaches to cultivation and husbandry. He also sees the importance of differences in soil and is perhaps the first to introduce the idea of "habitat suitability" or, in other words, "soil and climate vocation".

He also makes a distinction between different propagation techniques (each species having its own). Recounting his experiences with grafting he even goes as far as to write, with a poetic licence that pushes him beyond the bounds of reality, "And often we see one tree's branches harmlessly/ given over to another's, a pear altered, or mutated, to carry grafted apples, and stony cornelian cherries blushing on a plum"—"*Et saepe alterius ramos inpume videmus/ vertere in alterius mutatamque insita mala/ ferre pirum et prunis lapidosa rubescere corna*". His definition of juvenility is also rather original, "The tree that raises itself from scattered seed, grows slowly, creating shade for our descendants, its fruits degenerate, losing their former savour, and the vine bears sad clusters, a prize for the birds" ("*Jam quae seminibus iactis se sustuiit arbos, tarda venit seris factura nepotibus umbram, pomaque degenerant sucos oblita priores et turpes avibus praedam fert uva racemos*").

## *Cato (Marcii Porcii Catonis) (234 BC–149 BC)*

According to Columella, Cato (or Catonis, in Latin) the Elder was the first person (Fig. 38.3), in the second century BC, between the second and third Punic wars, to write about agricultural science in Latin. Before this, there were only Greek sources, none of whom Cato trusted. In fact he wanted only to write about his personal experiences which followed the traditions of his homeland. His "*Liber de agri cultura*" ("On Farming") (Fig. 38.4) (Catonis 1964) is a practical guide, a notebook of instructions and advice written primarily for overseers (stewards) and slaves. According to E.V. Marmorale (1949) any attempt to find a shimmer of love for the earth similar to that found in Virgil will be in vain His writing was purely utilitarian. His writing did however force a change in mentality and marks a great step forward along the road which made the Greeks and the Romans the wise and observant writers of great popular works of information. The Greeks laid the scientific foundations and developed the theories of agriculture. The Romans developed the methods employing them according to economic principles to accumulate assets and money thereby improving the quality of life.

Cato the Censor, to give him his Roman *cognomen*, forged a great political career and became a "persecutor of luxury", an intransigent upholder of tradition. Not a moralist but rather a defender of the ancient *Civitas romanorum*. He became an

**Fig. 38.3** Statue of Cato (Museo Lateranense, Rome)

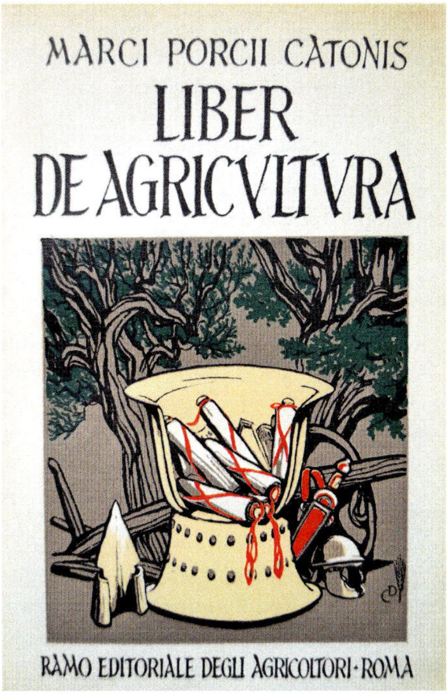


**Fig. 38.4** Frontispiece of M.P. Cato, "*Liber de agricultura*", REDA, Roma

excellent farmer. "From the very beginning I toiled my entire adolescence in parsimony, hardness and work, cultivating the stony Sabine hills, and working the stony ground and planting trees" (extract from oration no. 128).

Cato gave a detailed description of rural life in republican Rome before it came to know the scourge of the landed estates of *latifundium* and the ruinous upheavals of social strife. Unlike Columella, Cato described an agriculture fixed in time, empirical and still imbued with sorcery and sacred rites to the gods. He gave rules for the propagation and planting of olive trees and vines. He advocated the practice of green-manuring (with broad beans and vetch) to "fatten" the soil. He explained how to make wine and press olives but not without forgetting to propitiate the gods on feast days. His prose is somewhat meagre. In fruit farming he only mentions varieties of apple and pear, but this is not surprising given that the cherry, apricot and peach had still to arrive from the Orient (Lucullus brought these to Rome at the end of the first century BC). Cato dwells at length on "the cabbage and its digestive virtues" and offers advice on cultivating asparagus. He talks a lot about people. The overseer (or steward), for example, an important figure in the management of the estate and the farm. He proposes the outlines of contracts used to recruit labourers to thresh the grain and for all the other work in the fields such as picking olives and grapes. He also dedicates much attention to the housekeeper, a woman usually given in marriage by the master or *pater familias* to the overseer. The housekeeper had to keep the stores well supplied with dried apples and pears, sorbs, figs, raisins, sorbs in must, bunches of grapes in barrels or in grape pulp in buried pots, quinces and fresh Praenestine nuts also preserved in buried pots. The woman had to know how to spin, weave and sew clothes. She was assigned numerous tasks such as the "keeper of the hearth" and the provider of ritual devotions and offers but was dissuaded from performing acts of witchcraft. Only the master could order the sacrifices to be offered to the gods.

Some of his precepts were of doubtful value even at the time they first appeared; his reasoning was unconvincing and lacked credibility. This probably stemmed from the author's desire to remain faithful to popular traditions and beliefs. Cato indicated, for example, a remedy "… to prevent chafing: When you set out on a journey, keep a small branch of Pontic wormwood under the anus." (chapter CLIX). This sounds more like the advice given to boy scouts before they start a camping trip.

Here it is enough to quote his "propitiatory rite for purifying the land", an event which took place in spring and involved the sacrificing of a suckling pig, a lamb and a calf to Mars. Thus, the master sought the favour of the gods while trying to govern his people with "equity and honesty" (*bonus dominus*). In the final analysis, Cato favours a humanism based on life in the countryside, something which he exalts and believes is exemplary in maintaining rural peace and community because, as he puts it, "it is from the farming class that the bravest men and the sturdiest soldiers come". Today, there are those who still think like this.

Cato did however possess a "sour character" (according to Livio, the Roman writer) and was unyielding as a politician. Not surprisingly it was he who wanted to destroy Carthage (censeo Carthaginem *esse delendam!)* but died before Carthage burnt.

Generally speaking, the period of greatest splendour for fruit arboriculture during the Roman Empire started with the fall of Carthage and lasted until the fall of the Empire (Third century AD). The Romans were followers of the cult of the fruit tree. Pliny the Elder, the Roman author and writer of a natural history encyclopaedia (First century AD) proclaims the fruit tree "as the greatest gift given by the gods to man". Varro (Marcus Terentius Varro, 116–27 BC) wrote at 80 years the treatise "De re rustica", a very admirated and popular book in the Latin world for the high consideration of the Romans for the agriculture and horticulture. The Italian peninsula, he stated, was covered from fruit trees (*Tota pomarium*). Italy, at that time, had become "an orchard cultivated with art and science", so much so that around the Gulf of Naples the earnings from fruit exceeded those produced by vineyards and olive groves, otherwise the two most important forms of cultivation in the area.

The walls of Pompeii testify that the Romans already knew of the varieties available, many of which had already been described by Columella (First century AD) and Palladio (Fourth century AD). It was Palladio, in particular, who had described better than anyone else the cultivation of citrus fruit.

## *Theophrastus (371 BC–287 BC)*

The classical Greeks were the first to write about agriculture. They wrote mainly about the olive, a tree to which they assigned divine honours and strived to spread throughout their colonies. One of those who merits special attention is Theophrastus, who was to become for several centuries a point of reference for Italian authors.

Theophrastus was a pupil of Aristotle and therefore a philosopher, but one whose interests were far-ranging and who excelled at most things, dominating in most things "like an eagle that flies". His passion for natural science led him to study plants just in the same way that Aristotle had studied animals. He was not interested in superficial knowledge or mere facts as his two treatises demonstrate. One was a work of biology, before its time, where he wanted to delve deeply into many topics and find a synthesis. He succeeded in this and in doing something which no one else was able to do for more than a thousand years still to come. He lived in the Fourth century BC and died in 297 BC. His first work, a contribution to botanical science, was "Enquiry into plants" (in nine books). The second, "On the causes of plants" (in six books), was a treatise on plant physiology and husbandry.

Theophrastus, like Virgil and other Latin writers, had a very limited view of the validity of the reproduction of plants by seeds, not corresponding to the uses and usefulness of propagation practices. He believed that plants derived from seeds were deformed compared to their parents, perhaps even "degenerate"; hence the deduction that, for the proto-historic propagation of fruit trees, only vegetative propagation was able to reduce risks and maintain the identity of the tree, but according to Genesis, a sacred book *par excellence*, first drafted in the Fourth century BC, on the third day God said: "Let the earth bring forth grass, the herb yielding seed, and the fruit tree yielding fruit," (1.11). "And the earth brought forth grass, and herb yielding seed after his kind, and the tree yielding fruit, whose seed was in itself, after his kind; and God saw that it was good." (1.12) According to the historian, Antonio Saltini (2012), this is a better definition than the one provided by Theophrastus and others after him.

He also wrote about metaphysics and mineralogy. Some of his writings on naturalist philosophical doctrines appear to anticipate certain themes of today's environmentalist movements. Worthy of a separate mention is his book "The Characters". Here, Theophrastus outlines 30 'moral types' in an analysis of character profiles which highlights 'moral faults'. The analysis, seen from the ethical peripatetic viewpoint specifies through empirical observation the vices and virtues of human nature as already noted by Plato and taken up again later by Aristotle.

## 2nd Part—Middle Age

### *Ibn Al Awwam (Abū Zakariyā Yaḥyā ibn Muḥammad)*

Greek and Latin authors dominated in the writings about agriculture for centuries to come. After them, with the decline of the Roman Empire, the Byzantine era and the barbarian invasions, Europe did not produce anything of historical relevance in the world of agriculture. We have to wait until medieval times, after the year 1,000, to discover the contribution of the great scholar of agronomy *Abū Zakariyā Yaḥyā ibn Muḥammad* known as *Ibn al-Awwām* (end of the 12th century). At this time, the south of the Spanish peninsula and Sicily had already been settled by Arabs for several centuries. Ibn al-Awwām lived in Seville around the twelvth century and wrote a treatise on agriculture *Kitab al-felahah* (16 Chaps) later on translated into Spanish and French (several editions till nineteenth century). In the opinion of A. Saltini (2012), the scientific approach adopted in *Kitab al-felahah* makes this work the second real cornerstone of agronomy after *De re rustica* of Columella.

Let us try and outline the salient points that Ibn al-Awwām has left us in his writings about horticulture. First though, we need to say a few things about his cultural background. His knowledge referred back to that of the Nabataeans, an ancient Arab peoples whose cultural centre was the city of Petra, south of the Dead Sea. During the second century AD, Nabataean horticulturists were famed far and wide for their innovative methods. They were particularly known for their advice on to how to create and manage vegetable gardens and the soil and the different practices for cultivating each type of the numerous species of cultivated vegetable. The author lists many vegetables—garlic, onion, thistle, rape, melon, water melon and courgette. These were all species unknown in Europe but numerous varieties were already known and available in the Arab world. Ibn al-Awwām spread knowledge of the technical and scientific conquests in the world of Islamic agriculture. He states the principles to be followed when designing vegetable gardens and describes the irrigation techniques needed to produce a good result. He also had very clear ideas about the relationship between biological diversity and the different methods for treating each of the cultivated species individually. Mixings were to be avoided because he thought they were incompatible. Saltini reports that Ibn al-Awwām wrote, "The trees should not be planted at random. One should, on the contrary, bring together related types, so as to avoid the more vigorous species absorbing all the nutrient juices of the soil and leaving the more delicate species unprovided for". Virgil and Columella had already touched on this subject but the Arab botanists had gone to the heart of the matter in understanding the negative consequences of badly managing the mixing of various species; their understanding was such that their ideas, as Saltini writes, "are at the base of the modern theories on crop rotation".

Ibn al-Awwām also proposes a method for designing an orchard or a plantation to match the species grown and the end use of the product, fruit or wood. He writes, "The distance from one tree to another should be decided according to the nature of the soil and its vitality". For ornamental trees and trees used for their wood, he says that the species requiring longer growing times produce a wood which is harder, more durable and more suited for various uses.

For propagation, he lists the possibilities as using the seed (or the kernel), parts of branches, buds or root suckers. However, he does not advise grafting because he thinks this method does not correctly reproduce the tree but rather "modifies it characteristics because of the use of a scion belonging to another species or variety". True, but as we know grafting has other benefits. In effect, as Saltini shows, Ibn al-Awwām makes the same mistake as Theophrastus and Virgil.

Numerous other pages in Chapter 5 are dedicated to the methods for building a nursery and transplanting trees. Clearly, he recommends using well-tilled soil, fertilised with manure. To obtain a transplant, he recommends digging well around the tree taking care not to damage the roots and leaving the earth attached to the roots; this is a precursor of the current method of planting trees with the root ball in a container and the soil of the root ball intact. He provides much advice about sowing, growing and transplanting the saplings. Saplings should be transplanted two or three years from birth without cutting the roots. The biological precept he deduces from this is that "each species reproduced from seed produces a new tree which is similar to the tree of origin".

At that time, the present day idea of plant uniformity obtained by cloning had not fully emerged but this was already one step forward in comparison with Theophrastus, as A. Saltini (2012) noted. Theophrastus maintained that plant seedlings could degenerate, "producing plants which are different to those from which they take their origin". Ibn al-Awwām, on the other hand, found in practice that a seedling "does not produce another identical tree but rather one that is only similar to its parent". He goes on to explain that this is what makes it possible to obtain a new variety.

When discussing the temporary potting of plants obtained from air layering (a method he thought suitable for propagating myrtle, blackberry, pear, citrus fruit and other species), he recommends keeping plants in pots for no longer than a year before passing them temporarily to the nursery for cultivation until they were ready for final transplanting in the open field.

The climate of Andalusia must have been as arid in those times as it is today and not surprisingly he repeatedly recommends keeping the root system, in the pot and in the ground, well watered until it is well formed. Worth noting is the priority he

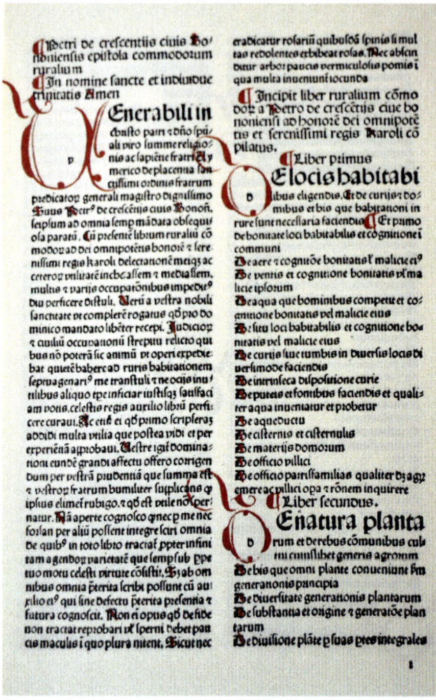

**Fig. 38.5** "Incipit page" of the first Latin issue of De Crescentis "*Opus ruralium commodorum*" (1486)

gives to preserving the roots rather than the foliage. When repotting, for example, he advised breaking the terracotta pot so as to avoid crumbling and breaking up the ball of earth protecting the roots.

## *Pietro de' Crescenzi (1233–1320)*

The Middle Ages left us numerous works of philosophy, literature, religion and naturalism, which the authors intended as compendia of all human knowledge on the subject, according to an ideal, unifying model.

Pier de' Crescenzi, a contemporary of Dante, wrote a Latin compendium on agriculture "*Opus ruralium commendorum*" (1486) (Fig. 38.5), which was very successful in Europe and translated into many languages.

The book is Aristotelian (the theory of the recombination of physical properties: hot, cold, dry, moist, and the four elements—water, air, earth and fire) but also includes the teachings of the Arabs, with thousands years of experience in horticulture. All living things, wrote de' Crescenzi, were a form of balance between the four basic elements. Imbalance was a "corruption of the humours" (the equivalent of sap today). Agriculture was the effort to maintain or restore balance to plants and animals.

The most interesting volume is the second, "Della natura delle piante …". Pier de' Crescenzi describes the ancients' understanding of natural phenomena within

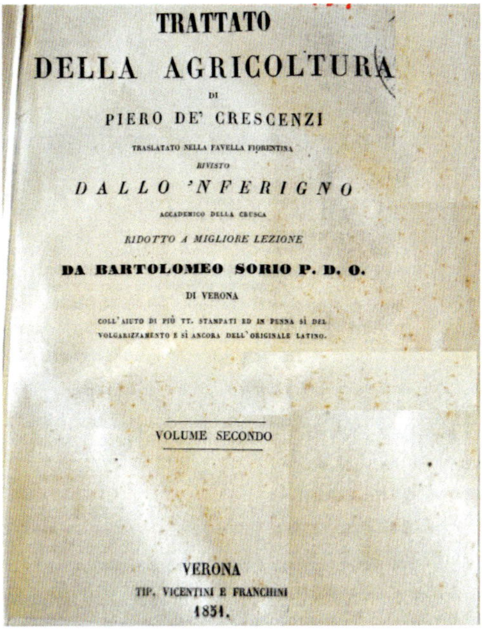


**Fig. 38.6** Title page of P. De Crescenzi's "Treatise on agriculture" (Trattato della Agricoltura), published in Verona (1851)

a philosophical framework (Saltini 1989). This inverted the theoretical hierarchy which put theology first, and hence the teachings of the Church; whereas de' Crescenzi had a scientific approach based on the material needs of man (Saltini 1989), which was out of step with the thinking of the time.

His work is an important source for the practices and understanding of agriculture in the Middle Ages. For example, propagation, a subject debated for centuries, is described on a biological basis, sometimes erroneously (in the light of today's knowledge), because authors tended to pass on what they had read in their source books, some of which went back to the Ancients.

The "Trattato dell'agricoltura" ("Treatise on agriculture") (Fig. 38.6) started in 1305 (1st edition) published in 1471, was consulted for at least five centuries (with translations into various languages). The author has consulted an Italian edition (De Crescenzi 1851), published in Verona.

De' Crescenzi cited many authorities. For the genesis of trees he used Aristotle, for whom their origin was from seeds or plants. In Chapter VIII he becomes a little obscure when he describes "the change of one tree into another, where it occurs, in the woods, is the fault of the roots" which have "narrow and closed pores" preventing the creation of new plants "of other species". Hence, "even fruit trees when cut and grafted with like trees will produce different fruit". Prunes and cherries grafted on willows trees would give fruit without stones. Vines grafted on cherry and pear trees "will produce grapes that ripen in the period of cherries and pears". Today, these statements raise a smile.

These errors were the result of a vivid imagination, and purely "virtual" experiments without any basis in fact, probably copied from a source. The author ignored the problems of grafting incompatibility between different species of trees. It is difficult to understand how such ideas were passed on for centuries, without being challenged and without correction in subsequent editions. Perhaps, no one noticed the oddities.

The book also contains however many valid and current ideas: a great deal of attention is paid by de' Crescenzi to agronomic practices like ploughing in the springtime, to be repeated twice in the summer (June and August) if the soil is dry. This was a recommendation of Palladium "clay earth must be ploughed deeper than healthy loose, sandy earth".

In Chapter XXII, de' Crescenzi left also good statements. He says that propagation by seed is dangerous "because it requires too long hoping" and the result would be a wild and not a domestic plant, whilst planting a shoot segment of a domestic plant (cutting technique) prevents wildness from spreading and hence he recommended this method (vegetative propagation) for vines, pomegranates and quince, and propagation by seed for stronger species such as walnut, almond, chestnut and peach; he also thought propagation by rooted sprouts (of shoots) valid for fruit trees bearing late, but forgetting the use of this practice for olive trees.

Chapter XXIII is dedicated to grafts; like Varro, he thought it an excellent method for propagating domestic but not wild trees, and quotes Columella's quaint idea that vines could be grafted onto elms which is recognized today as an impossibility.

In Europe, the end of the Middle Ages and the start of the Renaissance in the 1400s saw the fruit tree leaving the confines of cloistered convents and the walled gardens of patrician villas. Fruit trees were rediscovered, for their beauty, symbology and pleasing appearance. Fruit was now more widely grown and was now consumed by the wider population to whom it had previously been unavailable. This created a widespread, gradually increasing interest in the cultivation of fruiting varieties, but for aesthetic rather than purely commercial reasons. Landowners started to plant trees and discovered fruit and vegetables. The 1400s and the 1500s were the Renaissance period when the great schools of Flemish, German and French art produced the famous masterpieces of still life. Science, or rather the renewed study of plants, stimulated the publication of many books throughout Europe. Most of these were derived from Latin works or were encyclopaedias. Very few were true horticultural studies. We have chosen two of the most significant works of the time, written by Ulisse Aldrovandi (1522–1605) and Agostino Gallo (1499–1570) in Italy and Olivier de Serres (1539–1619) in France.

## 3rd Part: The Renaissance

### *Ulisse Aldrovandi, at the Dawn of Pomology (1522–1605)*

Dawn of Pomology was the statement used by E. Baldini (2008), illustrious professor of the Alma Mater Studiorum of Bologna, to describe the work of Ulisse Aldrovandi (1522–1605), written in Latin. Aldrovandi was an important naturalist,

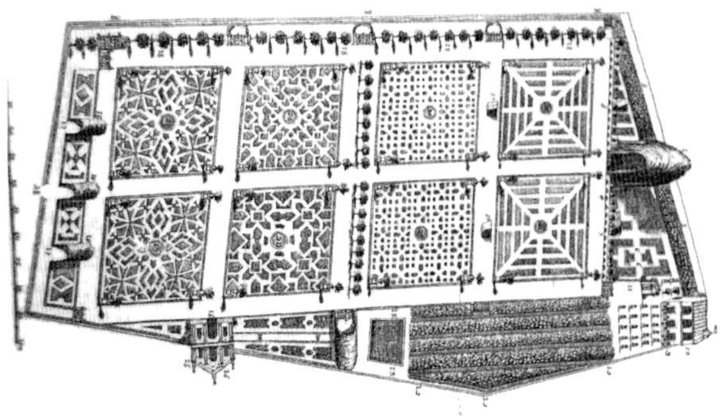

**Fig.38.7** Drawing of the "Giardino dei Semplici", one of the first botanical gardens for medicinal plants (Pisa, Italy, 1723)

Professor of the "Semplici" (plant varieties collection with medicinal properties) and founder of the first botanical gardens at Bologna University. See a dwarfing of that of Pisa (Fig. 38.7). His "*Hortus Pictus*" was, according to Baldini and Tagliaferri (1998), the first book of genuine pomological taxonomy. Aldrovandi left just under 2,000 drawings, the work of artists he paid himself, reproduced *in folio* in 10 volumes; they accurately represented all the varieties of fruit grown at the time, drawn and painted so precisely the varieties can be distinguished.

For example, the books illustrate ten varieties of apples 200 years before Linneo's botanical classification: white apples (*Malus alba*), red apples (*Malus rubra*); 'rusty' apples (*Malus ferrugineum*), ribbed apples with lengthened form (*Malus angulosa*), today not cultivated but known as 'snout apples'. Aldrovandi explained the name: "*Musabò vulgo dicta, quia rictus bovini in figuram habent*".

He illustrated the "Angelica" pear and listed and described the shapes of other varieties still common in various regions of Italy today: *Pyra moscatella augustan*, coinciding with cv Moscatella, *Pyra zucchella*, *Pyra viridia* (acidulous) and other varieties.

By correspondence he collected exotic fruit such as the exotic avocado, which he obtained from a Venetian traveller together with a drawing, included in his book.

He distinguished *Persica lutea* and *Persica alba*, i.e. peaches with yellow or white flesh, and described the pulp as more or less adherent to the stone, "*quorum es a pulpa separatum*" (cling peaches and freestones).

The tables include the very beautiful figure of a cedar (*Citrum piriforme*) and other citrus fruits. He categorized some "prodigious or monstrous" fruit, i.e. chimeras, mutations or the results of unknown grafts; for example 'bearded' grapes (*monstrifica barbis insignita*), then considered teratological, with a figure taken from the work of a German artist, C. Wolfhart (1557). Baldini thinks they were epiphytic threads of the dodder (*Cuscutum epythimum*), which may have affected vineyards, causing the appearance of the bunches illustrated.

Aldrovandi published a posthumous "dendrology" (1668) (*Dendrologiae naturalis scilicet, arborum historiae. Libri duo, sylva glandaria, acinosumq. Pomarium. Ubi, eruditiones omnium generum una cum botanicis doctrinis ingenia quaecunque non parum iuvant, et oblectant.* By *Ovidius Montalbanus*), included in a work of O. Montalbani who 'borrowed' material left by Aldrovandi. The book

**Fig. 38.8** Frontispiece of U. Aldrovandi's "*Dendrologiae, naturalis scilicet, arborum historiae. Libri duo, sylva glandaria, acinosumq. Pomarium. Ubi, eruditiones omnium generum una cum botanicis doctrinis ingenia quaecunque non parum iuvant, et oblectant.* (By *Ovidius Montalbanus*" edited by O. Montalbani (1667))

was translated into German and published in Frankfurt in 1690, with subsequent re-editions (Fig. 38.8).

A near contemporary of Aldrovandi, a Jesuit priest, Gian Battista Ferrari (Siena 1584–1655) wrote a treatise on citrus fruit with excellent illustrations, a forerunner of cytological studies, "*Hesperides sive de malorum aureorum cultura et usu*".

## *Agostino Gallo (1499–1570)*

Agostino Gallo was the author who probably exerted the greatest influence in the 1500s on the development of fruit growing in Europe. His best known work was "The ten days of real agriculture and pleasures of the villa" (written in Italian Gallo 1565). In practice, Gallo became the foremost advocate of vegetable and fruit gardens in patrician villas. Not only had the gardens to meet the new aesthetic and architectural canons of the time but they had also to provide for the pleasures of the table and the health of the gardener. They should produce medicinal herbs, hay and fruit from trees. Gallo possessed great experience, and the accuracy of his descriptions gives the impression that he could have been the designer of the things he described. He recommended the use of fertile soil and the importance of constant irrigation. He warned against overwatering because this would lead to an imbalance in the growth (root suffering) and ripening of a fruit. Fruit which was too large because it had been overwatered would have less taste. This principle still holds true today. The same can be said for his recommendation not to plant trees too densely, "If the plants are close together, when the trees are mature they will not only crowd upon each other but will cost upon the grow beneath so that it will yield little or nothing at all". Gallo also had much experience in planting methods and grafting. Clearly he was also superstitious, even to the point of suggesting that trees for transplanting should be removed from the ground towards evening and then only on the days when there was a new moon. Much debated by Gallo was the choice of rootstock. The alternatives were to use young wild plants from woods and uncultivated areas or the seeds of seedlings of cultivated varieties. He favoured the latter solution by far because it gave better, larger fruit. He even goes as far as to suggest regrafting (i.e. top working) the trees obtained in this way with scions from the same plant, for the purpose, as always, of obtaining still better and larger fruit. As Saltini shows, this is the first time that 'top working' is described in the literature, even though the purpose was different from that of today. Today, top working is performed on adult plants in order to change variety. Gallo, accurately, describes what types of grafting to use and how to perform them, using terms which are different from those of today: cleft, awl, budding and pipegraft.

Gallo is also worth quoting from another of his works, a pomological catalogue of apples and pears. In the introduction he writes that, "Pears are more delicate, sweeter and juicer than apples. They like sandy, gravelly soil; soil that is both dry and well-drained. … However, apples have a pulp that is more robust by nature … and they grow fine and large in rich, soft and humid soil".

Alessandro Saltini in his "*Storia delle Scienze Agrarie*—A history of agricultural science" (1984, 1987, 1989) writes "The use of grafting enabled the farmer to enrich his range of fruit trees, adding new varieties that he had encountered during his travels and from which he had been able to obtain scions. The result of patient research and courteous exchanges of species, the wealth of variety in his orchard was a source of pride for the landowner, who from late spring to autumn would thus be able to provide his guests with apples and pears straight from his orchards, the species that kept better were saved to serve both culinary and medicinal purposes during the winter months (in the sixteenth century, apples and pears figured not only in innumerable recipes but also as part of the recommended diet of those suffering from a wide range of illnesses). There were more varieties than those familiar to the modern supermarket shopper, whose choice is restricted to the few types that are to be found throughout the fruit-growing areas of the world".

Gallo goes on to describe citrus fruits, calling on his experiences in the areas overlooking Lake Garda in northern Italy. Gallo gave citrons, oranges, lemons and key limes (later Gallesio was to consider the key lime as a probable hybrid between the orange and the lemon) the generic name of Adam's pome. Gallo said they had two uses, first as food and second for the extraction of perfumes.

Gallo indicates also the species and varieties suitable for cultivation in protected struttures, (alias greenhouses). Here he describes how to build the protective structure with a mobile roof covering. He recommends removing the roof covering on sunny days but advises against uncovering too early at the end of winter because of the risk of frost damage. In his pomology, he complains that too often he had seen farmers lose their entire crop to the cold because they did not exercise sufficient care in the management of their "greenhouses". Probably they relied too much on the fact that the land around the shores of Lake Garda, in Italy and remained relatively untouched by winter cold.

## *Olivier de Serres (1539–1619)*

Olivier de Serres became famous in France towards the end of the 1500s for creating a model fruit farm on which his timeless work, "*Theâtre d'agriculture et mesnage des champs*" (De Serres 1600) (Agriculture and the cultivation of fields), is based. He introduced various species to France including the madder (used to make red dyes), the hop and the mulberry. He also introduced numerous varieties of pear, the best known of which bears his name and has survived to the present day. However, it is no longer cultivated and is only to be found in germplasm collections.

De Serres wrote how to build and manage gardens but not simply as a design exercise for the benefit of aristocratic families in the way that Gallo had done. For De Serres there was a second purpose—the sale of produce grown just outside cities (and therefore for sale in the suburbs) or grown close to waterways which enabled the rapid transport of produce to the marketplace. The sixth book of his *Theâtre* is entirely dedicated to the horticultural garden. There are four types: the "*jardin potager*" or vegetable garden; the "*bouquetier*" or flower garden; the "*medicinal*" or medicinal herb garden; the "*jardin fruitier*" or the fruit garden or orchard. The design of a garden, the form and layout of its trees, the layout of flowering plants and ornamental shrubs had above all to satisfy the wishes of the *mesnager* (the cultivator) but at the same time had to be striking in its beauty and effect.

De Serres, in particular, had a detailed knowledge of variability in genetic phenotypes, their varieties and species. He collected and catalogued these and incorporated this deep knowledge into his garden designs which often became a *potpourri* of colours, shapes and heights. His landscaping designs and the techniques he employed were widely recognised for their excellence. His reputation was such that his principal invention, the growing of trees on a frame flat against a south-facing wall, became the passport to success for his spiritual heir, Jean de la Quintinye (1626–1688). De la Quintinye, who was Louis XIV's consultant and administrator in Paris, later wrote a treatise on the techniques for growing and pruning of trees to obtain certain shapes. This became a guide followed for at least 200 years afterwards. It was a defining moment, clearly marking the transition from the tree growing techniques typical of Renaissance social aristocracy to the techniques which were to become widespread during the socialisation and development of agriculture for economic ends in the nineteenth and twentieth centuries.

A. Saltini describes how Olivier de Serres differed from his two predecessors, Estienne and Liébault, by abandoning what had until that time been the dominant tendency to see the garden as a "single harmonic design". Rather he preferred to offer a wealth of knowledge about the botany, pharmacology, comestibility and pomology of the single species cultivated. Medicinal plants, flowers, fruits, herbs and vegetables all had their proper place in the model vegetable garden, a place defined according to function. In describing these model gardens, De Serres makes reference to the large projects underway throughout Europe. He talks, for example, about the system of greenhouses and covers built in France and also at Heidelberg in the Palatinate. Mention is also made to the medicinal herbs cultivated in the *Giardini dei Semplici* botanical garden in Florence or Pisa and also to the herbs grown in Languedoc both before and after the time of French reign. When deciding the dimensions of a garden and calculating the potential amount of produce, De Serres characteristically advises that one should always take into account market demand. Obviously the need to provide water and fertiliser was assigned primary importance.

Saltini writes once again that Olivier de Serres was truly "the first French agronomist to draw up an overall survey of the experience that had been acquired—and was still being acquired—thanks to the passionate interest of French landowners in gardening". We can mention here at least two technical approaches which give us the ideas of how he improved the technical knowledge at that time:

1. First of all, to make a new orchard he wrote: "Many take still wild trees from the plant nursery and transplant them in the orchard, where they are grafted. Others, more skilful in this art, do the grafting in the nursery itself before replanting, so that the tree is *franc* when it is bedded in and the grower does not have to graft it at a later date. Indeed, going even further, they do not settle for grafting them a single time but come back repeatedly, to make them produce fine and valuable fruit trees". In this sentence there are two concepts: the *franc* plants (i.e. the seedlings) are grown from seeds and are to be grafted with the necessary scion in the orchard and not those that have "undergone no grafting at all". The second is the role, at that time, of 'top working' continuously used for the same tree, in a manner that each tree can change variety any number of times and at any time.
2. But the most important contribution of Olivier de Serres was on the tree forms (or shapes) and the related way of training the trees. In chapter XIX he discusses the *espalier* form as follows: "*Espalier* or *Palisade* [today "hedgerow"] is the name for that layout of orchards in which the trees, planted in rows, become intertwined with each other regardless of species, their branches, blossom and fruit growing in complete freedom, right up to the height that one decides upon … Reason argues and experience demonstrates that the fruit from an espalier is finer and better than that from other trees. This is because the trees planted in this manner have a much more abundant system of roots than of branches, given that in regulating the *espalier* one often trims back the vertical shoots … [This fact] works to the advantage of the fruit, which is more fully and freely nourished [by the roots and the foliage] and thus has a better quality. Though his description suggests an arabesque of different fruit trees intended primarily as a visual delight, De Serre's account does give precise reasons why *espaliers* produce better fruit than free-standing trees".

Today, as the reader knows, this theory it is no longer accepted as a reason for preferring *espaliered* trees (hedgerows) to full-foliage, free-standing trees planted at low densities. The taste of fruit and crop yields depends on orchard design and a host of pre-harvest and environmental factors and not just on the shape of the tree itself.

Probably, at the time De Serres was writing, the *espalier* produced better fruit simply because the *espaliered* tree was protected by the backing wall (no longer used today) which created warmer temperatures and thus higher photosynthetic activity of the foliage.

One last point the author wishes to emphasise about the pomologist De Serres is his description of how to build and manage an orangery dedicated entirely to citrus fruit growing in northern European climates. This time he did not copy Agostino Gallo who had previously described the coverings made in the Lake Garda area in Italy. He described a covering consisting entirely of glass, canvas, rigid pillars and brick walls (no plastic, of course). He made many recommendations to prevent cold damage. He confused the names of several species like lemons (limon), limes and the ornamental species *Poncirus trifoliata*. In his favour, though, it should be said that at that time botanists had probably still not fully classified all the numerous *Citrus* species being cultivated.

## *Jean Baptiste De La Quintinye (1626–1688)*

"Agriculture is certainly a noble art, able to pass nobility on to those who practice it as a profession… all can see and admire their work." This maxim by the ancient Greek Xenophon is one of the many cited by De La Quintinye to demonstrate the importance of the historical roots of horticulture, of which hc which he considered himself an expert practitioner.

De La Quintinye, who is pictured in Fig. 38.9, wrote his treatise "Instructions pour les jardins fruitiers et potagers" (Fig. 38.10 1746) after seeing the "lack of knowledge and understanding in so many of the books written before [his own], down through the centuries, in many languages and reaching very different conclusions". He appreciated very few, one of his eminent predecessors being the "Curé d'Enonville", who also wrote on fruit tree culture.

Another reason for his treatise was the existence of "too many botched orchards and gardens, where they have tried to imitate and apply my principles, without success… For example they have not understood why in some cases I cut the branches short and in others I leave them to grow long…But the principal reason I write is that I wish for 'my maxims' to be properly understood and not applied only partially, because in that case it would not be my responsibility if the results were poor."

De La Quintinye has the merit of describing how to design and look after gardens and orchards, including nurturing, soil and tree management according to the practices of the time, based not on economics but on aesthetics, and which could be appreciated at a glance by any visitor; an aim—he often repeated—entirely compatible with obtaining long-lasting fruit of excellent quality.

It is no accident that he begins his cited book by stating: "Whoever wishes to grow fruit trees or vegetables should have sufficient knowledge about how to do so" (an indication of the opinion he had of his readers). As the Director of Fruit and Vegetable Gardens under Louis XIV (Roi Soleil), he felt the serious responsibility of maintaining the gardens of Versailles as an example of French supremacy (Fig. 38.11).

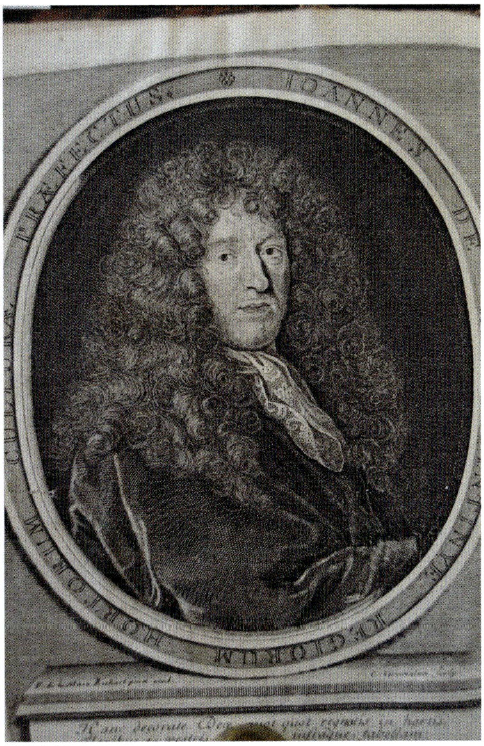

**Fig. 38.9** Jean Baptiste De La Quintinye (1626–1688)

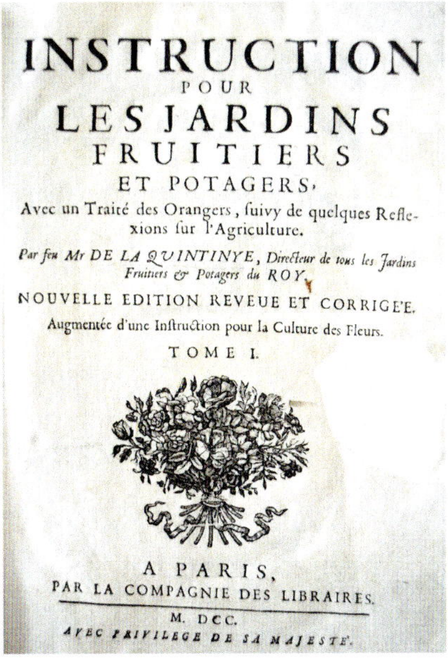

INSTRUCTION
POUR
LES JARDINS
FRUITIERS
ET POTAGERS,
Avec un Traité des Orangers, ſuivy de quelques Reflexions ſur l'Agriculture.
*Par feu Mr DE LA QUINTINYE, Directeur de tous les Jardins Fruitiers & Potagers du ROY.*
NOUVELLE EDITION REVEUE ET CORRIGE'E.
Augmentée d'une Inſtruction pour la Culture des Fleurs.
TOME I.
A PARIS,
PAR LA COMPAGNIE DES LIBRAIRES.
M. DCC.
*AVEC PRIVILEGE DE SA MAJESTE'.*

**Fig. 38.10** Title page of J.B. De la Quintinye's "Instruction pour les jardins fruitiers et potagers" (Paris, De la Quintinye 1746)

QUATRIEME PARTIE.
DES
JARDINS FRUITIERS
ET POTAGERS.

De la taille des Arbres Fruitiers.

PREFACE.

GENERALEMENT parlant tailler les Arbres c'est y couper des branches ; ainsi on dit pour l'ordinaire, qu'un Arbre est taillé, quand on y voit beaucoup de marques de branches coupées : On dit qu'un Jardinier taille, quand la Serpette à la main on le voit couper quelques branches à ses Arbres ; De tout temps cette taille a passé parmy les curieux d'Arbres fruitiers pour le chef-d'œuvre du jardinage. En effet ce n'est pas seule-
A ij

**Fig. 38.11** Incipit page with the drawing of Versailles jardins in De la Quintinye's Book (see Fig. 38.10)

These gardens were certainly of fundamental importance in building up knowledge on the cultivation of fruit trees, flowers and vegetables, based on direct experience, breaking away from the common place and principles of Latin authors until the eighth century (the so-called "Age of Enlightenment" in the agrarian field as in many others). Nonetheless, the book says little except how to prune trees, the various techniques being classified and illustrated, and how to plant and manage the trees. These two main topics have little in common and the book is far from an organic whole, despite some final thoughts on roots, sap, the movement of sap and grafts (for example, vascular transfer was described as possessing a network of "pores" that guided the sap within the fruit trees). The author made only a limited attempt to deal with all the factors contributing to success or failure. The author therefore dedicate more attention to a French author of the next century, M.Alphonse Du Breuil (1811–1870), whose merits in the field of horticulture include summarizing contemporary thinking, so that the understanding of tree culture at the time is very clear.

It is surprising that, along with the other defects of the book, La Quintinye should have written in such a prescriptive and self-referential fashion; at the end of the book there is even an Appendix praising the author, written by the famous writer, Charles Perrault, of the French Academy.

These brief remarks on La Quintinye would not be complete without his *alter ego* at the Gardens of Versailles, the architect André Le Nôtre.

Four hundred years after the birth of André Le Nôtre (1613–1700) Paris is dedicating a series of exceptional events to commemorate the great seventeenth century architect and inventor of the "French" garden, which replaced the Italian model of the previous century. Le Nôtre was in the service of Louis XIV for over fifty years, and created the Royal Gardens of Versailles. This masterpiece inspired park designers throughout Europe, striving for a balance between architecture and geography, geometry and design, to achieve aesthetic harmony with a touch of artistic imagination. He was the first to espouse the principle of inter-disciplinarity—what today we would call interaction—between agrarian knowledge (gardening) and city planning, architecture and engineering, hydraulics and art/sculpture. In Versailles, the park that rivals Kew Garden in London for the highest number of visitors in Europe is shaped by a "grand canal" in the form of a cross at the sides, from which tree-lined avenues extend, with sculptures, flowers beds, bushes, lawns, fountains and pools, all opening onto the city. Versailles highlights the aesthetic force of nature modeled by Le Nôtre, and hence the victory of culture and order over ragged and savage nature, as well as the triumph of the modern over the ancient.

## 4th Part: Take Shape the Whole Discipline of Horticulture. Birth of the "Pomologies"

### *John Evelyn (1620–1706)*

John Evelyn was the greatest and most colourful Englishman of the seventeenth century in this field. He was a prolific writer whose motto was "*Omnia explorate, meliora retinete*" (Explore everything, keep what is better). He travelled widely, above all to Italy, to avoid the English Civil War (1642 to 1651) only to become a Commissioner during the Second Anglo-Dutch War (1665 to 1667).

He was a great lover of trees, and wrote "Sylva: a discourse of forest-trees", with an Annex entitled Pomona (1664), reprinted as far as into the nineteenth century. He also denounced air pollution in London and was a member of the group that founded the Royal Society in London in 1660. He learnt horticulture from French literature and in 1658 published "The French Gardener: instructing how to cultivate all sorts of fruit-trees", translated from the French of N. de Bonnefons. In 1676 he published "A philosophical discourse of Earth", reprinted later (1693) with the title "Terra, the Compleat Gardener", a translation of the work by J. B. De La Quintinye.

Although not a professional horticulturist himself, he had great influence in disseminating knowledge of horticulture in the Anglo-Saxon world. Truly a man of his age, after his death he was celebrated and remembered for achievements in a wide range of intellectual endeavours.

### *Lajos Mitterpacher (1734–1814)*

This Hungarian author, a Professor at the University of Budapest, published a treatise on agriculture that was highly influential in Europe and translated into many languages. The author has the Italian edition, "Elementi di agricoltura" (Fig. 38.12)

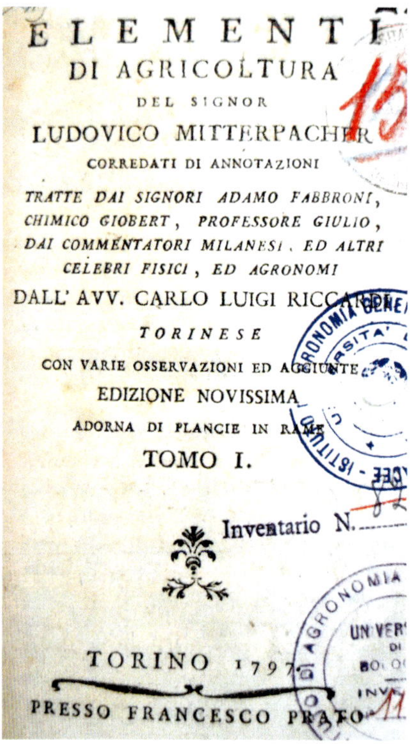
ELEMENTI
DI AGRICOLTURA
DEL SIGNOR
LUDOVICO MITTERPACHER
CORREDATI DI ANNOTAZIONI
TRATTE DAI SIGNORI ADAMO FABBRONI,
CHIMICO GIOBERT, PROFESSORE GIULIO,
DAI COMMENTATORI MILANESI, ED ALTRI
CELEBRI FISICI, ED AGRONOMI
DALL' AVV. CARLO LUIGI RICCARDI
TORINESE
CON VARIE OSSERVAZIONI ED AGGIUNTE
EDIZIONE NOVISSIMA
ADORNA DI PLANCIE IN RAME
TOMO I.
Inventario N.
TORINO 1797.
PRESSO FRANCESCO PRATO

**Fig. 38.12** Title page of the Italian issue "Elementi di Agricoltura" of L. Mitterparcher, integrated by G.L. Riccardi (Turin 1797)

in six volumes, published in Turin (Mitterparcher 1797), adapted and integrated by G.L. Riccardi. The author stated that many books on agriculture were then available and that: "Mine summarizes the ideas of others (and hence "I do not wish to be accused of plagiarism") mediated and filtered through personal experience".

Among the vegetable volume, he says that "none of the various species has a better taste and more fragrance than wild mountain strawberries" (*Fragaria vesca*). "Strawberries are advantageous every month, producing flowers throughout the year except in winter". Evidently, the diploid "reflowering plants" of *F. vesca* were not corresponding to today's "day neutral" everbearing strawberry (productive for several consecutive months). They are both cultivated. But the current strawberry belong to another 8-plopid species, *Fragaria* x *ananassa* (with big fruit), which through hybridization with another Chilean type, *Virginiana glauca* (carried out in California around 1950) could introduce the "day neutral" trait. This last type is grown in California and South Europe to produce strawberries continuously for 4 to 8 months. During the Mitterpacher age, strawberry beds lasted 3 to 4 years and were also used to produce runners for propagation. Today, in most plantations annual and propagation bed are separate.

Volume three is dedicated to vineyards; it describes the best choice of environment and soil, and lists the moist common Italian and French vines and the wines produced from them. Most of the book deals with the propagation, grafting and pruning of vines. For the former, the author recommends self-rooting through layering, with the production of plants in own field rather than the purchase of cuttings or rooted plants, because "they involve more risks" (evidently at the time nurseries were unreliable, unlike today's certified producers). He offers many tips on pruning. For example, Mitterpacher recommended shortening shoots during the year when the harvest was bountiful and leaving more fertile buds when it was less so.

Fruit trees are given less space (just 80 pages) than vines (over 200 pages); this was partly because, at that time, growers needed to defend themselves against thieves and partly because of the widespread belief that "fruits obstacle the fields and vines". Certainly this was the belief in Piedmont, as the integrated edition in Italian makes clear. The author complains of the poor quality of fruit in that region.

Mitterpacher recommended cultivating only grafted plants, obtaining the wild seedlings from nurseries (seedbeds), a system not used much in Hungary (where, as elsewhere, nurseries had a bad reputation). For grafting, he recommended the techniques of Latin authors, distinguishing between those "with a segment of budded shoot, between bark and wood", or "with cut and cleft of the stem" (in winter), or summer mono-bud shield (today all the same, with few variants). His description is valid today; he indicates a series of precepts, above all for crown grafting. He clearly describes offshoots and layering (as for vines) and illustrates transplant techniques. When pruning, he rightly says there is a difference between fruitful and non-fruitful branches (and shows the various types).

Some tips seem suited to today. For example, "Sometimes there is a top branch that flourishes more than others and this should be cut for the benefit of the whole tree". Cutting is considered more important (and essential) in some countries (e.g. France) he says, whilst in Italy "too much old wood is left". Strangely, he also deals with the tools used for pruning (numerous and with different shapes then as now).

He briefly examines the needs (above all environmental) of various species and distinguishes between growing tall trees and isolated trees against a wall (espalier), suitable mostly for stone fruit which can be harmed by the cold during springtime. He recommends cling peaches and the green pruning of peach trees, and the thinning out of fruit, advice which is still valid today.

In Europe, in the eighteenth and until the mid-nineteenth century, books flourished illustrating varieties of fruit, a discovery of the natural beauty of the world and hence an aesthetic, artistic and natural symbol for the bourgeoisie of the time. In almost all countries, under the influence of the French Enlightenment and the new discoveries of science, many "Pomologies" were published, often crossing over national borders.

Significantly, the Pomologies became a response to developments in art, where the "still life" painting of various naturalistic schools, the Flemish, Dutch, German and French, turned their attention away from overtly sacred works and towards the realistic representation of vegetables, fruit, meat, and game, with such precision that sometimes the specific species of the fruit can be identified (Sansavini 2004).

The iconography that made individual varieties perceptible and recognizable, as enshrined in the Pomologies, was often the work of professional artists familiar with still life painting, able to paint or draw with a high degree of precision.

POMONA
ITALIANA
OSSIA
*TRATTATO*
DEGLI ALBERI FRUTTIFERI

CONTENENTE LA DESCRIZIONE DELLE MIGLIORI VARIETA' DEI FRUTTI COLTIVATI IN ITALIA, COLLA LORO CLASSIFICAZIONE, LA LORO SINONIMIA E LA LORO COLTURA, ACCOMPAGNATO DI FIGURE DISEGNATE, E COLORITE SUL VERO, E PRECEDUTO DA UN TRATTATO ELEMENTARE DI POMOLOGIA.

OPERA
DI GIORGIO GALLESIO
AUTORE DEL TRATTATO SUL CITRUS, E DELLA TEORIA DELLA RIPRODUZIONE VEGETALE

*DISPENSA* 41

PISA
PRESSO NICCOLÓ CAPURRO
CO' CARATTERI DE' FF. AMORETTI
MDCCCXVIII

**Fig. 38.13** Title page of "Pomona italiana". (By G. Gallesio, Pisa (1817–1839))

One of the most famous illustrators of fruit varieties in the seventeenth century was Bartolomeo Bimbi, Court painter to the Grand Duchy of Tuscany, who left an unrivalled series of paintings illustrating pomology, including all species of fruit, totaling hundreds of different varieties, all of them immediately recognizable (via the copies in cartouches). They are currently on display in the Florentine museums of the Casate dei Medici (Baldini 1982).

## *Giorgio Gallesio (1772–1839)*

At the beginning of the nineteenth century, an Italian Pomologist, Giorgio Gallesio, with the help of several painters involved, estimated as a total of 10–15 drawers which documented and influenced strongly the work of fruit variety description and illustration.

The alliance between painters/drawers and pomologists was particularly evident in the masterpiece "Pomona italiana, ossia trattato degli alberi fruttiferi" (G. Gallesio 1839) (Fig. 38.13) where the author, a self-taught pomologist "sui generis" worked with unknown artists of immense talent to create the plates. Gallesio became a pomologist late in life, when serving as a Magistrate with some political connections internationally. He wrote a number of volumes, some left unfinished. They include "Traité du Citrus" which he published in Paris in 1811 and "Pomona italiana", published in Pisa over a period of 20 years, from 1817 to 1839 the date of his death.

**Fig. 38.14** Giorgio Gallesio (1772–1839)

Gallesio (Fig. 38.14) astonishes first and foremost because he was the standard-bearer for "periodicals". His work comprises forty slender volumes (compared to the planned 45), published over the years and mailed to well of 1,000 subscribers. The work was singular and was, in fact, singled out for its beauty and the fact that Gallesio had financed the project entirely through sales, without a sponsor.

For each variety discussed, Gallesio provides a wealth of "reasoned" description, a plate illustrating the variety in colour, perfect even today, bearing some similarity to hyper-realist art or artistic photography. It is somewhat strange to note that Gallesio the researcher did not keep a profile cards for the varieties he discussed but made informal notes along with his travel diaries (descriptions of what he had seen), which he wrote every day. He wanted to preserve his memory of things for posterity. He then made scientific descriptions of the varieties with the help of the technicians and specialists he met during his travels.

One of the reasons he was willing to get into debt to publish his work (his son accused him of squandering the family fortunes) was his outrage at the way current knowledge of orchard varieties was being lost in Italy, and the lack of appreciation of biodiversity. So he decided to get to know as many varieties of fruit as possible (one of the reasons it took him 20 years to complete his work). He began with citrus fruit and figs, then went on to apples, pears, peaches and lesser stone fruit, ending with grapes for the table and for wine. He made comparisons in an effort to settle questions of synonyms and homonyms. As before, in the study of pomology, he was not an expert linguist and earned himself a severe reprimand in 1820 for his use of foreign expressions, meaning non-Florentine words, in his case from Liguria; "in

**Fig. 38.15** Drawing of the apple cv "Astracan" included in Gallesio's "Pomona italiana" (Pisa 1839)

Tuscany—the criticism read—we appreciate the grace of the language, not defects and negligence". Some of the fruit he described is still grown today, albeit in a niche ecological market; they include Angelica and Spadona pears, Brogiotto and Dottato figs, Astracan (Fig. 38.15), Renetta and Carla apples, Marzemino, Brachetto and Nebbiolo grapes, and Claudia plums (Baldini 2004).

Gallesio wasn't only an important pomologist; he was a biologist ante-litteram, as shown by his "Traité du Citrus", the first chapter of which discusses plant reproduction. It was a controversial attempt (which met with some resistance) to establish the taxonomy and origin of citrus fruit in an effort to correct the mistaken and incomplete classification of Linneo on citrus fruit. In the book, however, Gallesio widens his horizons and attacks certain empirical prejudices, in particular the belief that "the nature of can be modified by cultivation, climate and soil", as many proclaimed. The error came from the fact that the best fruit was not wild or obtained from a seed, but was the result of grafting (Baldini and Tosi 1994).

Gallesio was the first to use the concept of "cloning" and vegetative material i.e. clonal propagation, as we now call individuals derived from "a single genetic source" (the mother plant) "multiplied by grafting over multiple feets" (rootstocks) or by offshoots "without being subject to change". This dismantles the old idea of conventional and ancient agronomy, according to which grafting was "able to transform the characteristics of fruit species". Since he had personally crossed fruit and created hybrid varieties, he was able to show that the genetic "oddity" of a citrus plant producing more than one species was a "grafting chimera".

In this connection, Gallesio wrote: "generations vary to an infinite degree but individuals do not change … nature has established the forms of living things". "The seed perpetuates the species, but is the source of the varieties … even where the characteristics are poorer than in the mother plant, these phenomena were observed in seeds from plantations with a mixture of species and varieties".

The key to the phenomena of reproduction was therefore to be found in the "law of identity" together with the "law of recombination", as he defined them. Gallesio suggested that "two different principles must be at work in the reproduction of all organized beings".

Darwin was an enthusiastic reader of Gallesio's work more than 50 years later, attributing to him the accolade of "precursor of the new conception of reproduction" (Saltini 1989).

## *M. Alphonse Du Breuil (1811–1890)*

The knowledge acquired by De la Quintinye, the arch-strategist of pruning, was passed on throughout Europe, on the basis of his reputation and position as Director of Gardens for Louis XIV (1643–1715) and, in France, was inherited a century later by Professor M.A. Du Breuil, who wrote several books, particularly a treatise on the "Cultures des Arbres et Abrisseaux à fruit de Table—Cours d'arboriculture" in many editions, including a volume on "Les vignobles et les arbres à fruit à cidre" (Fig. 38.16), used in universities for courses on tree culture and reprinted so many times that it became a sort of standard textbook in Europe for many years.

And, truly, Du Breuil succeeded in making" tree culture" (or arboriculture) a separate discipline. For the first time, his book described not only trees fruits and their classification (pomology) but also the concept of tree culture as the interaction of external soil and climate conditions (soil, temperature, sunlight, exposure), whose influence can be seen often at the level of a single species, as the outcome of knowledge and direct observation.

He studied trees not only as the producer of fruit but also for other functions (for wood and ornamentation). He also considered the vocation of the local environment and hence the reason why each species requires a particular kind of environment. He understood the importance of trees for ecology, their positive influence on the environment and their contributions to the health and wellbeing of mankind.

In relation to the discipline of fruit tree culture, Du Breuil (1863) begins by considering organography, which he believed was as necessary to understand as vegetable anatomy and morphology, in order to grow trees. This was because organography taught the "functions" of the tree, which were important to know in order to manage the tree, utilizing information from another botanical science, vegetable physiology, then a nascent discipline, able to help those interested in the cultivation of trees. For example, he wrote: "how can a tree be properly replanted if we do not understand the functions of the roots?" or "how can we prune a tree grown in trellis formation if we do not understand the importance of the leaves for the nutrition of

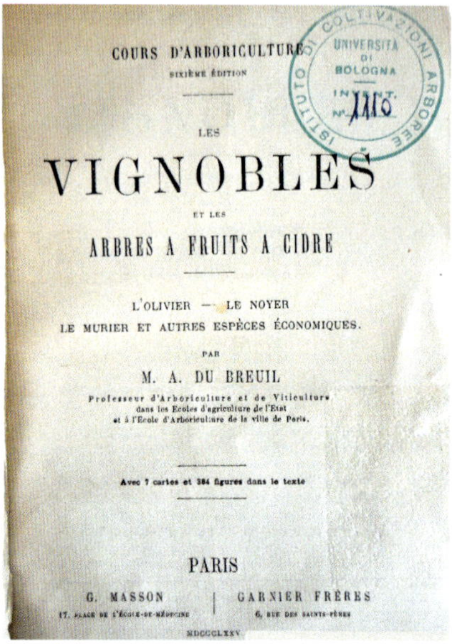
COURS D'ARBORICULTURE
SIXIÈME ÉDITION

LES
VIGNOBLES
ET LES
ARBRES A FRUITS A CIDRE

L'OLIVIER — LE NOYER
LE MURIER ET AUTRES ESPÈCES ÉCONOMIQUES.

PAR
M. A. DU BREUIL
Professeur d'Arboriculture et de Viticulture
dans les Ecoles d'agriculture de l'Etat
et à l'Ecole d'Arboriculture de la ville de Paris.

Avec 7 cartes et 384 figures dans le texte

PARIS
G. MASSON | GARNIER FRÈRES
17, PLACE DE L'ÉCOLE-DE-MÉDECINE | 6, RUE DES SAINTS-PÈRES
MDCCCLXXV



**Fig. 38.16** Title page of M.A. du Breuil's "Cultures des Arbres et Arbrisseux à fruit de table", vol. III. Vignobles et les arbres à fruits à cidre, published in Paris, 1875

the fruit trees?", and hence the need to restrict their number close to harvesting time in order to ensure that the fruit achieves a good colour.

The work of Du Breuil can also be appreciated for its beautiful drawings illustrating the object of study, almost at the level of photographic perfection (photography not having yet been invented). He also provided excellent drawings of laboratory experiments illustrating the physiology of trees (for example how roots absorb water, endosmosis, sap pressure, and growth) and field drawings of grafts, pruning and tree management.

In relation to reproduction, Du Breuil introduced data about "inter-specific hybrids". A number of pages describe the ripening of fruit, specifying the role of the organs, and contemporary practice for improving the quality of fruit is fully described, including "ring incision" of the branches (an idea of Lancry in 1776).

However, in the field of the propagation and arrangement of trees, the treatise astounds even today for its description, in physiological terms, of the way trees react to various kinds of treatment. For each main species (apple, pear and peach) he provided such detailed information as to comprise a gardener's handbook. In many cases, the techniques described are valid today, for example where he recommends planting pre-formed trees in nurseries if they are over a year old and have the proper rootstocks, a practice which is maintained today, although at that time there was no selection process (e.g. using quince for pear trees, paradise apple for apple trees, peach seedling, wild species and plum for the peach trees, and *P. mahaleb* for cherry trees). He also recommended replanting young trees in the nursery at times when the roots could not be affected by cold, winds or dryness.

He also set out numerous rules for designing orchards and the distance between plants. The principles he used for pruning are now out of date, with branches reduced to a few buds, making the process very lengthy indeed. He provided tables for each variety recommending where to plant (full wind, trellis, recommending a Verrier formation) and the exposure of the supporting walls or trees to the east, west or both. The geometries of tree skeletons (frame of the canopy) were extremely accurate and can be seen today in living museums, especially for the pear tree (a very plastic species) in for example East Malling (UK), Bologna (Italy), and Skierniewice (Poland). The chapters dedicated to defending the trees are highly detailed, with instantly recognizable figures of insects and diseases (e.g. pear clothes-moths and apple tree leaf borers). Defence was solely biological such as the harvesting and distribution of hibernating forms, collection on the ground and burning of the affected leaves, to prevent or reduce subsequent attacks and protect future generations.

The recommendation to set up small storehouses after the harvesting of the fruit is somewhat strange, with entire rooms in large villas given over to the ripening of fruit to prolong its shelf-life. This was based on ventilation, the creation of space between the lattices, the circulation of air in the room, and so on. Du Breuil also discusses harvesting, post-harvesting and packaging according to the type of fruit, hard or soft. He was really a great scientific innovator.

## 5th Part: The Modern Age

### *Hugo de Vries (1848–1935)*

A Dutch botanist, he was considered one of the most eminent naturalists of his age, earning him in some quarters the nickname of "Darwin's successor" (Hus 1906) and, indeed, he was famous for his studies of heredity and the "theory of mutations", integrating Darwin's idea of natural selection.

He was an *enfant prodige* in the study of natural sciences, carried out first at the University of Leiden, where he specialized in plant physiology (inventing the measurement of osmotic pressure via plasmolysis). After graduating in 1870, he studied in Germany (Heidelberg and Wurzburg), publishing a work that made him famous on the "mechanical causes of cell stretching". He became Professor of Botany at the Amsterdam Institute and even to give lectures abroad including University of Berkeley, California, in the United States.

To understand the man, suffice it to say that when given a lifetime achievement award he used the money to build a greenhouse.

In his book "Die Mutations Theorie" (vol. I and II, Leipzig, 1901/1903), 50 years after Darwin's theory of evolution, he set out to contradict popular opinion that species slowly evolve into "new types", arguing for the importance of mutation according to which "species and new varieties originate in pre-existing forms by sudden jumps. The original type remains unchanged, but can repeatedly give rise to new forms, simultaneously and in groups or separately in periods that may be close or distant". The following considerations are related to his book "Specie e varietà e loro origine per mutazione" (1909).

De Vries intended his work to be "entirely in agreement with the principles of Darwin", perfecting Darwin's much vaguer intuitions regarding the role of mutation in natural selection. A contemporary of de Vries, T. Hunt Morgan (1866–1945), another important researcher, published "Evolution and adaptation" (1903), which provided an accurate critique of the speculations concerning the theory of heritability. He starts from the premise that, in nature, species such as apple and pear trees have a huge genetic variability which could be used to create new varieties. He immediately discarded the hypothesis of De Candolle, Darwin and others, according to which varieties "owe their origin to the direct influence of cultivation", espousing the theory of the Belgian researcher J.B. Van Mons (1765–1842) (see "Arbres fruitiers ou Pomonomie belge"), a breeder himself, who demonstrated how over several generations of cross-breeding, new varieties could be created suitable for cultivation. He crossed varieties of apple trees already cultivated with several interesting forms of apple tree from the wild (which he chose in the Ardennes, which is in the south-east of Belgium) for features such as a non-woody pulp, aromatic scent, good taste and the size of the apple. For apple trees he thought two or three generations would be enough. In the USA, another formidable researcher, L.H. Bailey (1858–1954) (who left us the definition of variety), confirmed his ideas, with the seedlings of hundreds and thousands of seeds of cultivated or wild apple trees, originating varieties such as the "Wealthy" in the United States.

For pears he thought the same procedure could be used but that it would take more time, at least 6 generations to improve a new variety. It was still commonly believed that all the characteristics of the new varieties were already present in nature.

The true discoverer of the selective method by the choice of parental lines was Luigi Vilmorin who, towards the middle of the nineteenth century (working first with beetroot), applied selection procedures as they had been developed for animals. Virgil already believed that selection was necessary to keep cereals pure and prevent degeneration. In practice, what was being done in Europe and the USA was what today is called "mass selection".

Hugo de Vries believed the difference between "elementary species" (subspecies) and "varieties" was very distinct, because the latter "propagate by seed and are of pure, not hybrid, origin". We only knew the latter because they were cultivated. Where de Vries failed to fully understand mutation was in his belief that individual variations (now called sport) identified by seedling were or could be "a sudden transformation in the structure of the species and not a hereditary characteristic". Despite the many observations he had made and experiments he had carried out, de Vries was not immune from spurious theories such as the notion that some characteristics, for example the spinescence of apple trees or the florigenic differentiation of buds was influenced or determined by the climate ("thorns are lost in moist air"). These interpretations, which today we would call "ecotypes" were never confirmed for the agamic propagation of fruit trees. Nonetheless, de Vries, despite reservations about forms that exhibited different kinds of variability, or the influence of the "vicinity" on "retrograde varieties"—theories that have been disproved—distinguished between "mutations" from "phenotypical variants" manifest in hybrids and more generally the "variability manifested with the observed segregation in seedlings derived from crossbreeding". In other words, he didn't confuse mutants and the results of crossbreeding and hence the segregation of hereditary characteristics.

## *Luther Burbank (1849–1920) and Girolamo Molon (1860–1937)*

The author would like to present two authors together, the first American, the second Italian, who distinguished themselves in different but complementary fields: breeding and pomology. They are L. Burbank (1849–1926) and G. Molon (1860–1937).

Luther Burbank, the only important American breeder represented here, was a pioneer in the introduction of new varieties, which radically changed fruit growing. He was the most important twentieth century breeder in the USA, with almost legendary status. Before him, there were hundreds of "sowers" who selected plants at the time of fruiting. For example, from the millions of anonymous seedlings of the apple trees in the USA, varieties were selected which are still common today such as Golden Delicious and Red Delicious. Burbank, however, went much further, going back to the crossbreeding piloting the choice of parents. Luther Burbank lived and worked in Santa Rosa, California, near Sebastopol, the name of one of his famous plum varieties with red pulp, still grown in Europe and the USA due to its unrivalled taste and aroma.

We can see Burbank through the eyes of the Italian pomologist Girolamo Molon, his potential opposite number in the form of his own records of a journey and meeting between the two, which took place in the summer of 1912, which became a momentous occasion.

Molon described Burbank graphically as: "a short man, lean, with a lively, restless look in his eye. He was dressed modestly and had on a huge wide-brimmed hat hiding white hair ... he lives in a small chalet ... opposite which there is an office next to a small nursery ... all his work is kept there ... my conversation was brief, vague and inconclusive... He sang the praises of his *Opuntie* (prickly pears) without thorns ("I saw and tasted many"), his new plums and a very early sweet cherry... He spoke of the immense task he was carrying forward, publishing studies and discoveries about fruit growing (he has reached the 12th vol. of a work I could have bought in a shop not far from his chalet)... At the end of the visit and fruit tasting, he shook my hand and bustled off hastily".

Burbank left a practically unrivalled heritage. He created hundreds or varieties (about 800, including 200 fruit varieties), mainly of flowering plants and was the first to experiment with hybrids of the *Prunus* species, today documented on the web (Stansfield 2006)

In 1960, his nursery was turned into a "memorial park" dedicated to his memory, a testament to his formidable contribution. Dozens of schools in the USA bear his name. In 2004, Professor Jules Janick, wrote in the World Book Encyclopedia, that "Burbank cannot be considered a scientist in the academic sense" i.e. his results were based on a strong intuition and imagination, coupled with hard work. Janick surveyed Burbank's publications, from 1893 with a catalogue of his new varieties, "New creations in fruits and flowers", followed by a series of works in 12 vols, "Methods and discoveries and their practical application", inspired partly by

Charles Darwin on domestication and by Gregor Mendel on interspecies hybridization, which he developed and verified in person. He also knew of the experiments of G. Harrison Shull, who in 1908 had discovered and described the "hybrid vigour" derived from heterosis (still used today for the creation of hybrid maize, increasing yields from 1,000 to 10,000 kg per ha along the century). Burbank was among the first to understand the principles of genetics, which became applicable in the next century, including the possibility of the segregation in hereditary populations of certain characteristics of strength, tolerance to disease or abiotic stress.

The species he manipulated most was *Prunus* (he created 113 plums and prunes), followed by "fruiting cacti" (35), small fruit (16 blackberries and 13 raspberries), nuts (69), and hybrid plumcots (11). Sadly, he died four years before the law recognized the rights of the creator (Plant Patent Act 1930).

He was a profound nature lover, writing: "What a joy life is when you have made a close working partnership with nature, helping her to produce for the benefit of mankind new forms, colours and perfumes in flowers which were never known before; fruit in form, size and flavour never before seen on this globe … new food for all the world's untold millions for all time to come".

Girolamo Molon was won over by Burbank and the meeting with him, which he described in a publication 6 years later which was delayed by the Great War of 1914–1918). In the meantime he had collected a large quantity of statistical and commercial data on his travels in America (1912), which began in New York. He gave detailed descriptions of his visits to markets and cities, as well as research stations, that were responsible for the birth of the new American fruit industry. He was particularly struck by the aims of responding to market stimuli, almost unknown in Italy and the rest of Europe. Molon wrote that in Italy and Europe, at that time agriculture was "in a primordial state".

Molon was already a distinguished professor at Milan University when he visited America for several months in order to compare fruit growing on the two continents (first he visited Amsterdam, London and Liverpool). His international report was requested by the Italian government, which wanted to understand how the Americans had made such rapid progress, for example in creating specialist industrial crops, the importance of processing after harvesting and how the fruit industry, was so much more competitive than in Europe. Molon was impressed by the research centre in Geneva, New York (created in 1880) where the great pomologist U.P. Hedrick had 230 acres at his disposal, with a huge collection of the germplasm of temperate species, vines, apple and pear trees and all the varieties of *Prunus,* down to small fruit, for which pomological data was compiled each year, becoming the profiles in the centre's volumes of the Pomona series. In Geneva, Molon wrote, they are "fervent recorders of data" for all the development phases of morpho-phenological characteristics (including production quantities), employing a staff of fifty. "All the information gathered will be used to improve and transform horticultural production in many regions of North America."

Today, the words sound like a prophetic criticism of the failings of southern Europe, particularly Italy, a founder member of the European Union (1950s). Molon wrote: "The Italian governing class is convinced that such is the natural environment of the country, with its extraordinary potential for the production of food characteristic of the Mediterranean climate....that its typical products (olive oil, wine, fruit and vegetables, cheese, sweetmeats) are of such quality (so think Italian politicians and economists) that a bright future is assured in international markets, if not now then later." But at the beginning of the twentieth century, European agriculture and its fruit industry was already facing a crisis from international competition and globalization (as many specialist European journals warned), so much so that in 1902, Molon published a long article in the journal "The Chrysanthemum" praising American technology and its commercial power. He wrote: "competition with the US fruit industry may soon led to conflict which Europe is destined to lose" (at the end of the nineteenth century, the USA already exported 30,000 t of apples, prunes and dried fruit, the equivalent of 120,000 t of fresh fruit, leading Molon to believe the USA was the future world leader in the production and marketing of fruit.

Molon's contribution as a pomologist with an understanding of the varieties present in Italy began after 1892–1893, when the Ministry of Agriculture commissioned the dissemination of a list of pears and apples recommended for northern Italy. In 1901, Molon, then Professor of vine, fruit and vegetable growing at the Regia Scuola Superiore of Agriculture in Milan, published his first edition (in the Hoepli series of handbooks) of Pomology (Pomologia, descrizione delle migliori varietà di albicocchi, ciliegi, meli, peri, peschi) (1901) (Fig. 38.17). For each species (apricot, cherry, apple, pear, peach) he gave a brief botanical-taxonomic description, and then described the varieties from the most well-known areas of European pomology, with tables on nomenclature, classifications, ripening periods and observations on accession. It was intended as a long list of fruit varieties, and the book became an invaluable source of what was being grown in Europe at the beginning of the twentieth century. The descriptions were valid for the period and the terminology used was that of a professional pomologist. Molon does not touch on cultivation techniques unless it is to mention, criticize or praise what he had seen with his own eyes.

It is appropriate to underline that in the same age of Molon a new representative way of fruit collection was the wax and material for modelling. One of the best, which helps to recognize the single varieties was that of Garnier—Valletti (Fig. 38.18), still visible at the University Museum in Milan.

## *Ulysses Prentiss Hedrick (1870–1951)*

Ulysses Prentiss Hedrick, a native of Michigan, in the USA, was both a botanist and horticulturist. He acquired a culture in pomology in the 10 years he spent at three American Agricultural Colleges (Oregon, Utah and Michigan) before moving on to the New York State Agricultural Experiment Station, New york where he was awarded a PhD (Hobart College) from the Michigan Agricultural College and where he worked uninterruptedly until the end of his career (1937). He was also Director of the Geneva Station in the last 27 years.

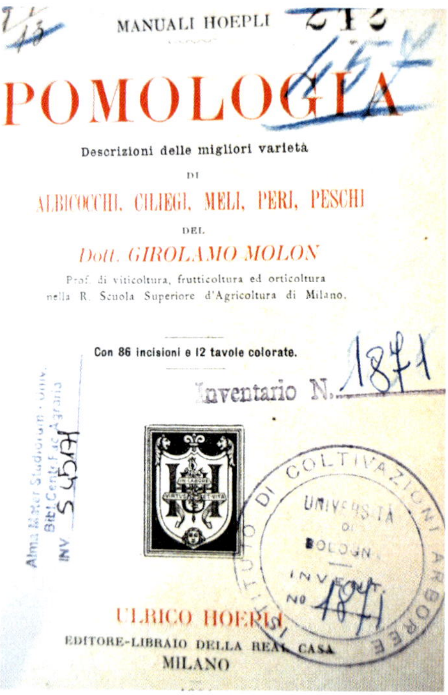


**Fig. 38.17** Title page of G. Molon "Pomologia" (Molon 1901)

He was the first and greatest of the American pomologists. The pomologists, American or European, who came before him were self-taught, whereas he had the right scientific background and a huge collection of species and accessions from the world over.

Hedrick was a sort of ideal "pomologist", motivated by a single mission: to disseminate the systematic classification of fruit trees, with the final aim of the recognition of individual varieties from not only by specialists but by the general public.

Hedrick wrote: "systematic classification must be a means not an end". For him, descriptive pomology needed to begin with the botanical understanding of each species, and reach not only their morphology but physiology as well (fruition habitus, resistance to cold and biotic adversities etc.). The ultimate aim was to encourage the cultivation of the best varieties and to improve the understanding of interactions with the environment, leading to genetic improvements and the appropriate choice of growing practices.

His six volumes, dedicated to six species ("Grapes of New York", "Plums", "Cherries", "Peaches", "Pears", "Small fruits", all with the New York suffix) took him seventeen years to complete, from 1908 to 1925. They are documents of immense historical importance. The monumental tome on pears (640 pages), in particular, became an indispensible text book for researchers throughout the twentieth century.

When Hedrick published "Pears of New York" (1921) (Fig. 38.19), the *Pyrus* species included both pears and apples, although botanists already differentiated.

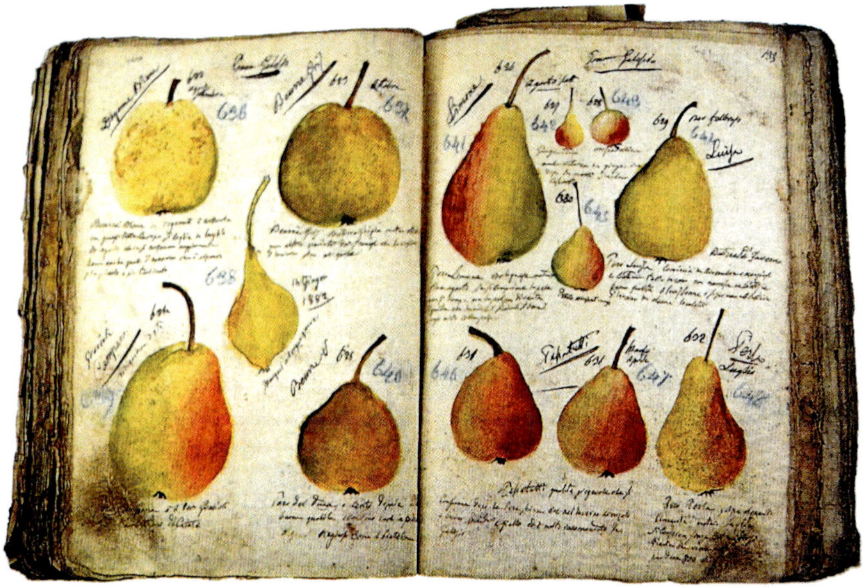

**Fig. 38.18** Garnier—Valletti drawing (1891) of pear varieties utilized to sculpt a collection of wax fruits (University of Milan's Museum)

Describing the *P. communis* as cultivated pears, Hedrick cited other well-known pear species, notably the *P. betulaefolia* (rootstock) and *P. serotina*, considered by Alfred Rehder (1863–1949), a famous European botanist, Hedrick's contemporary, the leading Oriental pear species (Lindley thought Oriental pears in the *P. sinensis* species were no longer cultivated). Rehder also distinguished two botanical varieties of the *P. serotina*, *stapfiana* (pear-shaped) and *culta* (apple-shaped). Hedrick described the latter in detail; he believed it had come from Japan but originated in China, where they were called Sand pears. The current name "Nashi", of the *P. serotina* varieties, very common in the United States, was not used a century ago. Hedrick describes the differences to European pears, in particular the fact that the calyx is deciduous. He lists three hybrid varieties of *P. serotina* and *P. communis*—"Kieffer", "Le Conte" and "Garber", all commercially available in the USA.

From the historical perspective it is interesting to note that unlike apples, where Europe is in debt to the USA for the many varieties obtained (Red Delicious, Golden Delicious), the 80 varieties of European pear described by Hedrick were mostly brought to America from Europe and often have their original French names (Beurré Bosc, Doyenné du Comice, Dr. J. Guyot), Belgian names (Beurré d'Anjou), English names (Bartlett or Bon Chretienne William), and some with American names (e.g. Seckel), all of these still in common use, particularly Bartlett.

Hedrick described all the temperate deciduous species, pome fruit, stone fruit, vines, strawberries and other small fruits, adopting an annotation system from his own observations. In his collection, in Geneva, he brought together the most well-known varieties and the newer ones introduced from Europe and the rest of the

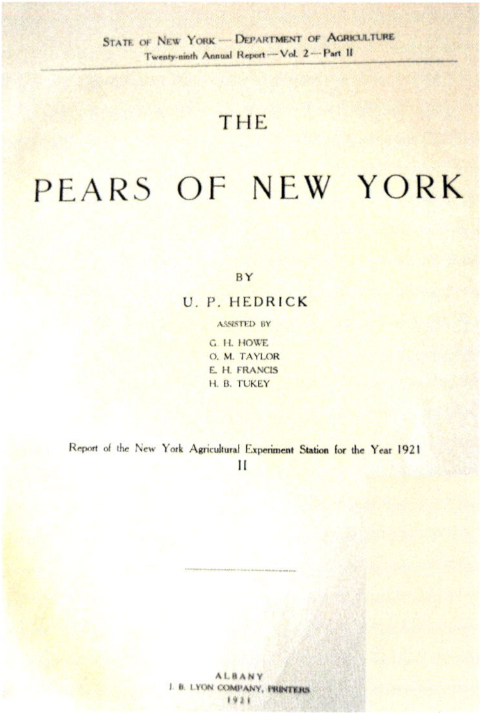


**Fig. 38.19** Title page of U.P. Hedrick's book "Pears of New York" (Hedrick 1921)

world. He noted that the work of a pomologist, irrespective of its scientific value, often failed to reach Colleges, Universities and practitioners as he would have liked. He therefore decided to disseminate the information himself. In 1925, he published "Systematic pomology" (Fig. 38.20), reworking a book he had published in 1900, which he hoped would become a text book in schools like the current Handbooks for laboratories and for practical exercises. He also wrote a "Cyclopedia of hardy fruits", more or less a compendium of the six volumes of "Fruits of the State of New York".

He wrote in a clear, concise and didactic style. Naturally, for each variety he discusses the pomological profile of each variety but not according to today's standard method but rather summarizes the different behaviour of each variety in different environments. For example, describing the pulp of the apple and pear he distinguishes between coarse and fine; tender and tough, crisp, breaking, melting and—for buttery pears—juicy or dry, mealy as undesirable, and—for other pears—granular or gritty. This classification, like that of flavour, aroma and quality, is still valid today.

Take, for example, the English pear "Williams' Bon Chretien", introduced into the USA in 1797 (or 1799) and shortly afterwards launched in Massachusetts by E. Bartlett with its new name. Hedrick's description is as invigorating today as it was when he wrote it, listing qualities and defects (e.g. vulnerability to fireblight, a contagious disease affecting apples, pears and some other members of the Rosaceae

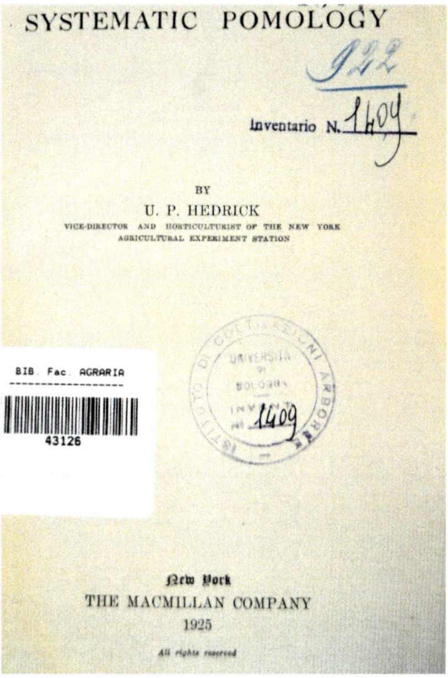


**Fig. 38.20** Cover of U.P. Hedrick's "Systematic Pomology" (Hedrick 1925)

family, to the cold winters, the need for good pollination, etc.), and commenting that it was the "most desired of all pears by the canning trade". Two hundred years later, it is still the most suitable if not the only variety of pear used for producing "Bartlett canned pears".

## *Luigi Savastano (1853–1937)*

The author would like to finish this summary of the great historical figures of horticulture at the end of the nineteenth and beginning of the twentieth century with a brief description of a relatively unknown Italian author, Luigi Savastano, who quickly learnt the lessons from France, chiefly through Du Breuil and his disciplined organization of horticulture, adding his experience and the results of his own experiments in agronomy, which had become fairly widespread throughout Europe, thanks to various agricultural academies, set up at the end of the eighteenth and throughout the nineteenth century.

Luigi Savastano was Professor at the Regia Scuola Superiore of Agriculture in Naples (Portici) and lived in the beautiful peninsula of Sorrento an area "covered with rich vegetation including citrus fruit, chestnuts, beech and fir trees". In the opening pages of the edition (1914) of his 1903 "Arboricoltura", "Treatise on tree growing" (Fig. 38.21) he summarized 20 years of research into trees, based on both theory and practice, because "it may not be difficult to grow well but it is certainly difficult to sell well" (a slogan still used by economists who underestimate the importance of technique and the organization difficulties of production). He wrote that

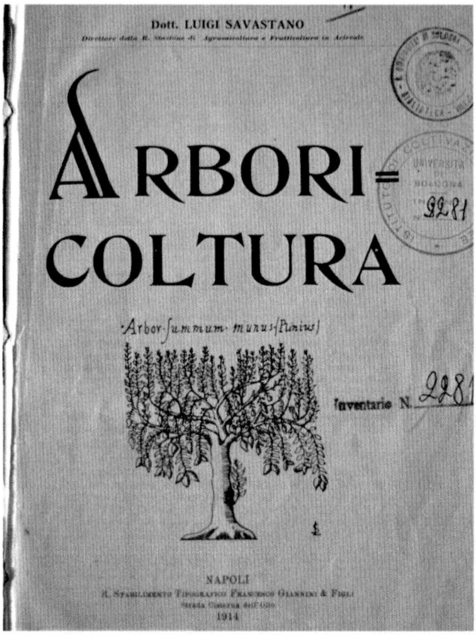


**Fig. 38.21** Title page of L. Savastano's "Arboricoltura", Naples, 1914

the "famous disagreement between theory and practice" between science and art does not exist; "they are two good sisters; whereas there are disagreements between theoreticians and practical growers, because each lacks the knowledge of the other whilst assuming that they do not". Both may be guilty of presumption.

Savastano then pays homage to two important Italian researchers and declares himself their follower of G. Gallesio, whose profile has been discussed, and G. Inzenga, for many years the editor of the "Annual of Sicilian Agriculture", whose technical studies brought new knowledge and innovative techniques to agriculture.

"Italy's geographical position makes it the fruit orchard of Europe … and so it must accept its civic duty to produce fruit for Nordic countries as well" in order to "satisfy not so much the taste of the wealthy as the needs of the poor". He concluded that modern tree growers must have a proper understanding of "agronomy, physiology, botany, chemistry and pathology". Savastano certainly understood that tree growing was not a science in itself but needed the understanding of other disciplines and a synthesis of this knowledge; this concept, as he wrote, had inspired a group of Italian researchers who published a new treatise on Italian tree growing "Arboricoltura generale" (2012). According to Savastano, the "students, more than the tree growers themselves, must take their understanding from up-to-date scientific sources, no longer standing on their own or derived straight from rational growing, but filtered, mediated and adapted to the specific territorial needs of the various branches of study in the fields of horticulture and tree growing".

After giving a history of tree industry and various civilizations down through the ages, Savastano describes the morphology of trees and new organs (with synoptic tables covering hundreds of species), making a classification of growing environments and stressing the role of "acclimatation".

Two chapters are dedicated to the origin of varieties, the hereditary nature of the variations described by Darwin, but the author had not yet read Mendel (1866), whose work was yet to be rediscovered and disseminated at the beginning of the twentieth century. Certain basic concepts of fruit growing were still misguided, lacking a proper understanding of genetics. For example, Savastano wrote: "Once obtained, a variety can be blocked by two processes: grafting or selection". Hence his recommendation of the method developed by the Belgian Van Mons, to "carry out the process of selection through subsequent generations of trees derived from controlled sowing".

The chapters on grafting and pruning are excellent, as is his description of the various forms of plantation, although the principles of selection were still unduly influenced by aesthetic considerations and the "geometry" of trees, irrelevant for production and quality purposes, preferred today. Many pages are dedicated to soil management, fertilization (especially organic) and irrigation.

For example, Savastano reminds tree growers that trees are organisms that respond to the wishes of the grower (always too demanding) only if properly guided (in proportion to their ability) but which "rebel against force and succumb if weak". He stated: "Fertilization is the most important work on the tree after pruning".

Another example was that considerable space is dedicated to the pros and cons of adopting the technique of not cultivating the soil by tillage (hence no hoeing or in-depth work). He concludes that the positive results obtained in France by Ravaz cannot be reproduced further south; in his experience after the first few years of using this technique, vineyards and orchards begin to suffer and age precociously. Comparing tree growing on flat land or in mountain areas, he cites a Latin proverb: "*Bacchus amat colles apricos*" (for vines, it is better to choose sunny slopes). He adds: "fruit trees prefer mountain climes, as the saying goes: "Praise the plateau but keep to the mountains".

Many pages are dedicated in detail to growing small trees, to nurseries, transplants and keeping allotments. He provides an economic analysis, with costs and likely revenues. The treatise is full of quotations from foreign authors and hence comprises a sort of review of current scientific knowledge and fin de siècle practice. The iconography of the book is its weakest part.

Its great merit is that it summarizes the changes in industrial and specialist tree growing over an entire century.

Savastano enlivens the book with philosophical remarks, put into the mouth of a tree grower friends of his: "Every good tree grower is an expert of his orchard and the best ones pass on their knowledge from the field". "Trees speak clearly to us but we do not understand them". "I continue to love Latin writers of '*Scriptores rei rusticae*'". For them: "*Bene colere optimum, optime vere damnosum*" (leave well alone). Savastano wrote: "These words are still today the basis of agriculture". He also cited a proverb from Cicero: "If agriculture is worthy of the free man, tree growing is worthy of the independent man, because over time the tree makes his character stronger: it breaks but does not bend".

So far we have described the most important contributions in the development of pomology. But when did orchard management begin to have a scientific basis? It would be wrong to omit a brief discussion of the disciplines and leading figures that made major contributions to agriculture and hence also to horticulture. The Treatise "Arboricoltura Generale" (Sansavini et al. 2012), has an up-date description of the pomology subjects, including orchard magament practices and a detailed focusing of physiological, biochemical and molecular mechanisms controlling tree growth, rooting fruiting and related metabolic processes.

## 6th Part: Pioneers of Chemistry and Physics Applied to Horticulture

F. Haber, N. T. De Sassure, J. Liebig, J. H. Gilbert, J. B. Lawes, F. Malaguti, J. Priestley, F. Blackman, C. B. Van Niel, L. J. Briggs

### *The First Synthetic Fertilizers. The Role of Nitrogen*

As we have seen, the importance of the use of organic fertilizers was known way back, Ancient Greek and Latin populations made widespread use of manure and similar fertilizers, as well as green manure cropping from ad hoc plantations. No progress was made in this field until the discovery in South America (and importation in Europe) of mineralized organic substances such as "guano". Some researchers immediately understood the importance of this development, among them the German genius Fritz Haber (1868–1934) who was the first person to synthesize ammonia, making industrial production a possibility. Others had the idea before but had not been able to convert the idea into an industrial product. After Haber, the production of nitrate and ammonia fertilizers began in earnest, along with calcium cyanamide, the latter being the brainchild of A. Frank and N. Caro who in 1895 synthesized calcium carbide and nitrogen in an electric oven.

Haber (a physical chemist) managed to produce ammonia (with a yield of 5 % in volume) after numerous experiments via an exothermic reaction, combining H and N at a pressure of 200 atmospheres and at a temperature between 300 to 600 °C, in the presence of catalysts (osmium and uranium). Over twenty thousand experiments were required to obtain the industrial patent granted in 1910, acquired by Basf, the process being perfected later by the chemist C. Bosch, Nobel laureate in 1931 together with German chemist Friedrich Bergius, and were rewarded by Haber gaining the Nobel Prize in 1918. But his discovery—which some have called the most important of the century—was bound to cause controversy, despite the benefits to agriculture and the pharmaceutical industry, because the processing of nitric acid was also used for the development of explosives. Haber supported the use of poison gas (yprite, aka mustard gas) in the Fist World War, provoking the suicide of his wife, a chemist, after the strategic attack on Ypres. Haber became wealthy and famous when in 1929, 40 % of the nitrogen produced in the world was synthetic ammonia, which is still produced today using the Haber/Bosch process in quantities of about 120 million t a year. At the beginning of the twentieth century in Europe, nitrogen-based fertilizers were used massively, the first step in the green revolution, before the contribution of genetics, leading to the doubling and tripling of the yields of maize and corn in the 1930s. Horticulture also had huge advantages, especially fruit trees and among them, particularly peach trees. With Haber and Bosch, humanity no longer depended on the biological synthesis of ammonia. It has been calculated that the chemical synthesis of ammonia saved the lives of about 2 billion people by making nitrogen, and hence sources of vegetable and animal protein, more available (Taddia 2010).

Nevertheless, the acquisition of the concept of the *fertilization of the soil* was by no means immediate, despite the work of many researchers and scientists. It was agricultural chemistry that first responded to the population growth during the twentieth century (from 1.6 billion in 1900 to nearly 6 billion at the end of the century). Without fertilizers and more rational soil management techniques, the world would not have been able to produce enough food. Today, in the Western world, many would like to return to more natural agriculture (e.g. biological cultivation), accusing the use of chemicals of terrible consequences for human health and the ecosystem.

At the same time as people are becoming aware of the dangers, chemicals are increasingly man's ally, especially in fruit production, for the affirmation of cleaner and more ecological agricultural production, benefiting man and the environment. The author gives just one example from peach farming. In the mid twentieth century peach farmers were unaware of the amount of nitrogen needed by trees and ignorant of solubilization processes, soil leaching and the loss of nitrogen in the water table, and therefore the more active farmers distributed annually up to 250 to 300 units/ha of N, i.e. two to three times the amount required, with adverse effects on the ecology. Now, thanks to research, the amount has been reduced to just 120 to170 units/ha.

Two other people stand out as fathers of chemical fertilization and, more generally, agricultural chemistry: the Swiss N.T. De Sassure (1767–1845) and the German J. Von Liebig (1803–1874). The first is remembered for his book "Recherches chimiques sur la vegetation" (1804), which for the first time dealt with the influence of the exchange with the atmosphere of the oxygen and carbon dioxide on vegetative processes. De Sassure's final statement was: "The presence or elaboration of CO2 (named as *gas carbonique acid*) is fundamental to the growth of the green plants to the sun; they die in dark conditions".

Von Liebig, at the then University of Giessen, then Munich, launched industrial fertilizers, convincing scientists and growers that synthetic fertilizers were very useful. He cancelled the belief that carbon in plants was taken from the humus arising from the degradation of organic substances in the soil, replacing it with the absorption of carbon dioxide in line with the discoveries of others. But he made the mistake of including nitrogen—considered important for nutrition—among the elements absorbed from the atmosphere by plants. He was the first researcher to experimentally demonstrate the role of inorganic mineral compounds in nutrition, through the analysis of vegetable ash.

Whilst Von Liebig understood how to integrate or replace organic soil fertilizers (various types of manure) with synthetic fertilizers, the true merit for this change lies with two English scientists, Joseph Henry Gilbert (1871–1901) and John Bennet Lawes (1814–1900) who, in strong disagreement with Von Liebig, demonstrated with experiments over many years carried out in Rothamsted, today one of the world's oldest experimental stations investigating the principles and practices associated with no tillage, arable land and crops. These practices gradually increased the administration of ammonium nitrogen to increase yield, in contrast to the Von Liebig's principles, which were based on even dosing, without excess. He claimed that the English scientists obtained their results partly due to the contribution of phosphorous rendered soluble by ammonium sulfate or nitrate. He believed that in 1844 the needs of agriculture (wheat and beets), under certain conditions, were best met with natural nitrogen from rainwater for good yields.

The disagreement went on for over 40 years, with the two English scientists and their correction of Von Liebig's "mineral theory" finally emerging victorious in 1893. They had always believed that it was nitrogen that boosted production above all for wheat, as they had shown with tests involving a mixture of nitrogenized salts compared to Von Liebig's synthetic fertilizer, that obtained higher yields. The scientific world accepted the results of the Rothamsted station rejected by Von Liebig.

Among the many researchers who made contributions to the understanding of agricultural chemistry, the Italian (later naturalized French, during exile) Faustino Malaguti (1802–1878), a University Professor, wrote numerous books, including "Chemie appliquée à l'agriculture", towards the middle of the nineteenth century. He introduced the concept of "chemical balance" between various elements, described their distribution in plants and carried out pioneering research on the chemical effects of sunlight.

## *Chemical Compounds to Fight Disease*

There is no father of synthetic pesticides but the origin of this branch of chemistry is probably the nineteenth century practice of curing cereal seeds to prevent "black blight" (*Ustilago tritici*) then very widespread in Italy, using lime, calcium chloride, sodium sulfate and even copper sulfate. In Italy there is evidence of the use of sulfated seeds in Tuscany in 1835. Certainly, in 1834, in North America, vitriol (copper sulfate) was used to protect vines against downy mildew with the cryptogam agent *Plasmopara viticola*, which arrived in Europe towards the end of the nineteenth century destroying many vineyards, together with phylloxera (an aphid, *Daktulosphaira vitifoliae*). Copper sulfate was adopted in Europe too, and today the copper derivates oxychloride and copper hydroxide are used to combat downy mildew.

## Photosynthesis and the Assimilation of Carbon

In relation to the discovery of photosynthesis, it is interesting to note that the most important physiological process of plants took three centuries to understand in terms of its biochemical mechanisms. Studies began with a Belgian chemist and doctor,

J. Baptiste Van Helmont (1579–1644), who noticed that a willow grown in a vase for five years grew in weight almost identically with the amount of water given (he did not consider the air), whilst the weight of the soil was unvaried. Subsequently, J. Priestley (1733–1804), an English chemist and philosopher, carried out an experiment with a plant under a bell jar, from which he had removed oxygen by burning a candle under the jar. He discovered that the plant did not die and that after a few days a candle once again burnt under the jar. Hence the idea that the plant regenerated the air. Then De Sassure, mentioned above, and others, suggested that carbon in plants comes from carbon dioxide

A few decades later, we encounter F. Blackman (1866–1947), an English physiologist, who—in 1905—affirmed the principle of "limiting factors" in the determinism of photosynthesis. He established the importance of individual factors in chemical transformations, involved separately from others, particularly of sunlight and temperature. But it was not until the twentieth century that a Dutch biologist, C.B. Van Niel (1897–1985), living in the USA after taking his PhD in1928, demonstrated for the first time through the study of violet and green sulfur bacteria, that photosynthesis is a redox reaction, dependent on sunlight, in which the hydrogen derived from an oxidizable compound reduces carbon dioxide to "cellular material". It was the water molecule, however, and not carbon dioxide that was split. From this he deduced that water provides the hydrogen for photosynthesis in green plants, giving off oxygen.

## Water Needs and the Relationship Between Water, Soil and Plant

The use of methods of irrigation probably dates back to the beginning of agriculture: by examining artistic illustrations. Janick (2003) showed that ancient civilizations such has the Egyptians, Chinese, Persians and Assyro-Babylonian used rudimentary but efficient systems for the provision of water for crops, that included digging channels, carrying water manually, and the hoisting of water from wells by wheels or animals (Fig. 38.22). The Hanging Gardens of Babylon became legendary, and Mesopotamia, with the waters of the Tigris and Euphrates rivers, became the centre for fruit and vegetables many of which had been imported from China and India.

Literature tells us that one population, the Nabataeans (forefathers of the Arabs), the first nomadic farmers to settle south of the Black Sea from between the second and third centuries before the Christian era, had raised horticulture to the level of an art, especially by means of irrigation. The oldest treatise on arboriculture (Nabataean Agriculture, written in Aramaic) was the work of a Nabataean, Ibn-Vahschiah; it was a sort of encyclopedia for its time, cited many times over a 1,000 years later by the most important Arab-Spanish scientist Ibn Alò-Awwam.

Historically, there are plenty of sources of information about the importance and use of water as well as hydrological research and investigations into the relationship between crops, scientific studies of the physiology of water, research into the effects of drought or excess water on crops. Naturally, most of these sources also provided

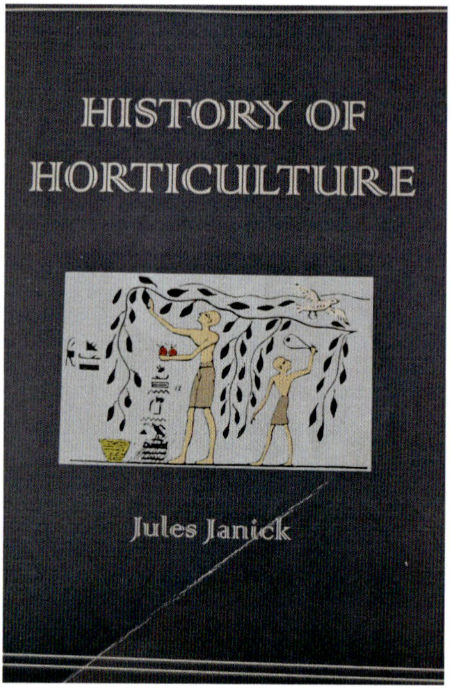


**Fig. 38.22** J. Janick's cover of the "History of Horticulture". (West Lafayette, 2003)

practical advice on the management and use of water, as well as describing the risks of too much or too little water.

For example, in Italy, Pietro de'Crescenzi (1230–1321) warned against excessive compaction and hardening of the earth "because it would not allow enough humidity to descend". "In particular, in the case of the olive tree, it is not necessary for the water to go deep, because its excess produces stagnation".

It wasn't until the end of the nineteenth and beginning of the twentieth century, however, that the fundamental role of water in the soil and in plant growth was understood and therefore could be properly managed. American research stations played an important part in these investigations, particularly Utah State University which published Bulletin 183 in 1922, in which researchers O.W. Israelsen and Frank L. West introduced the concept of "water availability in the soil" in their "field capacity" studies.

American irrigation technology, however, as generally applied for example in California and other fruit-growing states in the second half of the twentieth century presupposed the availability of enormous quantities of water, usually at no or little cost, with which the soil receives the water from furrows or was flooded, submerged each time it was irrigated.

During the same years, US researchers T.H. Kearney and L.J. Briggs carried out various studies to ascertain the relations between soil humidity and plant growth. They came up with a "wilting coefficient" which Australian scientist R.O. Slatyer, fine-tuned the relationship between soil salinity and autochthonous vegetation in

1957. Their research was carried out to compare results with and without irrigation. In 1898, one of the two, L.J. Briggs, described capillarity and the effect of the texturing of soil on water retention and movement.

At the University of California in Davis, the two American researchers continue to carry out in-depth investigations into water, soil and the growth of fruit trees, which they had chosen as a model. They pointed out the mistakes and uncertainties of their predecessors in "water relations". In particular, they showed that different soil conditions, including its moisture, caused different symptoms in the tree, including the term "water deficit" which was due to an imbalance between transpiration and available water for the plant. On the basis of the work carried out by Briggs and Shantz (USDA Bulletin 230 1912) they concluded that, for each soil, the wilting coefficient ("limit not to be exceeded for the leaf water deficit") is a constant for all species, whilst the amount of moisture in the soil can vary greatly before any observable "permanent wilting" of the tree can be attributed to the moisture of the soil. In addition "It is wrong to think that water flows easily in the soil from wet to dry, non-irrigated areas". In peach, apricot and plum trees planted in soil with good water availability, they observed stomatal leaf opening and water consumption compared to trees which had reached the wilting coefficient.

Fundamental were the studies of other two other American scientists, F.J. Veihmeyer and A.H. Hendrickson, who in 1927 discovered the physiological relations between water and trees (Veihmeyer and Hendrickson 1927).

Many other effects of water on plants had been investigated: for example, the larger leaf in peach trees being proportional to the increased water availability; but if the density of the plantation was increased this led to an earlier "decline" of the trees. They also noted the benefits of a "cover crop" to the orchard (not to be confused with today's grassing or meadow orchard) although it comprised only alfalfa. Evidently, they were looking at an orchard where too much water was given, by submersion or furrows.

It may seem strange but modern irrigation techniques, such as using sprinklers as a rain effect on foliage, began only in the second half of the twentieth century, and micro-irrigation (trickle, drip irrigation) only in the 1970s and 1980s, with "fertigation", at least in Europe, coming afterwards, followed by the use of "sub-surface drip irrigation", after the 1980s (Camp 1998).

These techniques have led to enormous savings in water, much less waste, a more rational approach to root systems and therefore a more efficient use of water, but are beyond the scope of this chapter. The most advanced research has been carried out in the USA, whilst Israel and Australia have pioneered methods of micro-irrigation in dry, semi-desert regions, adopting various solutions for "drip irrigation", unthinkable on the basis of traditional irrigation techniques.

Claims and rival claims have been made about who first invented these new techniques. For example, in Israel, Simcha Blan (1967) was a precursor of micro-irrigation and went on to found the Netafim company, a leader in the horticulture sector, whilst the first examples of sub-irrigation were the results of experiments carried out in SDI (Subsurface Drip Irrigation), according to F.R. Lamm et al. (Transactions of the ASABE, 2012, vol. 55:483–491) and dated back as far as about 1860, in Germany, using underground pipes both for irrigation and the drainage of water.

**Acknowledgments** The author is greatly indebted with Mrs. Clementina Forconi for her invaluable help in preparing the manuscript.

## References

Aldrovandi U (1667) Dendrologiae naturalis scilicet, arborum historiae. Libri duo, sylva glandaria, acinosumq. Pomarium. Ubi, eruditiones omnium generum una cum botanicis doctrinis ingenia quaecunque non parum iuvant, et oblectant. Ovidius Montalbanus. ex typographia Feroniana, Bononiae

Baldini E (1982) Agrumi, frutta e uve nella Firenze di Bartolomeo Bimbi, pittore mediceo. CNR Roma (ed) Grafiche Pizzetti, Firenze

Baldini E (2004) Cinque secoli di pomologia italiana. Dipartimento di Colture Arboree, Università di Bologna, Bologna

Baldini E (2008) Agli esordi della pomologia: Ulisse Aldrovandi. In: Baldini E (ed) Miti, arte e scienza nella pomologia italiana. CNR, Roma

Baldini E, Tosi A (1994) Scienza e arte nella Pomona Italiana di Giorgio Gallesio. Accademia dei Georgofili, Firenze

Baldini E, Tagliaferri C (1998) Complementi inediti della dendrologia di Ulisse Aldrovandi. Paper presented at Accademia delle Scienze dell'Istituto di Bologna. Classe di Scienze Fisiche, Bologna

Camp CR (1998) Subsurface drip irrigation: a review. Trans ASABE 41:1353–1637

Catonis MC (1964) Liber de agri-cultura. I classici dell'agricoltura. Reda, Roma

Columella LJM (1948) De re rustica: vol. I, III, IV, V, XII. In Calzecchi–Onesti RI classici dell'agricoltura. Reda, Roma

De Crescentiis P (1486) Opus Ruralium Commodorum. Biblioteca internazionale La Vigna, Vicenza (2010)

De' Crescenzi P (1851) Trattato della Agricoltura, 3 vol. Tip. Vicentini e Franchini, Verona

De la Quintinye JP (1746) Instruction pour les jardins fruitiers et potagers, vol I and II. Compagnie des Libraries, Paris

De Serres O (1600) Le théâtre d'agriculture et mesnage des champs. M.D.C., Paris

De Vries H (1909) Specie e varietà e loro origine per mutazione, 2 vol. Remo Sandron Editore, Napoli

Du Breuil MA (1863) Culture des arbres et arbrisseaux à fruits de table, 6th edn. G. Masson, Garnier Frères, Paris

Du Breuil MA (1875) Les vignobles et les Arbres à Fruit à Cidre, 6th ed. G. Masson, Garnier Frères, Paris

Gallesio G (1839) Pomona italiana, ossia trattato degli alberi fruttiferi, vol 2. N. Capurro, Firenze

Gallo MA (1565) Le dieci giornate di vera agricoltura e i piaceri della villa. D. Farri, Venezia

Hedrick UP (1921) The pears of New York. JB Lyon Company Printers, Albany

Hedrick UP (1925) Systematic pomology. The Macmillan Company, New York

Ibn al-ʿAwwām (early 1800) Kitāb al-filā-ḥah (Treatise on agriculture). Sevilla

Janick J (2003) History of horticulture. Purdue University, West Lafayette

Marmorale EV (1949) Cato Maior. G. Laterza & Figli, Bari

Mitterpacher L (1797) Elementi di agricoltura, 5 vol. Francesco Prato, Torino

Molon G (1901) Pomologia. U. Hoepli, Milano

Saltini A (1989) Storia delle Scienze Agrarie, vol 1–4. Edagricole, Bologna

Saltini A (2012) History of agrarian sciences, 1 vol., Museo Galileo, Istituto di Storia delle Scienze, Firenze

Sansavini S (2002) Un secolo e oltre di frutticoltura. In "L'agricoltura verso il terzo millennio attraverso i grandi mutamenti del XX secolo". Accademia Nazionale di Agricoltura, Bologna, pp 307–382

Sansavini S (2004) Presentazion de "Cinque secoli di pomologia italiana". Dipartimento di Colture Arboree, Università di Bologna, Bologna, pp 3–7
Sansavini S (2013) European horticultural challenges in a global economy: role of technological innovations. Chronica Hort 53(4):6–14
Sansavini S, Costa G, Gucci R, Inglese P, Ramina A, Xioloyannis C (2012) Arboricoltura Generale. Ed. Pàtron, Bologna
Savastano L (1914) Arboricoltura. Stab. Tip. Francesco Giannini & Figli, Napoli
Stansfield WD (2006) Luther Burbank: honorary member of the American Breeders' Association. J Heredity 97(2):95–99
Taddia M (2010) La chimica e l'agricoltura. CNS. La Chimica Nella Scuola 32(4):73–85
Veihmeyer FJ, Hendrickson AH (1927) Soil-moisture conditions in relation to plant growth. Plant Physiol 2(1):71–82
Virgilio Marone P (1908) Le Georgiche. G.C. Sansoni Ed., Firenze

# Chapter 39
# Gardening and Horticulture

**David Rae**

**Abstract** Gardening and horticulture are both activities concerned with the cultivation of plants. While there is much overlap between the two activities, the former refers to a leisure activity practiced by home or hobbyist gardeners, while the second refers to a scientifically underpinned, and highly specialised, professional occupation. Despite the undoubted similarities there are big differences in the techniques, technologies and scales of operation. Different types of cultivation, such as organic gardening, are used to highlight the differences in approach between gardening and horticulture, while garden styles or practices such as patio gardening or allotment gardening are used to show that while gardeners are the consumers of products and services, professional horticulturists are not only providers of the products and services, but have also developed the technology to make the style or practice possible. Commercial production and botanic gardens are examined as horticultural activities, while noting that some of the most skilled cultivators are, in the strict use of the word, amateurs, and are catered for by specialist societies. Techniques such as soil cultivation, propagation and pruning are examined for differences in approach and scale, and the chapter concludes with examples of the range of scientific endeavour underpinning plant breeding, glasshouse production and cultivation media.

**Keywords** Horticulture · Gardening · Cultivation · Horticultural technology

## Introduction

Gardening and horticulture are both activities concerned with the cultivation of plants but the terms are usually applied in slightly different ways. Gardening normally refers to hobbyists or home gardeners while horticulture is usually applied to professionals who earn a living from their work. The terms are not precise and are often interchanged. Many people who work in the industry, particularly those in parks or botanic gardens, would informally describe themselves as gardeners but if they were being formal about it, then as horticulturists while amateur gardeners

D. Rae (✉)
Royal Botanic Garden Edinburgh, 20A Inverleith Row,
Edinburgh, EH3 5LR, Scotland
e-mail: d.rae@rbge.org.uk

G. R. Dixon, D. E. Aldous (eds.), *Horticulture: Plants for People and Places, Volume 3*, 1307
DOI 10.1007/978-94-017-8560-0_39, © Springer Science+Business Media Dordrecht 2014

would virtually always describe themselves as gardeners and not horticulturists. In some dictionaries horticulture is defined as the art of gardening while gardening is defined as the act of cultivating or tending a garden, or the work or art of a gardener. This lack of precision is reflected in many societies and organisations that represent gardeners and horticulturists. The British Institute of Horticulture (Anon 2013a), the Horticultural Trades Association (Anon 2013b) and the International Society for Horticultural Science (Anon 2013c) very clearly represent professional horticulturists but so does the Professional Gardener's Guild (Anon 2013d) that represents paid, professional horticulturists who work in private gardens and estates. Likewise, the Worshipful Company of Gardeners represents professional gardeners (horticulturists), but uses the word 'gardeners', demonstrating that there are no clear established rules about how the words are used. First mentioned in City Corporation records in 1345, the Worshipful. Company of Gardeners is a survivor from the medieval craft guilds which exercised control over the practice of their particular 'mysteries' and ensured a proper training through the system of apprenticeship (Anon 2013e). In 1605, after existing for centuries as a "mystery" or "fellowship", the Guild was incorporated by a new Royal Charter. The Charter sets out the operations controlled by the Company: "The trade, crafte or misterie of gardening, planting, grafting, setting, sowing, cutting, arboring, rocking, mounting, covering, fencing and removing of plants, herbes, seedes, fruites, trees, stocks, setts, and of contryving the conveyances to the same belonging…". Apart from some old terminology and old fashioned spelling, these words are still a good description of what gardeners and some horticulturists do today, although the context within which each does it is very different.

The Royal Horticultural Society (RHS), one of the largest gardening societies and charities in the world, crosses the divide, with membership consisting mostly of amateur gardeners but with many professional horticulturists being members and the Society employs many horticulturists in their four gardens and its laboratories (Anon 2013f). The American Rhododendron Society is mainly populated by amateurs but has a considerable dimension of professionalism at which point it too, like many other similar organisation, bridges the divide (Anon 2013g). Many professional horticulturists are also passionate gardeners, with superb gardens at home, separate from their working lives. One characteristic that tends to prevail in this environment is a passion for growing plants and a very high level of job satisfaction meaning that many horticulturists continue to garden when they get home, often with little distinction between their day job and their preferred hobby or way of life. Likewise, many hobbyist gardeners develop their skills to such an extent that they can be more skilled than many trained, professional staff, spend considerable amounts of money on their enthusiasm, join specialist clubs and societies, take gardening holidays and attend gardening lectures, shows and events. Furthermore, while horticulture is regarded mostly as a science (but with art being an important component), gardening isn't considered a science, it's a leisure activity where one can garden for pleasure or garden at home, growing fruit and vegetables or derive pleasure from looking after the lawn, but it's not a profession for those people. In general all outdoor home care for plants can be brought together into gardening. The two terms are therefore very similar but if the terms are applied strictly then they are clearly different with horticulture referring to those who cultivate or manage plants as their paid employment and gardening being used to describe hobbyist or home gardeners. Botany, on the other hand is the science of plant life, covering aspects such as physiology, morphology and reproduction but dealing in the main with plants as ecological entities, although there is a branch subject embracing "economic botany".

The shear diversity of activities that can be listed under gardening and horticulture is immense. In this chapter a selection of these activities, such as turf care and propagation, are examined from both a gardening and horticultural perspective to demonstrate the differences between gardening and horticulture.

The entry for Horticulture in Wikipedia describes it as the science, technology, and business involved in intensive plant cultivation for human use and continues by providing a very good and comprehensive description of what is entailed in horticulture (Anon 2013h).

## The Diversity of Horticulture

Gardeners and horticulturists differ in the techniques and technologies they use and in the scale of their operations. There are also differences in philosophy when approaching gardening or horticulture (Hobhouse 2004; Taylor 2006). The forms of gardening selected for discussion may appear to simply be forms of garden design or style but they are actually different from these as they really emphasise the attitude, approach, background or thinking to garden making and plant cultivation and care.

Seven types of garden forms are described. The first two forms of gardening, organic gardening and wildlife gardening take a philosophical approach to gardening or the attitude taken by the gardener to horticulture while the second two describe allotment and patio gardening. These forms of gardening have largely been brought about by changing environmental, social, cultural and economic circumstances. The last three to be examined are botanic gardens, commercial horticulture and special interest gardening and this is where the enthusiast differs from the professional in the specific selection of plant material, their cultivation and financial motivation. All seven examples demonstrate both the differences and similarities between gardening and horticulture.

### *Organic Cultivation*

Organic gardening is not one precise activity and many people practice it in different forms and to varying degrees of strictness. Essentially it is plant cultivation without the use of artificial pesticides or fertilizers, but it is really much more than that, it is a philosophy and a way of life, with those practicing it also being interested in composting, recycling, the promotion of wildlife in the garden and adhering to sustainability standards. Organic cultivation has its strict adherents who avoid any artificial pesticides and who prefer to 'work with nature' in managing soil and controlling pests and diseases. Professional horticulturists who grow and sell products labelled as organic, or who have declared that their garden is organic, certainly have to operate within strict guidelines (although there is sometimes confusion about which guidelines are being followed), often established by government regulatory bodies. However, there are many home gardeners who generally work in this way, believing that it is 'a good thing' that not only benefits themselves and the environment, in producing wholesome, fresh and tasty produce but then occasionally revert to the use of some pesticides and herbicides when pests, diseases or weeds get out of hand.

Organic gardening has its roots back in the early to mid twenteeth century in tandem with the explosion in the use of artificial pesticides and fertilizers (Heckman 2005; Taylor 2006). Concern for the environmental and health effects of these materials, however, was only expressed by a minority of concerned individuals until people like Lawrence Hills, who is sometimes known as the father of organic gardening in Britain, started providing a scientific basis to what became known as 'the organic movement'. He and others were responsible for the foundation of the Henry Doubleday Research Association, HDRA, now known as HDRA- the organic organisation, which was established in 1954. The Soil Association, founded in 1946 and now with 27,000 members, is undoubtedly part of this story too. It is a membership charity campaigning for planet-friendly organic food and farming (Anon 2013i).

The organic movement is often associated with the social values and lifestyle of the 1960s with adherents often caricatured as long-haired, hippy, drop-outs and their produce thought of as small, badly shaped, pest and pathogen infested. But with concern for the environment becoming increasingly widespread from the 1980s onwards and coupled with good quality underpinning research, organic approaches to cultivation have become widely accepted in many parts of the world. Many important gardens are now run organically, for instance Prince Charles' garden at Highgrove (Wales and Lycett Green 2001) and many high profile celebrities such as Gwyneth Paltrow, Julia Roberts and Nicole Kidman also promote organic gardening.

While private organic gardeners are at liberty to adopt any degree of rigour they wish, professional horticulturists producing fruit, vegetables and ornamentals for sale have to adhere to strict preparatory and marketing guidelines. The problem is in trying to produce a set of guidelines as there is no legally binding definition of the word 'organic'. Likewise, there are many similar, but strictly different, issues which could be incorporated into any definition such as the cultivation and sale of genetically modified cultivars, fair trade and the question of air miles, in other words the distance that harvested crops are flown to markets, such as Peruvian asparagus being flown to Europe. Indeed, while the incorporation of animal manure into soil destined for organic cultivation would generally be regarded as good organic practice, the use of manure from intensively produced cows cannot be used. Instead there are various guidelines, produced by different organisations which are subject to change and updating as knowledge and techniques evolve. So, while the market for organic produce continues to grow, the search for helpful definitions and clearly labelled food can be argued to be hampering progress.

Despite these problems, organic gardening and organic cultivation which have at their core the care of the soil, absence of the use of artificial pesticides and fertilizers and environmental sustainability issues, continues to progress and is increasingly regarded as mainstream. Detailed information on organic gardening can be found from the following reference sources: Taylor (2006), Kruger and Pears (2001), Marshall et al. (2009) and Anon (2013j, hh).

## *Wildlife Gardening*

Wildlife gardening is mostly the preserve of home gardeners but many professional horticulturists working in large private gardens, botanic and heritage gardens often take a keen interest in promoting wildlife in and around their gardens too. Until the 1970s and 1980s the generally held view was that that the countryside was teaming with wildlife whereas the suburban landscape was devoid of many species except for a few common animals, birds and insects. Gradually, however, and supported by wildlife surveys it has become apparent that modern intensive agriculture with its monocultures, large-scale fields and heavy reliance on pesticides was not as full of wildlife as once thought (Carson 1963; Jameson 2012). In contrast, the mosaic of suburban gardens with their variety of flowers, shrubs, vegetables and fruit were surprisingly rich in wildlife.

Taken together gardens add up to a surprisingly large area of land and surveys of individual gardens have shown that they can contain large numbers of species. A 30 year study of a garden in Leicester, England, revealed that over 2,200 insect species had been recorded and it has been suggested that a typical garden probably exceeds 8,000 species of insect (Owen 2010). The single biggest factor in the value of gardens to wildlife is diversity, even on the very small scale. They don't have to be full of native species, but even a modest garden can accommodate a small tree, several shrubs, climbers, flowering plants, compost heaps, pond or water feature, vegetables, lawns, weeds and fruit and these, in turn provide shelter, nectar sources, natural habitats, food sources and water for different forms of wildlife. Gardens don't have to be left unkempt to encourage wildlife, as once thought, and even 'mini habitats' such as short runs of hedging, small ponds, natural stone walls and small piles of logs can support a diverse range of wildlife, while climbers provide nesting opportunities and shelter.

Encouraged by TV programmes, magazine articles, mass recording projects such as Big Bird Watch by the Royal Society for the Protection of Birds (RSPB) (Anon 2013k) and others, organic gardening, wildlife gardening, or wildlife-friendly gardening emerged as a legitimate type of gardening from the 1990s onwards. This does not mean that country gardens were not good for wildlife, (Fig. 39.1) or that all suburban gardens are good for wildlife or that many gardeners hadn't been wildlife friendly for many years. It simply meant that from the 1990s many urban and suburban gardeners have realised that their gardens could be important for wildlife and that with small adjustments, they could be increasingly rich in wildlife. In the professional environment it is no coincidence that many staff employed in both private

**Fig. 39.1** Wildlife bluebell garden, Dorset England

and public gardens are also interested in wildlife and are therefore keen to preserve and promote it in conjunction with the garden in their care. Furthermore, in the UK as in many other countries, businesses (which include public gardens), public and local authorities have a duty to preserve and promote wildlife (under the Biodiversity Duty which came into force on 1 October 2006), and may be required by law to protect species on their property listed by Biodiversity Action Plans and will be required to survey and protect species if they hold an accredited environmental standard such as ISO 14001 (Anon 2013l), the international standard for environmental sustainability, which includes a small section on duty of care for wildlife. Detailed information on wildlife gardening can be obtained from the following reference sources: Thompson (2007), Tait (2006) and Baines (2000).

## *Allotment Gardening*

Allotments or allotment gardens are plots of land set aside for non-commercial cultivation of fruit, vegetables and flowers and are therefore the preserve of amateur gardeners although the word 'amateur' in this context can be very misleading because many allotment gardeners are highly skilled, effective and respected growers. In North America they tend to be called community gardens but in Britain, such a term would mean a garden where members of a community came together on an individual plot of land to garden it whereas allotments tend to be gardened by individuals, families or, sometimes, a small group of friends. Allotments range in size from about 50–400 $m^2$ and are usually grouped together, sometimes totalling a hundred gardens or more. Allotments tend to be found in cities, or peri-urban areas, where many people don't have individual gardens and are located in all sorts of place such as corners of public parks or adjacent to railway lines. In some Scandinavian countries and parts of western Russia it is traditional for elaborate allotment gardens with small dwellings to be located out of town and used by flat dwellers at weekends to get out of town, reconnect with nature and grow fruit and vegetables.

Many allotments are gardened very intensively, producing large quantities of high quality fruit and vegetables but, like all forms of gardening, there are no limits on production. Likewise, the structures found in allotments vary enormously from production glasshouses, polythene tunnels and frames to tool sheds, summer houses with verandas and children's play equipment. Either way, they are recognised as providing important socio-cultural and economic functions over and above the opportunity to enjoy the growing of plants. They provide opportunities for exercise, relaxation, social bonding and play, for the unemployed the feeling of being useful and not excluded as well as a supply of fresh vegetables at minimum cost, for immigrant families the possibility of communication and better integration in their host country and for the retired and elderly a sense of routine and purpose.

The history of allotments can be traced back to 1809 in the Wiltshire village of Great Somerford where the Free Gardens were created following a letter from Rev Stephen Demainbray to King George III in which he asked the king to spare, in perpetuity, 6 acres from the Enclosure Acts for the benefit of the poor of the parish. Other well recorded allotments include St Ann's Allotments in Nottingham, created in the 1830s. A brief history of allotments in the UK can be found at Anon (2013m), while Anon 2013hh provides an overview of allotments in countries such as Finland, France, Germany and the Netherlands. At the end of the nineteenth century there were about 250,000 allotments in the UK but this number rose to about 1,500,000 by 1918. Numbers dropped after this but rose again during World War II with the need for more locally produced food and the 'Dig for Victory' campaign. Numbers dwindled to about 600,000 by the late 1960s but interest started to rise again in the 1970s in tandem with the interest in organic, locally sourced food and the rise in environmentalism. Despite this growing interest numbers continued to decline as land was sold for development, dropping to less than 500,000 in 1977 and 265,000 in 1997. In 2008 330,000 people held allotments whilst 100,000 were on waiting lists. So, while demand was at an all-time high, pressure from development only decreased the number available. However, while less land is set aside for allotments, there is an increasing understanding of their value not just for the socio-economic and cultural benefits for food, exercise and relaxation gained by allotment holders, but also for the broader issues of climate change, food security, the encouragement of biodiversity and environmental sustainability. The cultivation of fruit and vegetables is currently experiencing an upsurge in popularity and a recent England & Wales Department for Environment, Food and Rural Affairs (Defra) report has shown that many household foods which have been home grown in gardens or allotments have increased between 2007 and 2011, for instance beans, from 28 to 33 % and apples from 3 to 9 % over this period. Additionally, garden owners growing their own fruit and vegetables has increased from 34 % in 2007 to 43 % in 2011 (Anon 2011). Detailed information on allotment or community gardening can be obtained from the following reference sources: Anon (2013m), Akeroyd et al. (2010) and Clevely (2008).

## *Patio Gardening*

Some might consider that 'patio gardening' is hardly a type of gardening but it is interesting to consider how a combination of circumstances can combine to create what at least is a 'way' of gardening. In this comparison between professional horticulture and the amateur or home gardener it is interesting to see how this form of gardening has been created by a combination of the demands of the gardener, the garden centre and modified technological developments of the professional horticulturist. Today a visit to the garden centre can provide the home gardener with a range of tools and equipment that is suitable for the smaller garden.

The word patio comes from Spanish where it means a courtyard or forecourt and in general terms patios are paved areas adjoining or close to a house. The degree to which many people now live in and around and adorn such areas with plant containers, space heaters and barbecues is remarkable and demonstrates that patio gardening is at least a fashion, if not a style, born from a combination of late twentieth century influences.

One of the joys of garden making, whatever the size of the garden, is the creation of places to sit and enjoy the garden or a view outside of the garden. Finding the perfect spot out of the prevailing wind or to catch the last rays of the evening sun to enjoy morning coffee, al fresco eating or an early evening drink is undoubtedly one of the pleasures of gardening. These favourite spots, known affectionately in Scotland as 'sitooteries', range in sophistication from a couple of logs to elaborate terraces with ornate outdoor furniture, pots, urns and planters. With the spread of suburban housing schemes, smaller gardens, greater affluence and more leisure time coupled with the joys of outdoor eating in paved and decorated courtyards brought back from Mediterranean holidays, the patio as a defined garden space has undoubtedly grown in popularity.

Developments in the professional horticulture industry also fuelled this type of gardening because patio gardening relied on recreating that 'Mediterranean' atmosphere and called for bright annuals that grew well in containers. While the elaborate bedding schemes that peaked in the Edwardian era continued as popular features in urban parks right up to the 1970s, the labour costs associated with manual sowing, pricking out, growing on and then planting meant that gardening with half hardy annuals almost became an extinct style. Even home gardeners found that the time, skill and greenhouse space taken to produce these displays meant that they were becoming redundant. However, the advent of sowing machines, cellular or modular production and mass handling systems, which all brought down costs, allied to assured viability (that guaranteed no gaps in cellular trays), numerous new $F_1$ cultivars bred to be uniform, floriferous and shorter, and therefore suitable for patio containers, revived the cultivation of half hardy annuals, albeit in a different style. Hanging baskets, wall baskets, terracotta and other pots all leant themselves to patios and the requirement for regular watering and liquid nutrition was easily dealt with by adapting existing professional irrigation systems to the home gardening environment.

Patio gardening as a style or way of life has therefore been created from a combination of modern influences including package holidays to the Mediterranean and other idyllic locations, plant breeding, plant cultivation and handling systems, adapted intensive irrigation equipment, greater affluence, modern housing and the perennial desire to sit outside and enjoy the view.

## *Botanic Gardens*

The title of 'Botanic Garden' covers a multitude of activities and designations but usually infers that there is at least some scientific basis to the arrangement of the plants on show. There are about 3,000 botanic gardens in the world with the largest concentrations being in developed counties such as Germany, France, the UK and USA and fewest being located in developing counties, where, ironically, biodiversity is often greatest, and in greatest need of conservation and protection (Rae 1995). Their activities usually include research, conservation, education and amenity or display and at their best they include both serious research institutions with a broad range of schools and public education and high quality visitor attractions. At the other end of the scale, unfortunately, many botanic gardens languish with virtually no research, minimal public education and gardens that are poorly maintained. Designations range from government and university botanic gardens to private foundation and local authority gardens. In the last forty years it is probably university botanic gardens that had declined in number and quality the most due to the change in the focus of teaching and research from whole plant biology to molecular approaches meaning that associated botanic gardens that supported anatomy, morphology and physiology teaching were judged to be no longer required. Fortunately this trend may be beginning to reverse. Universities are finding that students judge them at least in part by the quality of their green environment. This can have a significant impact on student choice when deciding which university shall receive a student's fees (I. Park, Estates Manager Exeter University, pers. comm.).

Whatever their focus, all botanic gardens at least have gardens and therefore employ horticulturists to maintain and develop them. Numbers employed and levels of training vary enormously. Entebbe Botanic Gardens, Uganda, for instance, employs three horticultural staff whereas King's Park and Botanic Garden, Perth, Australia, employs 68 (Anon 2013n). Horticultural training for botanic garden staff can vary from one having little or no experience through to certificates, diplomas and botanic garden-validated courses, to undergraduate degrees and those with PhDs. All levels are valued in the establishment and maintenance of botanic gardens.

Botanic gardens in the modern era, meaning post Renaissance (there were gardens that could be termed 'botanic' before this period in countries such as Greece and Mexico) had their foundations in medicine, and therefore had a sound scientific basis. Generally described as physic gardens their purpose was to grow and distribute medicinal plants and train doctors and apothecaries in their use, so they also had an educational purpose. 'Classic' gardens of this period include Pisa, Padua, Bologna, Paris, Leyden and Montpellier, all founded before 1600. In the UK Oxford and Edinburgh Botanic Gardens and Chelsea Physic Garden were established in this tradition but were founded in 1620, 1670 and 1673 respectively. As plants became increasingly studied for interests beyond medicine, such as physiology, so botanic gardens adapted and supported teaching of these disciplines in universities. A good example of this is the research undertaken by Professor John Hope (1725–1786) at the Royal Botanic Garden Edinburgh in the latter part of the eighteenth century (the botanic garden having initially been founded as a physic garden in 1670). Again, there was a strong scientific basis for the garden and its collection of plants.

Botanic gardens have continued to evolve and develop reflecting the needs of society, with old gardens adopting new initiatives and new gardens being created for specific, new purposes. From those early days in medicine and physiology, botanic gardens have become involved in plant taxonomy and systematics, education, museum-like displays, acclimatisation of 'new' crops to different countries, plant discovery, conservation, biodiversity studies, interpretation and, most recently, well-being, social inclusion and food production. The most successful botanic gardens have maintained their effort in the most important traditional activities such as plant systematics but have also embraced the newer and frequently more public-focussed activities. The point is that for any of these activities, old or new, a plant collection maintained by trained horticultural staff is absolutely essential. This is why many botanic gardens concern themselves more with 'the collection', rather than the design or display of the plants, though the best gardens combine both successfully. It is the fundamental importance for such collections to be amassed and developed to support the scientific purposes of the garden, and for the requirement to have detailed and accurate records, that sets botanic gardens aside from purely ornamental gardens.

In selecting the most appropriate species to fulfil the needs of the collection, which are often documented in a Collections Policy, and ensuring that they've been correctly sampled from the wild and managed according to a plan we find the essence of curation, a word also used in museums and galleries. Few amateur gardeners would call themselves curators, but in the botanic garden world, the term is widely used and well understood and it is an important part of professional horticulture. Curators at the four gardens that make up the Royal Botanic Garden Edinburgh, for instance, follow the guidelines laid out in their Collections Policy for the Living Collection which includes sections covering national and international conventions and policies that apply, service provision to stakeholders, standards of information, targets, reviews, audits, presentation and design, collection types and acquisition (Rae et al. 2006). These and other management issues such as verification and the wild origin content are the very essence of their jobs and it is a branch of professional horticulture that links the selection and cultivation of plants to their use by a vast range of stakeholders from molecular biologists through to school children, artists and the general visiting public.

Detailed information on botanic gardens and their histories can be obtained from the following reference sources: Oldfield (2007), Oldfield (2010), Johnson and Medbury (2007), and Taylor (2006) and also the following web sites: Wikipedia on botanic gardens—Anon (2013o) and Botanic Gardens Conservation International Anon (2013p).

## *Production and Commercial Horticulture*

Both of these categories are exclusive to professional horticulturists rather than gardeners and, while they overlap considerably, the common link is they are business ventures with the intention of making money. In the UK the production horticulture industry consists of 7,700 businesses, with 95,000 people and is worth an estimated £3 billion (Anon 2013q). The industry splits into the distinct areas of ornamental plant, flower and tree production (including retail nursery outlets) and food production, also included are fruit, vegetable, salad, herb and potato production (Anon 2013q). Some might argue that the production of food crops is the preserve of agriculture. However, in this context the intensive cultivation of crops like fruit, vegetables, salads and herbs is regarded as horticulture while the more extensive production of crops such as wheat, barley, maize and soya beans is regarded as agriculture. While this division of terminology has been adopted for many years it is interesting to note that many vegetables, in particular, are now cultivated on such a scale that their cultivation is more akin to agriculture than horticulture. Pea growers in Lincolnshire and potato growers in East Lothian certainly consider themselves farmers rather than horticulturists, though the term 'grower' is a convenient way to avoid using either term. The precision required for the husbandry of many broad-acre agricultural crops now closely resembles horticultural practice. Consequently, distinctions between "farmers" and "growers" is becoming blurred.

Commercial horticulture would include production horticulture but to their numbers, in the UK, would be added the 172,000 people employed in the landscaping and sports turf industries which include those involved in landscape construction, from initial earthworks, through hard landscaping and on to final planting, grounds men who manage sports turf from racing tracks to golf courses, and those managing parks, botanic and historic gardens (Anon 2013q), see also the Commercial Horticulture Association website, (Anon 2013r).

While home gardeners certainly grow ornamental plants, fruit, vegetables and herbs and while they also landscape their gardens and manicure their lawns, the scale of operation is completely different, the range of machinery and equipment is different, the training is different and the motivation is different,- production and commercial horticulture being done for financial profit and home gardening being done, essentially, for pleasure.

The scale of many horticultural enterprises is considerable and at the upper end certainly eclipses anything that any enthusiast gardener could do. Three examples illustrate this point. Van Heyningen Brothers Ltd (VHB), a Sussex-based company growing fresh herbs and seedlings have 230 full time employees and produce 14 million fresh potted herbs and 12 million punnets of salad cress per year (Anon 2013s). The Green Pea Company is a co-operative formed in 2006 and covering a wide geographic area in East Yorkshire and North Lincolnshire. They include 230 growers, each of which have up to 1500 ha and collectively they grow nearly 10,000 ha of vining peas. The scale of the operation is vast, being the largest pea cooperative in the world, producing around 45,000 t of high quality peas per year. Pea viner harvesting machines can harvest 1 ha per h and cost £ 300k (Anon 2013t).

The third example demonstrates landscape construction and horticulture on a massive scale. The London Olympic Park, created for the venue of the 2012 Olympic Games was the largest new urban park to be developed in Europe for 150 years. The 100 ha site is situated in Stratford, East London and the master plan was developed by a consortium of the British landscape architecture practice, LDA Design, with the American landscape architecture practice, Hargreaves Associates. Professors James Hitchmough and Nigel Dunnett of the Department of Landscape, University of Sheffield were appointed in 2008 as principal horticultural and planting design consultants for the Olympic Park, working with LDA/Hargreaves. Their role has been to develop a whole-site planting strategy, and to produce concepts and detailed proposals for the herbaceous vegetation in the park. The planting approach is highly ambitious and revolutionary for a major UK urban park, being driven by biodiversity and sustainability objectives, whilst also providing for an outstanding aesthetic experience. The Olympic Park comprises two different character areas: the North Park which has a more extensive and informal character, and the South Park, which includes the main Olympic Stadium and has a more urban character. Plantings in the North Park largely represent designed versions of native UK habitats and celebrate native biodiversity. They include species-rich meadows of different types; wetland plantings, including rain gardens and bioswales; woodland underplantings, and dramatic perennial 'lens plantings'. Plantings in South Park focus on visual drama and have a strong horticultural basis. They include the 2012 Gardens, Display Meadows and the 'Fantasticology' art installation (Anon 2013u).

Created over four years, work on the Park included extensive demolition and the decontamination of nearly two million tonnes of soil (the largest ever soil-washing operation in the UK), the creation of vast areas of concourse, spectator lawns and landscape features like the London 2012 Gardens and Great British Garden, the largest wildflower meadow ever planted in the UK, more than 4,000 semi-mature trees planted, wetland planting on a massive scale—more than 300,000 plants, including reeds, rushes and grasses grown from cuttings taken before construction began, created new wildlife habitats, regeneration of the rivers and canals that weave through the site and transformation of the River Lee into wetland, swales, wet woodland, dry woodland and meadow to form crucial sustainable flood defences.

**Fig. 39.2** Eden Project, relaxation and education

The Eden project in Cornwall could equally well illustrate landscape construction and horticulture on a massive scale (Anon 2013v) (Fig. 39.2).

While it is difficult to tease out the number of horticultural staff and purely horticultural costs from all the engineers and construction staff there is no doubt that this was a massive horticultural project measured by any parameter but landscaped by professional horticulturists.

## *Special Interest Garden Societies*

Many botanic gardens certainly hold large numbers of species (RBGE 13,300, Botanischer Garten, Berlin-Dahlen, 16,865, (Rae 2012)) and they often have staff with very specialist knowledge in particular plant groups such as succulents, ferns or carnivorous plants. However, when it comes to the minutely detailed cultivation of specialist groups of plants there is no doubt that talented enthusiasts can outperform even the best botanic gardens. In this example of contrasting professional horticulturists with 'amateur' (which is not a good word to use in this context) gardeners the latter are every bit as professional as the former, the only difference being that it's not their full time, paid job. Many of the clubs and societies representing these special interests are so professional-looking in the quality of their publications, shows and cultivation knowledge that many professional horticulturists are members. Good examples of this level of expertise can be found in specialist societies such as the Alpine Garden Society (AGS) (Anon 2013w), Carnivorous Plant Society (Anon 2013x), International Dendrology Society (Anon 2013y) and the National Vegetable Society (Anon 2013z).

Membership numbers and activities vary but most include web sites giving detailed help and advice, produce publications and newsletters, arrange garden tours and distribute seeds. Many also take an active interest of their particular plants in the wild and arrange tours to view them growing in their native habitats. However, when it comes to the skilled cultivation of plants no other activity compares with the competitive shows that most of these clubs and societies arrange. At these events individual entries are grouped together and judged by skilled judges who look for excellence and deduct points for the most minute of defects. The ensuing discussion, advice and swapping of cultivation details is where younger or less experienced members gain invaluable help in cultivating their plants. Garden tours to visit others members' gardens and collections also facilitates this knowledge exchange.

The Alpine Garden Society (AGS) is a good example of all of the activities mentioned above. Started in 1929, and now with 7,000 members, the AGS describes itself as an "international society for the cultivation, conservation and exploration of alpine and rock garden plants, small hardy herbaceous plants, hardy and half-hardy bulbs, hardy ferns and small shrubs". In pursuing these goals, they have a seed distribution scheme, very high quality journal, arrange numerous competitive shows with many entry categories and award a wide range of cups and medals, have local and special interest groups and a slide and reference library. They hold national and regional meetings and conferences and arrange workshops to demonstrate and discuss the cultivation of particular plant groups or cultivation techniques/structures such as alpine bulbs, saxifrages, scree gardens or alpine houses. They also arrange three or four tours a year to mountainous regions to see alpine plants in the wild and organise garden visits to both other members' gardens and botanic and heritage gardens.

While the AGS concentrates mostly on flowering plant species the National Vegetable Society promotes the cultivation of edible vegetables. They were founded in 1960 and have 3,000 members. Their objectives are two-fold: "to advance the education of the public in the cultivation and improvement of vegetables" and "to advance knowledge of and further public interest in vegetables by the publication of information, by exhibition, by stimulating research and experiment and by awarding prizes open to public competition". In pursuing these objectives they have numerous national and regional shows, a web site packed full of helpful advice, maintain a network of local clubs, have an on-line shop selling books, leaflets and DVDs and arrange tours and demonstrations. While showing vegetables is not as popular as it once was, many shows are still well supported with numerous entries for each category. Competitors regularly produce vegetables of the highest quality and experienced judges are needed to select the very best from a range of entries that would be the envy of the average vegetable grower.

These examples of different types of approach, philosophy and motivation to gardening and horticulture demonstrate both the similarities and differences of the two vocations.

## Horticultural Practices

The successful cultivation of plants, be they in a garden, orchard or glasshouse, requires a diverse range of skills or techniques. Considering the fact that wild plants have grown and survived unaided by human intervention for thousands of years, it is remarkable that so many techniques to manipulate plants and the environments in which they grow have been developed to obtain so many different outcomes. These manipulations at their simplest include soil cultivation, propagation, environmental adaptations and pruning and training, mulching, composting and plant protection from pests and diseases. Their inclusion here along with a small selection of applications on turf care are not intended to provide a text book description of the techniques, but a sufficiently inclusive account to demonstrate the different manipulations possible and the degree of difference between skilled craftsmanship delivered by gardeners and the professional approach developed by horticulturists and to further demonstrate the differences between horticulture and gardening.

## *Soil Cultivation*

Soil cultivation includes primary and secondary cultivation, soil amelioration, mulching and physical forms of weed control. The purpose of primary cultivation is to bury weeds by inverting the soil to provide a weed free (and therefore competition free) environment to promote successful seedling emergence. Ploughing is the main form of primary cultivation in horticulture and agriculture but has traditionally taken the form of hand digging in small scale gardening. While both may seem mundane to some, cultivation is almost an art form when executed skilfully. Ploughing with horses certainly required skill but so too does ploughing with powerful tractors and both traditions have their enthusiastic followers who take part in fiercely contested competitions. Digging certainly requires skill, not to mention muscle power but, done by an experienced gardener, can appear almost effortless, resulting in straight lines, a level surface, a well prepared seedbed and absence of green material. As well as inverting the soil to give a clean weed free surface, digging also provides the opportunity to incorporate organic matter such as garden compost or horse manure to improve structure, retain moisture and increase soil fertility. This is usually accomplished by double digging where the top spade's depth is inverted to one side leaving a trench into which the organic material is added and incorporated with a fork (Brickell 2007). While ploughing, organic matter can be incorporated if it has first been spread over the surface with a 'muck spreader' but it is never incorporated to the same depth or with the same degree of amelioration achieved with hand digging.

Traditionally, and especially with heavy clay soils, primary cultivations are undertaken in the autumn to allow the soil to settle over the winter and provide an opportunity for frost action to break down the lumps of soil making it easier to work with a 'frost tilth' later. In contrast, the use of powerful machinery and the rapid cycling of crops can usually render this traditional approach unnecessary or unaffordable in intensive horticulture.

The purpose of secondary cultivation is to break down the large lumps of soil left from primary cultivation into a finer tilth, ready for sowing. Again, there is a clear difference between intensive horticulture where powered and non-powered harrows are used and gardening where hoes and rakes are used. Either way, the purpose is to create a fine tilth so that seeds can more easily come into contact with water-providing soil particles and also so that they are evenly buried thereby ensuring even germination. Hoeing and harrowing are also used for mechanical weed control, leaving the soil weed free if done frequently enough although the mechanical action of both approaches can cause considerable damage to roots and stems if not done carefully.

## *Propagation*

Seed sowing is an ancient horticultural and agricultural technique going back to biblical times and beyond. To the untrained layperson the simple act of scattering seeds on the ground may seem basic but the opportunities for failure are great, ranging from factors such as poor seedbed preparation and wrong time of year to wrong sowing rate, depth and aftercare. Vegetable gardeners will quickly learn that one rule does not apply to all as some plants are more cold hardy than others and can therefore be sown earlier, some require deeper sowing than others, some are best grown in broad drills whilst other are best in single lines, some require thinning soon after emergence, and they all have different emergence rates and irrigation requirements and more. And that is just for direct sown vegetables and can well apply to the wider range of horticultural crops.

Again, the approach taken by horticulturists and gardeners is different and driven by scale, economics and training. Vegetables grown by a home or allotment gardener would always be sown by hand and with considerable time and trouble taken over sowing depth and density. The treatment of field grown vegetables is far more akin to agriculture with the use of precision sowing machines or planters pulled by tractors.

Half hardy ornamental annuals require sowing and early cultivation under protection before planting and, commercially, have undergone a revolution in production during the last 20 years. Traditionally, both in commercial horticulture (probably parks and heritage gardens in this example) and home gardening the technique was quite similar, the only difference being one of scale and operator speed. The task was often painstakingly undertaken by skilled gardeners who relished the challenges offered and attained deep satisfaction in sowing or planting their gardens. For commercial horticultural staff the task might not have been regarded as quite so special but it was none-the-less undertaken with skill and precision but probably at a much faster pace than home gardeners. For both sectors seeds would initially be sown into pots or containers, pricked out to allow growth and development followed by further growing under protection, hardening off to promote acclimatisation, and eventual planting out. Half hardy bedding displays in parks and heritage gardens almost ceased at the end of the last century due to the high costs associated with the process. In home gardening their popularity declined due to the sheer amount of work required or the costs involved in purchasing appropriate seedlings from a nursery. Half hardy bedding plant displays are once again popular due to a combination of factors including smaller home gardens, so-called 'patio' or container gardening but most of all because the cost has reduced both for public gardens and also for home gardeners. This is because production techniques have become so automated that home gardeners can now buy relatively cheap half hardy bedding plants from garden centres or supermarkets, and parks and heritage gardens now buy in such plants from specialist professional growers or nurseries rather than producing them themselves. The reason for this is that the whole process has been streamlined by precision sowing, modular production and automated handling and irrigation. This, in turn, has been made possible by improvements in plant breeding which have increased uniformity of size, not only in visual attributes, but also in seed and seedling viability.

For plants that don't produce a lot of seeds or where germination is slow or erratic or for cultivars with complex percentage, then vegetative propagation techniques, which maintain parental genetic integrity, are used. These techniques, typically cuttings, grafting or layering, are ideal for woody shrubs, trees and half hardy perennials such as *Pelargonium* cultivars. Simple division or splitting is easiest for herbaceous plants. There are sound scientific principles underpinning propagation by cuttings and grafting including an understanding of callus production, rooting hormones, transpiration control, propagule maturity and root zone heating. The approaches adopted by gardeners and horticulturists do not differ greatly in their principles, but certainly differ in their scale, use of equipment and speed of propagation. Propagation by cuttings requires skill and dexterity to accomplish the task carefully and effectively but a gardener is likely to propagate fewer plants, taking a lot of time in the process, using pots, bought compost, domestic propagators and then relying on small domestic often poorly environmentally controlled glasshouses. The success of the task then relies on skill and judgement with experienced gardeners able to control temperature, humidity and the outbreaks of diseases while those without such skills often failing to root a single plant. Horticulturists, by contrast, deal in tens of thousands, combine skill with speed, use specialist modules and rely on climate controlled glasshouses where temperature and humidity are carefully controlled and where outbreaks of disease are rare because of the precautions taken and the attention paid to climate control. All aspects of the environment are likely to be carefully controlled, from air temperature and humidity, air circulation, root zone temperature and substrate moisture content (Hartman et al. 2010).

Grafting requires skilful knife craft, knowledge of the different techniques available and which is best for different species, time of year and aftercare experience to maximize the union of the scion with the rootstock. While simple in theory, competence only comes with long-term practice and experience and the names of the techniques used provides ample testament to the degree of 'craft' involved- whip and tongue, cleft and saddle grafting, budding, bridge grafting, awl, veneer and stub grafting (Garner and Bradley 2013; Alexander and Lewis 2008). While rootstock selection for ornamental plants such as *Sorbus* and *Rosa* is usually confined to seed-grown wild species, that is not the case with cropping apples (*Malus*) and pears (*Pyrus*) where there has been a huge amount of research into the effect of rootstocks on the scion, particularly as it affects vigour (Garner and Bradley 2013). The same scion material can be grafted onto numerous different rootstocks which control the plant's vigour in varying degrees from dwarfing to vigorous. While some enthusi-

**Fig. 39.3** Espalier trained rose

astic gardeners may occasionally graft a few apples, *Sorbus* cultivars, or bud some rose (*Rosa*) cultivars, this is an area of propagation that is predominantly the preserve of horticultural specialists. Detailed information on propagation can be obtained from the following reference sources: Hartman et al. 2010, Macdonald 2000, Toogood 2006 as well as the International Plant Propagator's Society (IPPS) web site at Anon (2013aa).

## *Pruning and Training*

Pruning has a sound scientific basis although for some it seems to be shrouded in myth and mystery. Pruning controls and balances plant growth regulators within the plant. In the cultivation of fruit trees such as apples and pears, pruning both promotes flower-bearing shoots and regulates the amount of flower buds (and hence fruit) on stems (Fig. 39.3). This avoids the otherwise typical boom and bust cycles where trees flower and fruit heavily one year followed by a year of poor fruiting. Removal of section of stem removes flower buds and thus regulates fruiting. In flowering shrubs, such as *Philadelphus* and *Deutzia*, pruning after flowering pro-

**Fig. 39.4** Arboriculture

motes the production of new, flower-bearing shoots for the following year whereas pruning roses (which flower on shoots produced in the current year) well before flowering stimulates vigorous new growth which supports good flowering. Severe pruning of all stems down to the ground, or stooling, on species such as coloured stem willows (*Salix*) and dogwood (*Cornus*) promote vigorous one year stems of a metre or more to sprout from the base. Since the colour, typically reds, oranges and yellows, are only produced on one year stems, this treatment produces thickets of coloured stems.

Pruning is also carried out on field grown trees or nursery stock to provide the necessary specification for a tree's eventual landscape purpose, such as standard trees for street planting or feathered or multi stem stock for more natural planting situations. It also affords the opportunity to remove diseased stems and, undertaken with an artistic eye, allows shrubs to be kept within their allotted space, keeping them compact and well balanced. Pruning can therefore be regarded as both an art and a science, with the art more the preserve of gardeners and the science more the preserve of horticulturists. Taken up a level and applied to trees, pruning becomes arboriculture with the opportunity for crown reduction to reduce wind drag and crown lifting to allow more light to penetrate beneath the canopy (Fig. 39.4). These arboricultural operations are almost exclusively the domain of skilled and highly trained horticultural arborists.

Training, which virtually always requires pruning to accomplish, refers to the practice of forcing plants into particular shapes. This is sometimes done to increase production as in the case of fruit trees grown against sunny walls, or to make the most of restricted space, such as training the stems of a grapevine up the inside of a glasshouse or conservatory or in espalier grown apples. On other occasions it can make protection from predating birds easier such as when cherries are grown against a wall to make netting easier, or because it's the most effective way to manage climbing plants such as *Wisteria*, climbing roses or *Clematis*. Often, plants are trained into particular shapes for design or aesthetic purposes such as cloud pruned box plants or climbing roses trained over a pergola. Sometimes art and science come together such as the traditional practice of growing espalier pears on the gable ends of cottages which is both scientifically valid and aesthetically pleasing. The same is the case with step over and cordon grown fruit trees. In each case the training provides for efficient production and is visually pleasing.

Plant training might therefore be considered purely an aesthetic pursuit and therefore a home gardening technique rather than an operation carried out by commercial horticulturists. Professional horticulturists employed in heritage or botanic gardens, however, spend considerable amounts of time training plants to create aesthetic displays and professional horticulturists at famous display gardens such as Longwood in Pennsylvania, USA (Anon 2013b) deploy considerable time and skill into training species from *Chrysanthemum* to *Wisteria* into dazzling shapes and spectacles. Going back in history, famous garden designers such as André le Notre (1613–1700), relied on training to achieve spectacular garden creations. Training is therefore the preserve of both gardeners and horticulturists with, again, scale of operation being the major divide. Detailed information on pruning and training can be obtained from the following reference sources: Brickell and Joyce (2011), Brown and Kirkham (2009) and Joyce and Lawson (1999).

## Composting

The process of organic matter decomposition and nutrient cycling (the carbon cycle) is a natural process in the environment and both gardeners and horticulturists have harnessed this biological activity for their benefit for generations. In the wild dead leaves and other plant and animal remains are colonised by fungi and bacteria resulting in the liberation of heat, water, carbon dioxide, a reduction in volume, the liberation or recycling of some nutrients, and leaving a by-product, humus, that slowly continues to further decompose and reduce in size. In the garden situation, any organic material such as weeds, leaves, vegetable peelings, mown grass clippings and unwanted herbaceous plants can be stacked in a heap and allowed to decompose. Undertaken with skill, the waste heat produced kills pathogens and weed seeds and the humus remaining (which will have reduced to about a quarter of its original bulk) can then be incorporated back into the garden where it increases fertility but, perhaps more importantly, helps improve soil structure making the soil easier to work and less prone to wind or water erosion.

Composting is frequently described as an art and it is certainly a craft, at least, with many gardeners lacking the necessary skills finding success difficult to achieve. Composting fits very well with organic and wildlife gardening and is very popular at the moment, often featuring very prominently in garden society lecture programmes and demonstrations.

In the professional horticulture environment, while composting has always been practiced to a greater or lesser extent over the last 50 years and more, the rise in popularity of organic production, along with public pressure and coupled with imposed environmental legislation has forced up the usage and importance of composting and, with it, the machinery necessary to handle large quantities of material. In the recent past many parks and public gardens would collect and burn leaves, or throw organic matter onto rubbish tips rather than compost heaps and there was certainly no thought that local authorities would collect domestic garden waste and compost it in municipal compost heaps. Today progressive large botanic gardens now feature their massive compost heaps as educational opportunities allowing the public to see the tractors turning the heaps and witness the clouds of steam rising into the air. Composting on this scale is subject to strict legislation to promote public safety, monitor transportation and guard against environmental damage such as pollution of ground water. This and the size and cost of the machinery involved contrasts clearly with the home gardening approach. Detailed information on composting can be obtained from the following reference sources: Scott (2010) as well as Anon (2013bb).

## *Environmental Manipulations*

While gardening itself is sometimes described as the art of plant manipulation, it is probably fair to say that gardeners manipulate the environments in which they cultivate plants almost as much as they manipulate plants themselves. Probably the most obvious way to benefit plants in northern temperate regions is to grow them on south facing slopes to benefit from the extra warming that provides as opposed to growing them on north facing slopes. The Romans understood this when they cultivated cherries in the Scottish Borders but it was well known long before that. On a smaller scale, Victorian gardeners would grow peaches and nectarines on south facing walls for the same reason and, apart from the benefit for fruit growing, gardeners growing ornamental plants know well the value of seeking out sunny spots for Mediterranean species while retaining cooler shaded spots for understory ferns and large leaved herbaceous plants. The use of microclimates in this way to benefit from extra warm or avoid frost pockets is all part of both the art and science of gardening.

Not only do gardeners seek to benefit from small differences in site microclimate but they invest in numerous types of structure and equipment, all in the name of gardening and the benefit, financial or for pleasure, in producing bigger, or earlier or more cold intolerant plants in places where they would not naturally grow. Different styles of greenhouses, cloches, cold frames, fleece, polythene, bell jars and heavy mulching are all ways to protect non hardy plants from frost, or to produce fruit such as tomatoes in places where they would not naturally ripen well. Wind breaks, both natural and synthetic provide shelter and tree tubes do this for individual plants aiding their establishment by reducing exposure. Mulching with both organic and inorganic materials smothers weeds, thereby reducing competition, and helping to retain moisture while all types and scales of irrigation equipment provide plants with extra water to help establishment or increase growth and yield.

The practices listed above are just a small insight into the craft or practice of gardening but demonstrate the intricate nature of what skilful plant cultivation at the amateur level can entail. Horticulturists employed in public gardens often adopt the same types of approaches and this is perhaps best seen in botanic gardens where a wide range of the world's biodiversity is being grown on a single location. Every opportunity to manipulate the environment so as to cultivate plants that naturally grow in deserts, shade, alkaline soils, waterlogged situations or on shallow mountain tops, for instance, are taken to successfully grow plants from other environments. The sophistication of professional horticulturists in adapting and controlling the environment is, perhaps, demonstrated to its extreme in glasshouse cropping. Here computer controlled environments manipulate day and night temperature, humidity, carbon dioxide levels, nutrient applications, pest and disease control, pollination, light levels and day length, while also maximising opportunities for environmental sustainability through rainwater harvesting, thermal insulation and renewable energy sources. Detailed information on environmental manipulations can be obtained from Cockshull et al. (1998).

## Mulching

Mulches, to smother weeds and retain moisture, are traditionally composed of organic matter such as garden compost, leaf mould or farm yard manure. Inorganic materials do much the same job and are often used in intensive crop production, even if they are aesthetically less pleasing. Black polythene, woven geotextile membranes, gravel and even decorative, coloured recycled glass aggregates can all be used but don't add to the soil's fertility or improve its structure. In collections-based gardens, such as botanic gardens, different types and depths of mulching can help diversify the range of species it's possible to cultivate at a particular site. For instance, heavy applications of moisture retaining garden compost allow the cultivation of rhododendrons on the east coast of Scotland where the annual rainfall amounts to only 650 mm whereas such applications are unnecessary on the west coast where precipitation can reach 3,000 mm.

## Pest, Weed and Pathogen Control

Pests of all sorts including insects, molluscs, pathogens and weeds have been the scourge of gardeners and horticulturists for as long as husbandry has been practiced. Killing, trapping, removing, scaring, shooting and burning have been traditional ways of trying to remove the source of the problem and there is even evidence of chemical control using sulphur, for example, as an insecticide and fertilizer going back more than 4,000 years. As a result of the agricultural revolution machinery became more effective and more widely adopted leading to massive changes in cultivation techniques which in turn led to more intensive cropping creating the conditions for some pests to become even more of a problem. Combating pests became consequently more serious with solutions being found in plant breeding and the production of resistant cultivars, cultivation techniques such as crop rotation and intercropping, more efficient chemical sprayers and machinery and, as the green revolution dawned, so the manufacture of more and more sophisticated and highly effective pesticides. Since the latter part of the twentieth century concerns regarding the damaging effect on human health and the environment have caused a gradual shift away from chemical treatment to other forms of control using biological agents, physical techniques such as mulching for weed control, organic pesticides, holistic approaches and integrated techniques.

Possibly nowhere else is the difference between home gardening and professional horticulture more widely exhibited than in the control of pests and diseases but there are not simply two scenarios at different ends of the scale but three:- home gardeners, public ornamental gardens, such as botanic and heritage gardens and commercial horticultural crops. Many gardeners and horticulturists adopt an Integrated Pest Management (IPM) approach which combines minimal use of pesticides with physical and cultural techniques of pest control (Radcliffe et al. 2008).

### *Turf Culture*

Possibly no other area of plant cultivation more amply demonstrates the divide between horticulturists and gardeners than turf culture or lawn care. While they might have been regarded as only of a different scale in the past, the science and technology now applied to sports turf in particular, now places them far apart, at least at their extremes. Modern sports turf cultivation now requires a thorough understanding of soil texture, irrigation, draining, cultivar selection, aeration, topdressing, plant nutrition and disease and weed control. The perfect-looking swards achieved at famous sporting locations such as Wimbledon, Pebble Beach, Lords or Old Trafford Cricket Grounds each have very specific characteristics required from each sport and are managed with skill, backed by science and technology. The equipment required to create and maintain these surfaces includes not only numerous cutting and irrigation types of machinery but also technique such as sand and vibro slitting, overseeding, scarification, topdressing, rolling and a means of alleviating compaction.

## The Science of Horticulture

Horticulture is both an art and a science. The art is expressed from the smallest urban garden right the way through to the grandest of designs such as Peterhof Palace, west of St Petersburg, Russia while the science underpins every stage of cultivation from plant selection right the way through to post harvest storage. Horticultural research has resulted in new cultivars, with greater pathogen resistance, greater colour range, longer cropping season, higher yield, better flavour, different sizes and shapes and more. Cultivation research has led to improvements in propagation and handling, irrigation techniques, soil cultivation, urban tree survival, organic cultivation, effective pest, disease and weed control, cost effective harvesting, and post harvesting procedures to extend shelf life. In addition there is considerable art and science in amenity horticulture when we are dealing with the design, construction and maintenance of different landscapes.

## *Science Associated with Growing Media*

Sometimes known as potting composts or just compost for short (but not to be confused with garden compost and composting), growing media describes the range of materials used to cultivate plants in containers, as opposed to plants growing in the open ground. Plants have been grown in pots or containers for many centuries, even if it was only to transfer them from one place to another. Using a predominantly mineral soil for this purpose is unsatisfactory because the frequent and heavy watering required when plants are containerised leads to a loss of structure. This in turn causes the loss of natural air spaces that would normally allow free drainage resulting in a material that does not drain leaving waterlogged conditions in which roots eventually die.

Up until the middle of the last century gardeners used mixtures of all sorts of materials to overcome this problem in the physical structure of growing media, along with a vast array of chemical additives to try and promote healthy plant grow. The main problem was that these materials had to withstand daily watering which is detrimental to soil structure and answers lay in a combination of organic matter and coarse sand or grit– the former have an extraordinary combination of high porosity, large surface area capable of retaining moisture and strong physical structure able to withstand the destruction of the hose or watering can. The latter add weight and improve porosity. Sources of organic matter usually consisted of garden compost or leaf mould. To these two basic material were usually added a certain amount of garden soil and a plethora of other ingredients such as charcoal, sawdust, broken up clay pots or river washed sand. Ingredients to provide plant nutrients consisted mostly of organic products such as ground bone meal, hoof and horn meal, dried blood and manure but also some mineral components like limestone. Mixed carefully and used skilfully there is no doubt that such materials could grow satisfactory crops and plants. The main problems, however, lay in procurement, consistency, sterility and the time taken in preparation. In commercial production and large municipal parks, in particular, it became difficult and time consuming to source the materials, although this was less of a problem in estates and private gardens. Consistency of product, which is so important in commercial production, was probably the main issue with the variable quality of each of the materials giving inconsistent results in terms of crop size, quality and harvest date. Finally, many of these materials contained sources of pathogens leading to high levels of pest and disease infestation and they took a long time to prepare.

Many of these problems were at least partially overcome following research undertaken in the 1930s by William Lawrence and John Newell working for The John Innes Horticultural Institution which demonstrated that a reasonable degree of uniformity could be obtained by using just three components– loam, peat and grit, supplemented by a standard combination of chemicals added at stated weights per volume and increasing in amount with type of plant and the likely length of time that they would remain in the growth media (Anon (2013cc). Interestingly, this uniform approach to making compost was initially developed to remove the variables from crop experiments at the institute and were only commercialised at a later date.

This was true also of the University of California standardised peat and sand mixes which were developed first for standardisation demanded from research and were then commercialised (Flegman and George 1979).

While these composts marked a substantial advance in the science of plant cultivation and were used by gardeners and professional horticulturists for many years (and are being revived today as part of moves to reduce the use of peat), the problems of procurement and standardisation remained, albeit at a reduced level, particularly with the use of loam which was supposed to be sourced from stacked turves, left to decompose 6 to12 months, then riddled to create an evenly sized product before steam sterilization to kill pests and pathogens. But, while the procurement, standardisation and use of the loam were problems, the procurement and use of peat was a revelation to many. While it was acid in reaction and devoid of any nutrients, its ease of procurement, superb physical composition and sterile nature made it an ideal constituent and led to further research on the possibility of using peat only, or peat with added grit for potting composts. More careful attention had to be paid to plant nutrition but these were easily added by readily available inorganic fertilizers, while the acid reaction was countered by the addition of ground limestone.

Peat-based composts were suitable for commercial and home use and were easily supplied in bulk to the former, particularly for its use in the nursery stock and garden centre industry, and conveniently in bags for home use. The product was clean, easily used and uniform, giving good, healthy growth and a uniform product. Many peat-based composts were only composed of peat with added nutrients but sometimes this was supplement with sand or grit to add weight or, for instance, perlite or vermiculite to increase aeration even more, particularly for use in propagation. Unfortunately, the supply of peat is not sustainable and there have been increasing pressures put on the use of peat due to concerns about the loss of valuable wildlife habitats. Since the turn of the century alternative materials, such as coir and bark, have been sought and have been the subject of research but it has proved to be very difficult to find a product as good as peat. Today a variety of materials are available on the market but are arguably not as easy to use as peat-based composts. However, with an understanding of their nutritional requirements and, especially, irrigation needs, good results can be obtained.

The 1960s and 1970s also saw experimentation of the cultivation of plants without soil at all, but in nutrient enriched water (Douglas 1976; Anon 2013dd). This generally took two forms, types of hydroponics, which found popularity in the commercial specimen pot plant industry, particularly for interior landscapes such as hotel lobbies, shopping malls and corporate board rooms (Manaker 1996). This type of cultivation involved containers with water reservoirs which could easily be topped up by unskilled staff and with plants supported by light expanded clay aggregate (LECA), a light material looking like gravel but with a pumice-looking internal structure. The other form of cultivation, commonly termed nutrient film technique (NFT), first found its outlet in the commercial production of tomatoes, aubergines and peppers but is now applied to many other vegetables produced for the supermarket such as lettuce and strawberries. Production involved plants growing in shallow channels of nutrient enriched water with the advantages of a very clean, soilless product which never experiences drought and in which the nutrient supply is constantly monitored and automatically topped up. Add to this an almost sterile environment with the floor covered with polythene, ideal environmental conditions in controlled glasshouses and biological pest control, adding up to a commercially successful growing 'package'.

Home gardeners have certainly benefitted from the research undertaken on composts and are easily able to buy bags of ready-mixed high quality compost from garden centres and supermarkets. To this extent they are using very similar materials, backed by the same research and development, as commercial horticulturists, the only difference being in the scale of delivery and use with gardeners buying modest sized bags and commercial users purchasing their compost in cubic metre sacks or in bulk by the trailer load. When it comes to fruit and vegetable production, while it is possible to by home-scale hydroponic kits, commercial scale production, along with its technology and equipment is very much the preserve of the commercial horticultural producer and is poles apart from home gardening.

## *Science Associated With Plant Breeding*

Forms of plant breeding have existed for as long as humankind has been domesticating agricultural and horticultural plants, probably for at least ten thousand years. The earliest application was probably the simple selection of useful characteristics such as size or disease resistance and the subsequent use of seeds from plants exhibiting these features. Over time such selections would have reduced the genetic variability within the species to a more uniform (but still genetically diverse) appearance leading, eventually, to recognised landraces. With an understanding (or at least observation) of how pollination resulted in seed production so the opportunity for deliberate hybridisation resulted in the possibility of combining desirable characteristics from two related species. However, this was still well before Gregor Mendel's (1822–1884) time and deliberate hybridisation was the preserve of farmers and growers who had no knowledge or understanding of genetics. It was Mendel's methodical approach to hybridising peas that eventually lead to the new science of genetics and an understanding that there was a rational underlying explanation of what was being observed (Mawer 2006).

Traditional plant breeding uses crossing and backcrossing to combine useful traits such as high yield, increased quality and disease resistance and these techniques were in use by commercial, mostly agricultural, seed producers in the latter part of the nineteenth century. In 1908 heterosis, the superiority of heterozygous genotypes with respect to one or more characters in comparison with the corresponding homozygous, the phenomenon of hybrid vigour (Rieger et al. 1991), was described and explained and this ability of the progeny of a particular cross to outperform both parents was used widely in crop development. However, while scientists were steadily working away understanding the science of genetics and applying this gradually to crop development and improvement, large numbers of Edwardian Head Gardeners and commercial nurserymen were busy hybridising and selecting a vast array of ornamental plants, fruit and vegetables. Everything from

**Fig. 39.5** Display of flowering plants for hobby gardeners at the Chelsea Flower Show, London

violets, antirrhinums, sweet peas and delphiniums to plums, pears, potatoes, cabbages and beans were the subject of breeding and selection and nursery catalogues bulged with vast lists of the cultivars available (Fig. 39.5).

The era when both the scientific and 'gifted amateur' approach to plant breeding and selection continued for many years with the former concentrating on more commercial crops and the latter on ornamentals. However, with the increasing application of technology such as protoplast fusion, mutagenesis, genetic modification and exploiting somaclonal variation to increase diversity that would not normally occur naturally, and the trend of chemical companies buying seed houses with the intention of 'harmonising' the cultivation of specific cultivars with the use of particular pesticide, so the day of the amateur plant breeder has declined, but has certainly not disappeared.

## *Science Associated with Glasshouse Production Technology*

The purpose of glasshouse production is normally to raise the growing temperature to permit the cultivation of warm temperate or tropical crops like tomatoes, aubergines and peppers in northern latitudes. This is not always the case, however, and in some tropical countries glasshouse temperatures are cooled to enable these same crops to grow successfully. Glasshouse, or protected, cropping does more than simply heat or cool the temperature, it makes possible the complete control of all growing parameters within an almost closed, and therefore controllable, environment. This, in turn, makes possible the cultivation of high volume, high quality crops matched to precise harvest and sale schedules. The costs associated with both the infrastructure and on-going production are considerable but, given the value and profits possible, are considered to be worth it. Like any industry though, to keep ahead of competition and to make a profit requires investment in new technology, and this new technology needs to be underpinned by sound research. More research

**Table 39.1** World greenhouse vegetable production area (ha) (2002)a

| Country | Production area (ha) |
|---|---|
| Canada | 876 |
| United States | 395 |
| Netherlands | 4,300 |
| Mexico | 1,520 |
| Spain | 70,000 |

Note: When comparing relative size of operations between countries the different production technologies should be taken into account. For example, production in Mexico and Spain consists of a variety of production systems ranging from low to high technology greenhouses. Spain consists mostly shade cloth production not glass production. 1 ha = 2.471 acres. Source: BC Vegetable Marketing Commission.

**Table 39.2** North American greenhouse vegetable production area (ha) of major crops (2002). (Source: BC Vegetable Marketing Commission)

| Production area (ha) | | | | |
|---|---|---|---|---|
| Crop | Canada | US | Mexico | Total North America |
| Tomatoes | 482 | 350 | 790 | 1,622 |
| Cucumbers | 199 | 25 | 118 | 342 |
| Bell Peppers | 174 | 20 | 210 | 404 |
| Total | 855 | 395 | 1,118 | 2,368 |

has probably been devoted to protected cropping than any other aspect of the horticultural industry.

Selected statistics demonstrate the scale of glasshouse production. Worldwide, the main greenhouse vegetable production areas include: Spain, the Netherlands, Mexico, Canada and the United States (Table 39.1). Production in Mexico and Spain consists of a variety of production systems ranging from low to high technology greenhouses. Spanish production consists mostly of shade cloth production, not glass production. Production in the Netherlands, Canada, and the United States consists primarily of high technology greenhouses with significantly higher yields. Table 39.2 shows the production area of the major greenhouse crops in North America (Anon 2013ee). The Netherlands has around 9,000 greenhouse enterprises that operate over 10,000 ha of greenhouses and employ 150,000 workers, efficiently producing € 4.5 billion worth of vegetables, fruit, plants, and flowers, 80 % of which is exported (Anon 2013ff). In the 1950s, the global flower trade was less than US$ 3 billion. By 1992, it had grown to US$ 100 billion. In recent years, the floral industry has grown six percent annually, while the global trade volume in 2003 was US$ 101.84 billion. While production has traditionally been centered around the main centres of population in North America and Europe and with Dutch growers leading production technology and volume, in recent years other countries such as Kenya and Ethiopia have been developing their capacity, based on cheap labour and efficient air transportation (Anon 2013gg).

Every aspect of infrastructure and cultivation technique impact on quality and yield and have been the subject of research and innovation. Many glasshouse companies, university departments and governments carry out this research and whole research institutes are or have supported the industry. In the UK the Glasshouse Crops Research Institute was created for this very purpose and, in its heyday, employed over 200 scientific staff. In the northern hemisphere a lot of research has been devoted to maximising light levels and reducing heat loss as these two factors influence heating costs, and therefore production costs, the most. Even a 1 to 2 % increase in light intensity can make a difference to profitability and roof and crop orientation are both significant factors in these studies (Nelson 2011). In terms of reducing heat loss, cladding material (e.g. glass, polythene or plastics such as polycarbonate sheets) side wall insulation and thermal blankets are all important while the source of heat (e.g. gas or oil), the efficiency of combustion and distribution systems all have a role to play in glasshouse efficiency (McCullagh 1978; Nelson 2011; Boodley 2008).

The cost of energy is such a major influence on cost and yield that even small deviations from ideal temperatures can have an impact. As a result electronic control systems and environmental management are important components of glasshouse cropping, while sustainable forms of heating such as ground source and air source heat pumps and the use of deep geothermal energy are being vigorously investigated. If $CO_2$ enrichment and supplementary lighting systems are used, the interplay of light intensity (not just from supplementary lighting but also dawn, dusk and ambient intensity), temperature and $CO_2$ can all be controlled for maximum effect. Temperature is affected naturally by outdoor temperature and light radiation penetrating the house but can be controlled by a combination of ventilation and water temperature within heating pipes.

When it comes to cultivation, crop layout to maximise space by minimising walkways or using mobile bench systems, ensures that as much of the expensive glasshouse space is used for cropping as possible. Breeding programmes to increase yield and improve disease resistance have also made a major contribution as has research into plant nutrition and cropping systems such as hydroponics, nutrient film technique (NFT) and modular systems of cultivation (Fig. 39.6).

## Conclusions

While gardening and horticulture are both concerned with cultivating plants, this chapter has served to show that the motivation, scale and technology between the two is different. Gardening is primarily a leisure activity while horticulture is a paid, professional occupation. While gardeners pursue their hobby in gardens and allotments, however large, horticulture is often on a large scale such as parks, botanic gardens and production nurseries. Finally, while gardeners however skilled, and many are exceptionally skilled, reply on domestic-based technology, horticulturists work in or with sophisticated machinery and equipment and in high-tech, highly controlled environments, backed by advanced scientific research.

**Fig. 39.6** Container grown trees for landscaping and amateur gardens

# References

Akeroyd S, Barter G, Draycott S, Hodge G (2010) The RHS allotment handbook: the expert guide for every fruit and Veg grower. Mitchell Beazley, London
Alexander D, Lewis WB (2008) Grafting and budding: a practical guide for Fruit and Nut plants and Ornamentals Plants. Landlinks Press, Collingwood
Anon (2011) Defra family food report. http://www.defra.gov.uk/statistics/foodfarm/food/family-food. Accessed Oct 2012
Anon (2013a) Institute of Horticulture. www.horticulture.org.uk. Accessed Feb 2013
Anon (2013b) Horticultural Trades Association. www.the-hta.org.uk. Accessed Feb 2013
Anon (2013c) International Society for Horticultural Science. www.ishs.org. Accessed Feb 2013
Anon (2013d) Professional Gardener's Guild. www.pgg.org.uk. Accessed Jan 2013
Anon (2013e) Worshipful Company of Gardeners. www.gardenerscompany.org.uk. Accessed Nov 2012
Anon (2013f) Royal Horticultural Society. www.rhs.org.uk. Accessed Oct 2012
Anon (2013g) American Rhododendron Society. www.rhododendron.org. Accessed Oct 2012
Anon (2013h) Horticulture Wikipedia. http:/Wikipedia.org/wiki/Horticulture. Accessed Feb 2013
Anon (2013i) Soil Association. www.soilassociation.org. Accessed Nov 2012
Anon (2013j) Wikipedia. http://en.wikipedia.org/wiki/Organic_farming. Accessed Feb 2013
Anon (2013k) Royal Society for the Protection of Birds. www.rspb.org.uk/birdwatch. Accessed Feb 2013

Anon (2013l) ISO14001. www.iso.org. Accessed Jan 2013
Anon (2013m) The Allotment Society. www.allotment.org.uk. Accessed Nov 2012
Anon (2013n) Botanical garden. http://www.bgci.org/garden_search.php. Accessed Feb 2013
Anon (2013o) Botanical garden. http://en.wikipedia.org/wiki/Botanical_garden. Accessed Feb 2013
Anon (2013p) Botanic Gardens Conservation International, Garden Search. http://www.bgci.org/garden_search.php. Accessed Dec 2012
Anon (2013q) LANTRA. http://www.lantra.co.uk/Industries/Production-Horticulture.aspx and http://www.lantra.co.uk/Industries/Horticulture-Landscaping-and-Sports-Turf.aspx
Anon (2013r) Commercial Horticulture Association. www.cha-hort.com. Accessed Oct 2012
Anon (2013s) VHB. www.vhbherbs.co.uk. Accessed Feb 2013
Anon (2013t) Green Pea Company. www.greenpea.co. Accessed Nov 2012
Anon (2013u) Nigel Dunnett. www.nigeldunnett. Accessed Oct 2012
Anon (2013v) Eden Project. www.edenproject.com. Accessed Feb 2013
Anon (2013w) Alpine Garden Society. www.alpinegardensociety.net. Accessed Oct 2012
Anon (2013x ) Carnivorous Plant Society. www.thecps.org.uk. Accessed Oct 2012
Anon (2013y) International Dendrology Society. www.ids.co.uk. Accessed Oct 2012
Anon (2013z) National Vegetable Society. www.nvsuk.org.uk. Accessed Oct 2012
Anon (2013aa) International Plant Propagator's Society. www.ipps.co.uk. Accessed Feb 2013
Anon (2013bb) Longwood gardens www.longwoodgardens.org. Accessed Feb 2013
Anon(2013cc)JohnInnesCompost.http://www.jic.ac.uk/corporate/media-and-public/compost.htm. Accessed Feb 2013
Anon (2013dd) Wikipedia. http://en.wikipedia.org/wiki/Hydroponics. Accessed Feb 2013
Anon (2013ee) North American Greenhouse Crops. http://www.al.gov.bc.ca/ghvegetable/publications/documents/industry_profile.pdf. Accessed Nov 2012
Anon (2013ff) Greenhouses. http://en.wikipedia.org/wiki/Greenhouse. Accessed Feb 2013
Anon (2013gg) Wikipedia. http://en.wikipedia.org/wiki/Floral_industry. Accessed Feb 2013
Anon (2013hh) Allotment. http://en.wikipedia.org/wiki/Allotment. Accessed Feb 2013
Baines C (2000) How to make a wildlife garden. Frances Lincoln, London
Boodley J (2008) The commercial greenhouse. 3rd edn. Delmar Publishers, Albany
Brickell C (Ed) (2007) Encyclopaedia of gardening. Dorling Kindersley, London
Brickell C, Joyce D (2011) RHS Pruning and training. Dorling Kindersley, London
Brown GE, Kirkham T (2009) The pruning of trees, shrubs and conifers. 2nd edn. Timber press, London
Carson R (1963) Silent spring. H. Hamilton, 304 pp
Clevely A (2008) The allotment book. Collins, London
Cockshull K, Gray D, Seymour GB, Thomas B (1998) Genetic and environmental manipulation of horticultural crops. CABI Publishing, Wallingford
Douglas JS (1976) Advanced guide to hydroponics. Pelham Books, London
Flegman AW, George RAT (1979) Soils and other growth media. MacMillan Press, London
Garner RJ, Bradley S (2013) Grafter's handbook. 6th edn. Octopus Publishing Group, London
Hartman HT, Kester DE, Davies FT, Geneve R (2010) Hartmann and Kester's plant propagation: principles and practices. 8th edn. Prentice Hall, London
Heckman J (2005) A history of organic farming: transitions from Sir Albert Howard's war in the soil to USDA National Organic Program
Hobhouse P (2004) The story of gardening. Dorling Kindersley, London, 468 pp
Jameson C (2012) Silent Spring Revisited. Bloomsbury Acad & PR, 320 pp
Johnson B, Medbury S (2007) Botanic gardens: a living history. Black Dog Publishing, London, 288 pp
Joyce D, Lawson A (1999) Topiary and the art of training plants. Frances Lincoln, London
Kruger A, Pears P (eds) (2001) Encyclopedia of organic gardening. Dorling Kindersley, London, 416 pp
Macdonald B (2000) Practical woody plant propagation for nursery growers. Timber Press, Portland

Manaker G (1996) Interior plantscapes: installation, maintenance and management. Prentice Hall, London

Marshall B, Barbara W, Phillipes E (eds) (2009) Rodale's ultimate encyclopedia of organic gardening. Rodale, Pennsylvania

Mawer S (2006) Gregor Mendel: planting the seeds of Genetics. Harry N. Abrams, New York

McCullagh JC (1978) The solar greenhouse book. Rodale, Emmaus

Nelson PV (2011) Greenhouse operation and management. 7th edn. Pearson Education, London

Oldfield S (2007) Great botanic gardens of the world. New Holland Publishers, London, 160 pp

Oldfield S, Botanic Gardens Conservation International (2010) Botanic gardens: modern day arks. New Holland Publishers, London

Owen J (2010) Wildlife of a garden: thirty year study. RMS Media, Peterborough

Radcliffe EB, Hutchison WD, Cancelado RE (eds) (2008) Integrated pest management: concepts, tactics, strategies and case studies. Cambridge University Press, New York

Rae DAH (1995) Botanic gardens and their live plant collections: present and future roles. PhD thesis. University of Edinburgh

Rae D et al (2006) Collection policy for the living collection. Royal Botanic Garden, Edinburgh, 69 pp

Rae D et al (2012) Catalogue of plants 2012. Royal Botanic Garden, Edinburgh, 771 pp

Reiger R, Michaelis A, Green MM (1991) Glossary of genetics: classical and molecular. 5th Edn. Springer, Berlin

Scott N (2010) How to make and use compost: the ultimate guide. Green Books, Totnes

Tait M (Ed) (2006) Wildlife Gardening for Everyone. RHS and the Wildlife Trusts

Taylor P (Ed) (2006) The Oxford companion to the garden. Oxford University Press, Oxford, 554 pp

Thompson K (2007) No nettles required. Eden Project Books, 183 pp

Toogood A (2006) Royal horticultural society propagating plants. Dorling Kindersley, London

Wales C, HRH The Prince of Wales, Lycett Green C (2001) The Garden at Highgrove. Weidenfeld and Nicolson, 171 pp

# Index




G. R. Dixon, D. E. Aldous (eds.), *Horticulture: Plants for People and Places, Volume 3*, DOI 10.1007/978-94-017-8560-0, 

**D**

E

**H**

## K



## L

**T**

## V